Landscape Ecology for Sustainable Environment and Culture

Bojie Fu · K. Bruce Jones
Editors

Landscape Ecology for Sustainable Environment and Culture

Springer

Editors
Bojie Fu
Research Center for Eco-Environmental Sciences
Chinese Academy of Sciences
Beijing
People's Republic of China

K. Bruce Jones
Desert Research Institute
University of Nevada
Las Vegas
USA

ISBN 978-94-007-6529-0 ISBN 978-94-007-6530-6 (eBook)
DOI 10.1007/978-94-007-6530-6
Springer Dordrecht Heidelberg New York London

Library of Congress Control Number: 2013932855

Printed on acid-free paper

Springer is part of Springer Science+Business Media (www.springer.com)

Preface

With increasing human interferences to the earth system, most landscapes are no longer natural but rather are an integration of both natural and culture systems. Landscape ecology, as an interdisciplinary science, is originated from the integration of geography and biology to study landscape structure and function, and has now evolved to seek relationships between humans and nature. Although the important role of humans on the environment and its sustainability has been recognized, their coevolving relationships have not been well connected yet. A more comprehensive understanding about the nature–culture entangled systems could be obtained by taking the advantages of the special theoretical and methodological frameworks of landscape ecology. For example, landscape services and biocultural diversity of landscapes are new research areas. At the same time, during the past years, landscape modeling has made advances by including pattern–process interactions, and the concepts and approaches of landscape ecology is more extensively applied in landscape planning, management, and policy.

There came an excellent chance to raise concerns and to identify opportunities related to the intensified relationships between nature and culture for landscape ecologists. The 8th IALE World Congress was held in the capital of the ancient oriental country—China on August 18–23, 2011, Beijing. The theme of the congress was "*Landscape Ecology for Sustainable Environmentand Culture*". The country has a rich cultural legacy in the long history and complex human-landscape relations. Especially in recent decades, it experienced dramatic economic development, which aggravated the conflicts between human and environment—similar problems exist in many other countries and regions globally. It is the first time that IALE held a meeting in a developing country and the congress was attended by about 1,000 participants from more than 47 countries or regions.

The theme of the congress is proposed as the title of this book. This volume includes a selection of papers presented in the 8th IALE World Congress but not excluding other interested groups or scholars. We hope that this book will replenish the existing literature and provide useful information to the scholars or students in the fields of landscape ecology, geography, ecology, environmental sciences, and sustainability studies. The chapters are grouped into three sessions with the main topics focusing on:

Part I. Concepts and Approaches
Part II. Landscape Modeling
Part III. Landscape Planning and Management

The congress would not be that successful without the support from many organizations including International Association of Landscape Ecology (IALE), Chinese Academy of Sciences (CAS), National Natural Science Foundation of China (NSFC), State Key laboratory of Urban and Regional Ecology, WWF China, Chinese Ecosystem Research Network (CERN), and Chinese Ecological Restoration Network (ER-China). We would like to thank the contributions of the International Partnership Program for Creative Research Teams of ''Ecosystem Processes and Services'' supported by Chinese Academy of Sciences and China's State Administration of Foreign Experts Affairs (CAS/SAFEA) and to Springer for considering publication of this book.

We also give thanks to the following reviewers who contributed their time and intelligence to review the book chapters. They are Jesper Brandt, Chansheng He, Hubert Gulinck, Billy Johnson, Nan Lu, Paul Opdam, Gloria Pungetti, Simon Swaffield, Zhi Wang, and Zhonglong Zhang.

We finally express our gratitude to all the authors of the book.

Bojie Fu
Kenneth Bruce Jones

Contents

Part II Landscape Modeling

Part III Landscape Planning and Management

Part I
Concepts and Approaches

Chapter 1
Coupling Landscape Patterns and Ecological Processes

Bojie Fu, Changhong Su and Yihe Lü

Abstract The relationship between landscape patterns and ecological processes lays the foundation of landscape ecology, and understanding the relationship between them is key to further promoting the study of landscape ecology. This chapter addresses the importance of coupling research on the relationships between landscape patterns and ecological processes. Direct measurement and model simulation are the two basic approaches. Being robust in tapping the in situ measurement data and integrating them with multiple spatial and temporal scale data, ecological models have gradually taken the lead in landscape ecological research. This chapter discusses methodology in analyzing driving, feedback, and coupling relationships between landscape patterns and ecological processes, and explores a 7-step coupling framework on landscape patterns and ecological processes. In particular, the links between landscape patterns and soil erosion processes are addressed at patch, slope, and watershed scales. Finally, we look into the future of the coupling research on landscape patterns and ecological processes as: (1) developing landscape pattern indices reflecting ecological processes; (2) exploring the scale dependence of the relationship between landscape patterns and ecological processes; (3) integrating landscape modeling with long-term ecological research; and (4) strengthening research on the effects of ecological processes on landscape patterns.

Keywords Landscape patterns · Ecological processes · Coupling · Model integration · Scale · Soil erosion

---

B. Fu (✉) · C. Su · Y. Lü
State Key Laboratory of Urban and Regional Ecology, Research Center for Eco-Environmental Sciences, Chinese Academy of Sciences, P.O. Box 2871, Beijing 100085, China
e-mail: bfu@rcees.ac.cn

B. Fu and K. B. Jones (eds.), *Landscape Ecology for Sustainable Environment and Culture*, DOI: 10.1007/978-94-007-6530-6_1,

1.1 Introduction

As an interdisciplinary science linking ecology and geography, landscape ecology has been applied widely for its theories and approaches. The link between landscape patterns and ecological processes forms the foundation of landscape ecology, understanding the relationship between them is key to further promoting the study of landscape ecology (Fortin and Agrawal 2005; Fu et al. 2001; Wu 2007; Wu and Hobbs 2002). Landscape pattern generally refers to the spatial structural characteristics of a landscape, i.e., the spatial arrangement and deployment of landscape units in various sizes and shapes. Ecological processes relate to the flow and transfer of matter, energy, water, biota, and information within or among ecosystems, with special emphasis on the dynamic characteristics of the occurrence and development of the ecological events or phenomena. Landscape patterns are associated closely with ecological processes; the interactions between them demonstrate some ecological functions tinted by scale dependence. Theoretically, landscape patterns and ecological processes are inseparable. To make the research easier, some researchers focus on the features of landscape patterns, while others focus on the dynamics of ecological processes.

Spatial statistical analysis, landscape pattern metrics, and dynamic models constitute the three primary methods for analyzing landscape patterns. As a basic method, spatial statistical analysis explores the areas/percentages and geometries of different landscapes and their variations at different stages (Odland 1988; Tobler 1970). Landscape pattern metrics are designed to quantify the landscape pattern by analyzing landscape structural composition and spatial configuration. Dynamic models of landscape pattern consist primarily of spatial Markov (Aaviksoo 1995; Li 1995), cellular automata (Wolfram 1984; Wu 2002), and agent-based models (Bithell and Brasington 2009; Ligtenberg et al. 2004; Matthews et al. 2007).

Landscape pattern metrics are used widely in analyzing landscape spatial configurations (Haines-Young and Chopping 1996). To date, a variety of indices have been developed, including patch area, edge/shape, proximity, diversity and lacunarity indices, aggregation, and location-weighted landscape contrast metrics (Chen et al. 2008). Unfortunately, most existing landscape metrics fail to capture the fundamental patterns that can be used to identify or explain ecosystem processes, due to lack of ecological implications; e.g., landscape metrics dealing with land cover composition and connectivity/fragmentation fail to capture the linear features or the hub-and-corridor spatial patterns (Jones et al. 2012). Current landscape analyses based on pattern metrics do not go any further than analyzing the geometrical characteristics of the land cover pattern, let alone linking them with relevant ecological processes (Chen et al. 2008; Gustafson 1998; Lü et al. 2007). Thus, landscape pattern models are stagnating in defining the land cover transfer rules and dynamic simulations by analyzing landscape transfer probability. The absence of ecological implications makes it difficult to explain the underlying mechanism of the landscape pattern change.

In situ observation and model simulation are the two major approaches to ecological process analysis. At finer spatial scales, in situ observation and experimentation are used widely in collecting data for ecological process research. As the scale increases, models play a more important role. In some cases, in situ measurement and modeling are complementary to each other; i.e., some large-scale models are built from and tested by measurement data (Carlisle et al. 2011; Herrick et al. 2010). Currently, models are used widely in the exploration of biomass analysis (Bergen and Dobson 1999; Zhang et al. 2008), carbon sequestration (Billen et al. 2009; Caldwell et al. 2007), nutrient flow (Kohlmeier and Ebenhöh 2007; Sogn and Abrahamsen 1997), climatic change effects (Jackson et al. 2011), and fire and human disturbances (Keane and Karau 2010). In most cases, the parameterization of process models is confined within small spatial scales, which may compromise the spatial heterogeneity of landscape patterns and hinder its application at broader landscape or regional scales (Xu et al. 2010).

As landscape patterns and ecological processes interact in many ways, ignoring either of them will make it difficult to grasp the panoramic dynamics of the landscape. In-depth probes of the relationship between landscape patterns and ecological processes are indispensable for sustainable ecosystem management, given the current situation in which humans are exerting influence on the planet at an ever-increasing rate.

1.2 Theoretical Framework and Approach to Coupling Landscape Patterns and Ecological Processes

1.2.1 Theoretical Framework

All ecological processes occur within a certain landscape space, and ecological processes and landscape space are intermingled, demonstrating complex spatial characteristics. In most cases, landscape has a macro-controlling effect on the ecological process. The focus of ecological process research varies greatly across different scales. At scales smaller than ecosystem, for example, the organizational scale, traditional ecological research composed the bulk of ecological process exploration, such as matter cycle, energy flow, population dynamics and interspecific relationship analysis, in which human activities are treated as external disturbances (Lü et al. 2007). Such traditional research at finer scales provides "bricks" for ecological process research at broader scales. At landscape (1–10^4 km^2) or regional scales (10^4–10^6 km^2), human activities are deemed as endogenous component of a landscape, i.e., human-related social, economic and cultural processes are integral components. "Land unit", an ecologically homogeneous tract of land in topography, soil, and vegetation, provides an effective stepping-stone for multi-dimensional process research across multiple scales, and serves as a means for landscape ecological research characterized by topology and land cover composition (Zonneveld 1989).

Hierarchical patch dynamics theory and scaling strategy are the key theoretical bases for the spatially-explicit hierarchical modeling of landscape patterns and ecological processes (Wu and David 2002). According to the hierarchy theory, complex systems are characterized by vertical organization and horizontal arrangement of sub-systems. These sub-systems at different levels interact with each other and demonstrate a relatively stable state. The hierarchical patch dynamics theory holds that the ecosystem is a dynamic patch-mosaic characterized by conspicuous general features formed by the cumulative effects of the interactions at patch level. Scaling is a key issue in landscape ecology. Coupling research on landscape ecology and ecological processes needs to balance the relationships among core scale and small-scale components and large-scale backgrounds. Integrating the studies of multiple scales can help reduce the uncertainties of research and reveal the coupling mechanisms between landscape patterns and ecological processes accurately.

A framework for coupling landscape patterns and ecological processes can be proposed as follow (Fig. 1.1). The details of the framework include: (1) determining the study area and defining research objectives, e.g., the sustainable management of crop production, habitat and wildlife conservation, control of soil and water loss; (2) delineating land units at specific resolutions—natural factors,

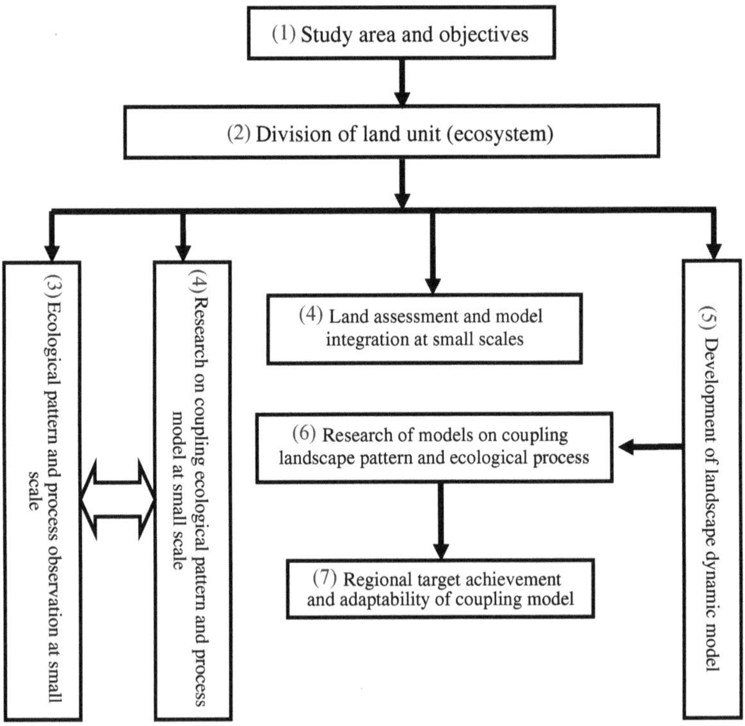

Fig. 1.1 Framework on coupling landscape pattern and ecological process (Lü et al. 2007)

such as those affecting crop production, habitat quality, and soil erosion—should be taken as the major classifying criteria; (3) measuring the patterns and processes within the typical land unit, as well as the reorganization of source or sink, e.g., the spatial configuration of crops, formation of crop production, wildlife habitat selection, transfer and cycling of water, soil, and nutrients; (4) constructing and validating ecological process models at the land unit scale; (5) developing landscape pattern dynamic models; (6) land assessment at the regional scale, designing modeling algorithms of the ecological processes at the land unit scale, and the coupling of regional landscape dynamic models, and (7) testing, validating, and applying the model system on coupling regional landscape patterns and ecological processes. In this framework, step (3) is a coupling analysis based on direct measurement—case study and data accumulation—that provides a basis for model analysis at small scales. Step (5) is relatively independent from steps (3) and (4). Step (6) conducts trans-scale studies. Steps (2) through (4) focus primarily on natural factors and processes, while steps (5) and (6) address the roles of human activities and socio-economic factors.

1.2.2 Coupling Approach

1.2.2.1 Coupling Based on Direct Measurement

Measurement-based coupling approach is effective at small spatial scales to establish quantitative relationships between landscape patterns and ecological processes by monitoring both the "snap shot" *status quo* and the chronical responses of the ecological process to the dynamics of landscape pattern. Major achievements include: the role of fire on forest landscape patterns (Kong et al. 2005); the species' establishment, growth and movement under specific landscape configurations (Berggren et al. 2001; Franken and Hik 2004); the land use configuration's influences on soil, water, and nutrient contents (Fu et al. 2000; Wang et al. 2009a; Zornoza et al. 2007); the effect of landscape pattern on runoff, soil erosion and nutrient transfer at plot or watershed scales (Bucur et al. 2007; Fu et al. 2009; Girmay et al. 2009; Wei et al. 2007); the role of vegetation on evapotranspiration (Sun et al. 2008) and the influence of land use change on soil carbon features (Boix-Fayos et al. 2009). Small-scale measurement can be controlled and is conducive to identify the quantitative relationships between pattern and process. Special attention need to be paid to sampling strategies and monitoring protocols to reduce uncertainties in data acquisition, and enhance the accuracy and representativeness of the results (Lü et al. 2007).

Because of scale dependencies, the measurement-based, small-scale coupling results cannot be taken directly as conclusions applicable at larger scales. This information, however, provides good references for large-scale research, e.g., the in situ measurement data can be used as either input or calibration data for large-scale models. In addition, as more factors are involved in coupling landscape patterns and ecological processes at large scales, systematic analysis and modeling approaches can make up the deficiencies of measurements and experiments, i.e., lack of accessibility, low replication and continuity.

1.2.2.2 Coupling Based on Systematic Analysis and Model Simulation

At larger scales, the landscape patterns and ecological processes involve multiple complex factors of ecological, socioeconomic and cultural categories. Two problems need to be solved in advance: (1) the location and magnitude of landscape pattern variation, and (2) the ecological effects of landscape pattern variation. Models are robust in revealing the underlying mechanisms and supporting scenario simulations at larger scales (Xu et al. 2010). So far, various types of models have been developed in exploring landscape pattern, e.g., systematic, statistical, cellular automata, and agent-based models. As single models are limited in coupling landscape patterns and ecological processes at larger scales, it is necessary to integrate multiple models by modularization, with due consideration of hierarchical structures. Model integration of landscape patterns and ecological processes have developed very rapidly, such as HILLS (Hesse Integrated Land Use Simulator; Schaldach and Alcamo 2006), PLM (Patuxent Landscape Unit Model; Binder et al. 2003; Costanza et al. 2002; Voinov et al. 1999), LANDIS (Land Information System; He et al. 1999), CLUE (Conversion of Land Use and its Effects) and TESim (Terrestrial Ecosystem Simulator Model; Xu et al. 2009).

The relationship between landscape patterns and hydrological processes is a key component in landscape ecology. Various models have been used widely to assess the effects of landscape patterns on hydrological processes. Brath et al. (2006) examined the effect of land use change caused by human activities on the frequency of floods in the Samoggia River basin in Italy from 1955 to 1992. Niehoff et al. (2002) developed land use change scenarios based on the LUCK (Land use Change Modeling Kit) model, and simulated flood events using a modified version of the hydrological physical model of WaSiM-ETH (Water Flow and Balance Simulation Model-Evapotransporation Hydrology). Savary et al. (2009) simulated the influence of land cover change on the water yield and runoff of the Chaudière River Basin in Canada over the last 30 years, and concluded that land cover change was a key factor in hydrological processes. The hydrological models also have been used in assessing the effects of geology, climate, human factors, and glaciation on river sediment transfer in the offshore watershed (Syvitski and Milliman 2007). On the other hand, hydrological processes also drive landscape pattern changes. For instance, soil erosion can change the landscape pattern by altering micro-topography, soil layer depth, and even land use policies (Bakker et al. 2005; Jimenéz-Hornero et al. 2009). Some hydrological processes can exert effects on biological processes or geochemical cycles by altering landscape patterns, e.g., soil erosion can influence carbon accumulation (Liu et al. 2003) and crop yield (Lu and Stocking 2000).

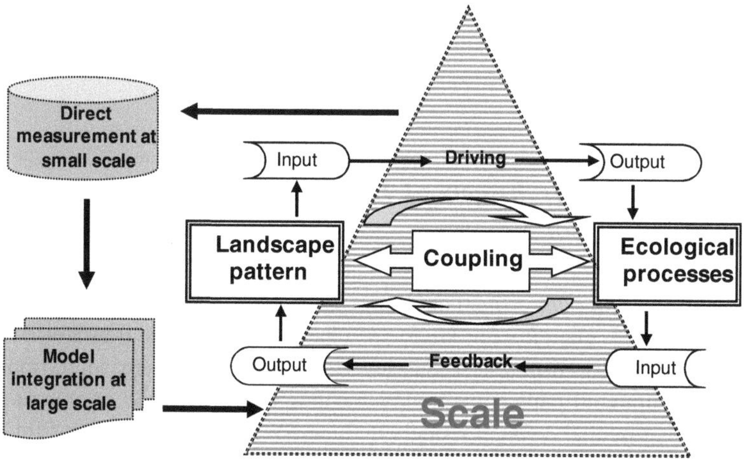

Fig. 1.2 Mechanisms and approach in coupling Landscape patterns and ecological processes-Ecological processes

Coupling models also have been used widely in hydrological simulation. Xu et al. (2009) coupled the hydro-ecological process based TESim and CLUE-based land use model in which soil organic matter and total nitrogen simulated by TESim were used as input for the CLUE model. These coupled models were used successfully in the temperate crop-pasture transitional area in northern China (CCPB) to analyze the interactions between land use and major ecosystem processes and functions.

To summarize, the coupling research on landscape pattern and ecological process mainly depend on in situ measurement and experimentation at small scales and spatial modeling backed up by ArcGIS software at large scales (Fig. 1.2). Assessing the mechanism between patterns and processes through measurement and modeling, and conducting coupling research by constructing integrative models have become the frontiers of landscape ecology research.

1.3 Categories of Models on Landscape Pattern and Ecological Processes

Landscape patterns affect ecological processes, and vise versa, which underpins the overall landscape dynamics. An ideal landscape dynamic model needs to combine both landscape patterns and ecological processes, which poses a daunting challenge to model development. According to the directions of the interactions between landscape patterns and ecological processes, the models generally fall into three categories: models that analyze the effects of landscape patterns on ecological processes (Hattermann et al. 2006; Li et al. 2007), models that analyze the influences of ecological processes on landscape patterns (Jeltsch et al. 1999; Paruelo et al. 2008), and models that couple landscape patterns and ecological processes (Fig. 1.2).

1.3.1 Models on the Effects of Landscape Patterns on Ecological Processes

To develop such models, the landscape pattern is taken as the input for the mechanism model. The effect of landscape changes on ecological processes can be tested by as setting different scenarios of landscape pattern alterations e.g., removing certain landscape components or changing the location or area of certain parts of the landscape. Typical models include SWIM (Soil and Water Integrated Model; Krysanova et al. 1998), THMB (Terrestrial Hydrology Model with Biogeochemistry; Li et al. 2007) and ANSWERS (Areal Non-point Source Watershed Environment Response Simulation; Beasley et al. 1980). The SWIM model is used to simulate water flow and nutrient retention processes in riparian and wetland ecosystems, the underground water and nutrient transfer of plants, as well as the influence of riparian zones and wetlands on hydrological processes. The THMB model is used to quantify the influence of forest and grassland on hydrological processes. The ANSWERS model can assess the influence of vegetation on runoff and sediment yields. The deficiency of such models lies in the fact that the pattern change is analyzed through scenario simulation, whereas the factors affecting the landscape pattern and the potential landscape evolution are not given due consideration, thus compromising their accuracy in predicting the landscape pattern change and ecological process responses.

1.3.2 Models on the Influences of Ecological Processes on Landscape Patterns

Modeling the effects of ecological processes on landscape pattern change can be conducted by assessing the corresponding landscape pattern changes after altering or removing some input data pertaining to the ecological process, i.e., landscape pattern is taken as an indicator to reflect the ecological process variation. The difficulties in establishing such models lie in the chronicity and subtlety of the inherent traits of ecological processes' effects on landscape patterns, the abundant driving factors underlying landscape patterns, and insufficiency in basic data and mechanism. Ambitious trials have been made for such models. For example, Jeltsch et al. (1999) simulated the effects of precipitation, fire, and grazing on tree distribution in savannas by using a spatially-explicit, grid-based model, and found that higher precipitation increased tree numbers characterized by an enhancement in tree clumping, whereas lower precipitation, fire, and grazing decreased tree density, characterized by a tendency towards random or even tree spacing. Schurr et al. (2004), using a series of integrated models, found that seed dispersal played a larger role than root competition in shaping the vegetation pattern in Karoo scrubland, South Africa. Bleher et al. (2002), using an individual-based, spatially-explicit model, found that the dispersal distance of seed and tree density are the major factors influencing tropical forest patterns.

1.3.3 Models on Coupling Landscape Patterns and Ecological Processes

The details of building the coupling models for landscape patterns and ecological processes are as follows: first, building model structure, writing codes, and applying modeling tools and GIS software to integrate the pattern and process models; and second, transferring or sharing the data across various models to extend the models' functions. Coupling models effectively tap the advantages of various specialized models, and validate and complement each component model through cross-checking. The major coupling models include the meta-population, vegetation dynamic, biogeochemical cycle, and hydrological models. Childress et al. (2002) employed LMS (Land Management System) to integrate the EDYS (Ecological Dynamics Simulation Model) model and the hydrological model of CASC2D (Cascade 2 dimensional sediment), embedding the hydrological model with the factors of climate, content of soil, water, and nutrients, as well as plant growth, fire, disturbance, and management measures. The PLM links topography, hydrology, nutrients, vegetation and land use, and simulates the ecological processes by use of Pat-GEM (Patuxent-General Ecosystem Model), which successfully quantifies the matter cycles among pixels in rasterized landscapes (Costanza et al. 2002). Similar coupling models include: IMAGE (Integrated Model to Assess the Global Environment; Alcamo et al. 1998) and LANDIS (Land Information System) pattern model (Mladenoff et al. 1996).

1.4 Landscape Patterns and Soil Erosion Processes

Links between land use/land cover and soil erosion provide an entry point to explore the relationship between landscape patterns and ecological processes. The complex mechanisms between landscape pattern and soil erosion at various scales have become a frontier in earth sciences (Fig. 1.3; Huang et al. 2003; Wang et al. 2009b). The pivotal driving factors on soil erosion vary greatly depending on the spatial scale (Table 1.1). Landscape pattern, represented by land use and vegetation structure, influences the runoff yield and soil erosion by affecting multiple factors, i.e., evapotranspiration, rainfall interception, soil water infiltration, and underground

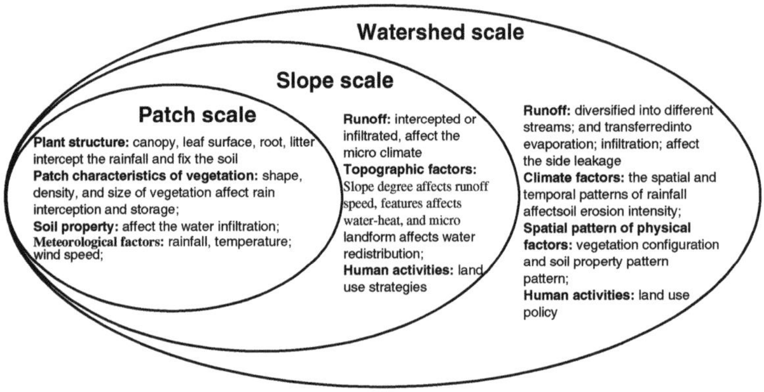

Fig. 1.3 Mechanisms of landscape patterns on runoff and soil erosion at various scales

Table 1.1 Leading factors affecting landscape patterns and soil erosion processes at different scales

Scale	Patch scale	Slope scale	Watershed scale
Impact factors	Plant characteristics	Vegetation and soil patch configuration	Vegetation and soil pattern
	Soil property	Topography	Land use pattern
	Meteorological factors	Meteorological factors	Climate
	Disturbance	Disturbance	Disturbance

water flow (Wang et al. 2006). On the other hand, soil erosion affects the vegetation pattern by altering water content and soil redistribution and accumulation, which further drive the dynamics of the landscape pattern (Wilcox et al. 2003).

1.4.1 Landscape Patterns and Soil Erosion Processes at Patch Scale

In semi-arid areas, landscape pattern at the patch scale is characterized by the spatial variabilities of vegetation and soil properties (Ludwig et al. 2005), while soil erosion is directly affected by runoff, infiltration, and interception. Vegetation plays an active role in reducing surface flow (Reid et al. 1999), lowering runoff speed, and increasing soil water content (Li et al. 1991), as well as influencing biomass (Noble et al. 1998) and bioactivity (Roth et al. 2003), and alleviating soil erosion (Ludwig et al. 2005). The canopy and leaf surfaces absorb the energy of rain drop splashes and effectively intercept the rainfall (Liu et al. 1994), while the root systems help fix the soil and improve its physical properties (Li et al. 1991). The litter not only enhances the soil's anti-erosion capacity (Hu and Shao 2001) and increases infiltration (Ludwig et al. 2005), but it also helps foster a sound environment for the growth of other organisms. Soil property is a key factor affecting erosion. High water content, infiltration rate, and surface roughness can increase sediment and nutrient retention, and greatly improve vegetation growth. Soil property, vegetation type and their interactions jointly regulate water cycling and micro-climate.

In addition to vegetation type and soil properties, other factors are also involved in the soil erosion process: (1) Patch characteristics of vegetation (shape, size, and density) play key roles in rainfall interception, runoff storage, and soil erosion processes. Ludwig et al. (1999, 2005) compared the capacity of different shapes of vegetation patches in intercepting runoff, sediment and nutrients, and found that banded patterns (stripes or strands) captured about 8 % more rainfall as soil water than did stippled patterns, and increased plant productivity by about 10 %. (2) Meteorological factors such as rainfall, temperature, radiation, wind speed, etc., significantly affect vegetation biomass (Hao et al. 2008) and water soil content (Huang et al. 2005) at patch scales. In semi-arid areas, water is a limiting factor for plant growth, and precipitation serves as the primary water source. Precipitation in semi-arid areas is characterized by "impulsiveness" at the patch scale; rainfall, rain intensity, and rain frequency exert profound influences on the magnitude and intensity of runoff and on the vegetation distribution in each patch. (3) The disturbances of human activities, such as crop growth and grazing, are important factors for soil erosion. Wilcox et al. (2003) found that the average annual runoff on disturbed vegetative slopes was almost double that on undisturbed slopes, while the average annual erosion on disturbed slopes was more than three times greater than on undisturbed slopes. McIvor et al. (1995) and Scanlan et al. (1996a, b) also found that soil erosion on grazed slope land was much higher than that on undisturbed slope land, while the aboveground biomass of long-term grazed slopes was significantly lower than that of undisturbed slopes.

1.4.2 Landscape Pattern and Soil Erosion at Slope Scale

Slope is composed of multiple landscape patches. Runoff has several possible destinations at the slope scale: being intercepted by plants, being infiltrated into deep soils, remaining as runoff, and transforming to evapotranspiration; furthermore, the evapotranspiration can have a bearing on the micro-climate at slope scale by exchanging water with the atmosphere. Factors affecting landscape pattern and soil erosion are more voluminous at slope scale than at patch scale. (1) Topographical factors, such as slope gradient, features (direction), and micro-topography play important roles in soil water movement. Specifically, slope gradient is the most important factor in affecting the runoff generation, i.e., runoff at low slope gradients has low kinetic energy, and causes less soil erosion. In addition, the ability of vegetation to intercept water and nutrients is more significant at low slope gradients. Slope aspect influences the water-heat conditions of the slope and consequently affects evapotranspiration. Generally, north-facing slopes have weaker evapotranspiration than south-facing slopes, which is favorable to soil water retention. Studies on the Loess Plateau in China indicated that soil water storage is highest on the north-facing slope, followed by west-facing and south-facing slopes (Chen and Shao 2003). Micro-topography drives water redistribution at the slope scale. Generally, the central and lower parts of a slope are more active in water exchange and have larger soil water storage capacity (Jiang 1997). (2) Human land use strategies affect vegetation patterns and soil properties strongly at the slope scale, leading to net loss of matter (Ludwig et al. 2005; Wilcox et al. 2003), and such damage results in severe disturbance over long periods of time. (3) The frequency of runoff and erosion varies depending on the spatio-temporal scales. Wilcox et al. (2003) found that the erosion happened more frequently at patch scale than slope scale.

1.4.3 Landscape Pattern and Soil Erosion Processes at the Watershed Scale

Watershed, as the basic unit of hydrological response, is an ideal scale to study landscape patterns and soil loss processes (Wang et al. 2001). Most watersheds used in ecosystem studies are relatively small, first order streams with drainage areas of 10–50 ha. The interaction between landscape patterns and most eco-hydrological processes (except the nutrient cycling or flooding, which is fairly straightforward) at watershed scales is complex and subject to multiple factors: (1) Climate, particularly the spatial/temporal pattern of precipitation, affects riparian water conditions, runoff, sediment, and vegetation patterns; (2) Land use strategies affect vegetation, water and nutrient cycling. For example, in the Loess Plateau in China, forest and grassland, as well as terraced fields, can effectively prevent soil loss, while bare land (particularly slope cropland) generates runoff. The arrangement of farmland, forest and grassland determines the generation or interception of runoff, and decides the extent of soil loss (Wang et al. 2000). (3) Configuration patterns of soil properties and vegetation at larger scales are important factors. In semi-arid areas characterized by over-infiltration runoff, barren land is the major generator of surface runoff, while vegetated areas absorb rainfall. The mosaic patterns of vegetation patch and soil properties affect the sediment concentration of the watershed, flood peak flow and runoff at the estuary (Zhang et al. 2006). (4) Human activities, particularly imprudent ones, often intensify the watershed's soil erosion by affecting the landscape pattern, resulting in soil loss and desertification, as well as altering river flow modes. For example, in Little Karoo, South Africa, inappropriate agricultural practices such as overgrazing, cultivation, and irrigation have led to the disruption of landscape linkages (e.g., hydrological flows, organic matter recycling), resulting in decreasing infiltration and increasing runoff yield). In addition, after each individual rainfall, the disturbed river has larger water flow and more frequent rip currents than the undisturbed river, which causes more severe erosion on the river banks (Le Maitre et al. 2007). The vegetation loss alters the evapotranspiration pattern of the watershed, leading to a dry and extreme micro-climate at watershed scale.

1.5 Perspectives on Landscape Pattern and Ecological Process Coupling Research

Pattern, process, function, and scale are the core components of landscape ecology. Although some progress has been made on the theories and methodology on these issues, the research on the whole is far from mature. The multi-scale coupling research on landscape patterns and ecological processes remains a big challenge for landscape ecology research. Future breakthroughs may lie in the following directions:

1.5.1 Developing Landscape Pattern Metrics that Reflect Ecological Processes

Landscape pattern metrics play a crucial role in analyzing the spatial configuration of landscapes. Along with the development of computer softwares, these metrics have promoted landscape ecology research greatly. However, given the rapid improvement of awareness and understanding of landscape ecology, the mere description of landscape patterns can no longer meet the needs of academic and applied pursuits. There is still much to be done to develop landscape pattern indices with significant ecological implications or to tap the potential ecological implications of existing pattern indices. Developing landscape pattern indices that reflect ecological processes has much theoretical and practical significance for the future of landscape ecology.

1.5.2 Exploring the Scale Dependence of the Relationships Between Landscape Patterns and Ecological Processes

Landscape pattern and ecological processes vary constantly and interact at multiple scales. Due to the complexity and abstract nature of ecological processes, most current studies are isolated and confined to small and medium scales; synthesis studies at large scales or cross-scale research are very rare. It is necessary to test the multi-scale relationships between landscape patterns and ecological processes, to reveal the characteristics and tendencies of ecological processes through scaling and spatial analysis.

1.5.3 Integrating Landscape Modeling with Long-Term Ecological Research

Long-term ecological research (LTER) is designed to monitor ecological factors and analyze their interactions and dynamics. LTER can reduce the uncertainties caused by disturbances, thus playing a significant role in dynamic research on ecological processes. One of the key objectives of landscape modelling is to quantify the relationship between landscape patterns and ecological processes. As models are derived from the real-world, validation techniques and sensitivity-testing play decisive roles in perfecting the ecological models. The functional relationship between landscape variables and parametric values of models needs to be determined by in situ measurement that necessitates the use of LTER data.

1.5.4 Strengthening Research on the Effects of Ecological Processes on Landscape Patterns

So far, research on the influences of landscape pattern on ecological processes is relatively mature. However, research on the effects of ecological processes on landscape patterns has not aroused sufficient attention. The reasons lie in the multiple complex driving factors underlying landscape pattern dynamics, and the inherent traits of chronicity and subtlety of landscape pattern variations. The limited studies in this regard are confined to the response of vegetation spatial patterns to climate change, fire, grazing and crop growth, propagule dissemination, and species competition. The key problems that are urgently need to be solved in the near future are: land use/land cover change caused by redistribution of surface water and soil water due to runoff, effects of erosion on soil spatial patterns and vegetation configurations, and the effect of nutrient cycling on the spatial patterns of organism growth. Such ecological processes are subtle and long-lasting, and call for long-term and unremitting study.

Acknowledgments This research was supported by the CAS/SAFEA International Partnership Program for Creative Research Teams of "Ecosystem Processes and Services", National Basic Research (973) Program (No. 2009CB421104), and National Natural Science Foundation of China (40930528).

References

Aaviksoo K. Simulating vegetation dynamics and land use in a mire landscape using a Markov model. Landscape Urban Plann. 1995;31(1/3):129–42.

Alcamo J, Leemans R, Kreileman E. Global change scenarios of the 21st century: Results from the IMAGE 2.1 model. London: Pergamon & Elsevier Science; 1998. p. 296.

Bakker MM, Govers G, Kosmas C, Vanacker V, van Oost K. Rounsevell M. Soil erosion as a driver of land-use change. Agric Ecosyst Environ. 2005;105(3):467–81.
Beasley DB, Huggins LF, Monke EJ. ANSWERS: a model for watershed planning. Trans ASAE. 1980;23(4):938–44.
Bergen KM, Dobson MC. Integration of remotely sensed radar imagery in modeling and mapping of forest biomass and net primary production. Ecol Model. 1999;122(3):257–74.
Berggren A, Carlson A, Kindvall O. The effect of landscape composition on colonization success, growth rate and dispersal in introduced bush-crickets Metrioptera roeseli. J Anim Ecol. 2001;70(4):663–70.
Billen N, Röder C, Gaiser T, Stahr K. Carbon sequestration in soils of SW-Germany as affected by agricultural management-Calibration of the EPIC model for regional simulations. Ecol Model. 2009;220(1):71–80.
Binder C, Boumans RM, Costanza R. Applying the Patuxent Landscape Unit Model to human dominated ecosystems: the case of agriculture. Ecol Model. 2003;159(2–3):161–77.
Bithell M, Brasington J. Coupling agent-based models of subsistence farming with individual based forest models and dynamic models of water distribution. Environ Model Softw. 2009;24(2):173–90.
Bleher B, Oberrath R, Bhning-Gaese K. Seed dispersal, breeding system, tree density and the spatial pattern of trees a simulation approach. Basic Appl Ecol. 2002;3(2):115–23.
Boix-Fayos C, de Vente J, Albaladejo J, Martínez-Mena M. Soil carbon erosion and stock as affected by land use changes at the catchment scale in Mediterranean ecosystems. Agric Ecosyst Environ. 2009;133(1–2):75–85.
Brath A, Montanari A, Moretti G. Assessing the effect on flood frequency of land use change via hydrological simulation (with uncertainty). J Hydrol. 2006;324(1–4):141–53.
Bucur D, Jitareanu G, Ailincai C, Tsadilas C, Ailincai D, Mercus A. Influence of soil erosion on water, soil, humus and nutrient losses in different crop systems in the Moldavian Plateau, Romania. J Food Agric Environ. 2007;5(2):261–4.
Caldwell IM, Maclaren VM, Chen JM, Ju WM, Zhou S, Yin Y, Boland A. An integrated assessment model of carbon sequestration benefits: a case study of Liping County China. J Environ Manage. 2007;85(3):757–73.
Carlisle DM, Wolock DM, Meador MR. Alteration of streamflow magnitudes and potential ecological consequences: a multiregional assessment. Front Ecol Environ. 2011;9(5):264–70.
Chen H, Shao M. Review on hillslope soil water movement and transformation mechanism on the Loess Plateau. Adv Water Sci. 2003;14(4):513–20 (in Chinese).
Chen L, Liu Y, Lv Y, Feng X, Fu B. Landscape pattern analysis in landscape ecology: current, challenges and future. Acta Ecologica Sinica. 2008;28(11):5521–31 (in Chinese).
Childress WM, Coldren CL, Mclendon T. Applying a complex, general ecosystem model (EDYS) in large-scale land management. Ecol Model. 2002;153(1/2):97–108.
Costanza R, Voinov A, Boumans R, Maxwell T, Villa F, Wainger L, Voinov H. Integrated ecological economic modeling of the Patuxent River watershed Maryland. Ecol Monogr. 2002;72(2):203–31.
Fortin M-J, Agrawal AA. Landscape ecology comes of age. Ecology. 2005;86(8):1965–6.
Franken RJ, Hik DS. Influence of habitat quality, patch size and connectivity on colonization and extinction dynamics of collared pikas Ochotona collaris. J Anim Ecol. 2004;73(5):889–96.
Fu BJ, Chen LD, Ma KM, Zhou HF, Wang J. The relationships between land use and soil conditions in the hilly area of the loess plateau in northern Shaanxi China. Catena. 2000;39(1):69–78.
Fu BJ, Chen LD, Ma KM, Wang YL. The principles and application on landscape ecology. Beijing: Science Press; 2001. p. 1–14 (in Chinese).
Fu BJ, Wang YF, Lu YH, He CS, Chen LD, Song CJ. The effects of land-use combinations on soil erosion: a case study in the Loess Plateau of China. Prog Phys Geogr. 2009;33(6):793–804.
Girmay G, Singh BR, Nyssen J, Borrosen T. Runoff and sediment-associated nutrient losses under different land uses in Tigray, Northern Ethiopia. J Hydrol. 2009;376(1–2):70–80.

Gustafson EJ. Quantifying landscape spatial pattern: what is the state of the art? Ecosystems. 1998;1(2):143–56.
Haines-Young R, Chopping M. Quantifying landscape structure: a review of landscape indices and their application to forested landscapes. Prog Phys Geogr. 1996;20(4):418–45.
Hao W, Chen C, Liang Z, Ma L. Research advances in vegetation biomass. J Northwest A & F Univ. 2008;36(2):175–82 (in Chinese).
Hattermann FF, Krysanova V, Habeck A, Bronstert A. Integrating wetlands and riparian zones in river basin modelling. Ecol Model. 2006;199(4):379–92.
He HS, Mladenoff DJ, Crow TR. Linking an ecosystem model and a landscape model to study forest species response to climate warming. Ecol Model. 1999;114(2–3):213–33.
Herrick JE, Lessard VC, Spaeth KE, Shaver PL, Dayton RS, Pyke DA, Jolley L, Goebel JJ. National ecosystem assessments supported by scientific and local knowledge. Front Ecol Environ. 2010;8(8):403–8.
Hu L, Shao M. Vegetation coverage index in soil and water loss studies. J Northwest For Univ. 2001;16(1):40–3 (in Chinese).
Huang Y, Chen L, Fu B, Wang Y. Spatial pattern of soil water and its influencing factors in a gully catchment of the Loess Plateau. J Nat Resour. 2005;20(4):483–92 (in Chinese).
Huang Y, Fu B, Chen L. Advances in ecohydrological process research. Acta Ecologica Sinica. 2003;23(3):580–7 (in Chinese).
Jackson CR, Meister R, Prudhomme C. Modelling the effects of climate change and its uncertainty on UK Chalk groundwater resources from an ensemble of global climate model projections. J Hydrol. 2011;399(1–2):12–28.
Jeltsch F, Moloney K, Milton SJ. Detecting process from snapshot pattern: lessons from tree spacing in the southern Kalahari. Oikos. 1999;85(3):451–66.
Jiang D. Soil erosion and control models in the Loess Plateau. Beijing: China Waterpower Press; 1997. p. 198–206 (in Chinese).
Jimenéz-Hornero FJ, Giráldez JV, Laguna AM, Jiménez-Hornero JE. An educational computer tool for simulating long-term soil erosion on agricultural landscapes. Comput Appl Eng Educ. 2009;17(3):253–62.
Jones KB, Zurlini G, Kienast F, Petrosillo I, Edwards T, Wade TG, Li BL, Zaccarelli N. 2012. Informing landscape planning and design for sustaining ecosystem services from existing spatial patterns and knowledge. Landscape Ecology. doi:10.1007/s10980-012-9794-4.
Keane RE, Karau E. Evaluating the ecological benefits of wildfire by integrating fire and ecosystem simulation models. Ecol Model. 2010;221(8):1162–72.
Kohlmeier C, Ebenhöh W. Modelling the ecosystem dynamics and nutrient cycling of the Spiekeroog back barrier system with a coupled Euler–Lagrange model on the base of ERSEM. Ecol Model. 2007;202(3–4):297–310.
Kong F, Li X, Yin H. The effect of fire intensity on the patterns of forest landscape in the north-slope of Da Hinggan Mountains. J Nanjing For Univ (Natural Sciences Edition). 2005;25(2): 33–7 (in Chinese).
Krysanova V, Müller-Wohlfeil DI, Becker A. Development and test of a spatially distributed hydrological: water quality model for mesoscale watersheds. Ecol Model. 1998;106(2–3): 261–89.
Le Maitre DC, Milton SJ, Jarmain C, Colvin CA, Saayman I, Vlok JHJ. Linking ecosystem services and water resources: landscape-scale hydrology of the Little Karoo. Front Ecol Environ. 2007;5(5):261–70.
Li BL. Stability analysis of a nonhomogeneous Markovian landscape model. Ecol Model. 1995; 82(3):247–56.
Li KY, Coe MT, Ramankutty N, De Jong R. Modeling the hydrological impact of land-use change in West Africa. J Hydrol. 2007;337(3–4):258–68.
Li Y, Zhu X, Tian J. Study on the effectiveness of soil anti-scourability by plant roots in Loess Plateau. Chin Sci Bull. 1991;36(12):935–8 (in Chinese).

Ligtenberg A, Wachowicz M, Bregt AK, Beulens A, Kettenis DL. A design and application of a multi-agent system for simulation of multi-actor spatial planning. J Environ Manage. 2004;72(1/2):33–55.
Liu SG, Bliss N, Sundquist E, Huntington TG. Modeling carbon dynamics in vegetation and soil under the impact of soil erosion and deposition. Global Biogeochem Cycles. 2003;17(2):24.
Liu X, Wu Q, Zhao H. The vertical interception function of forest vegetation and soil and water conservation. Res Soil Water Conserv. 1994;1(3):8–13 (in Chinese).
Lü Y, Chen L, Fu B. Analysis of the integrating approach on landscape pattern and ecological processes. Prog Geogr. 2007;26(3):1–10 (in Chinese).
Lu Y, Stocking M. Integrating biophysical and socio-economic aspects of soil conservation on the Loess Plateau, China. Part II. Productivity impact and economic costs of erosion. Land Degrad Dev. 2000;11(2):141–52.
Ludwig JA, Tongway DJ, Marsden SG. Stripes, strands or stipples: modelling the influence of three landscape banding patterns on resource capture and productivity in semi-arid woodlands, Australia. Catena. 1999;37(1–2):257–73.
Ludwig JA, Wilcox BP, Breshears DD, Tongway DJ, Imeson AC. Vegetation patches and runoff-erosion as interacting ecohydrological processes in semiarid landscapes. Ecology. 2005;86(2):288–97.
Matthews RB, Gilbert NG, Roach A, Polhill JG, Gotts NM. Agent-based land-use models: a review of applications. Landscape Ecol. 2007;22(10):1447–59.
McIvor JG, Williams J, Gardener CJ. Pasture management influences runoff and soil movement in the semiarid tropics. Aust J Exp Agric. 1995;35(1):55–65.
Mladenoff DJ, Host GE, Boeder J, Crow TR. LANDIS: a spaital model of forest landscape disturbance, succession, and management. In: Goodchild MF, Steyaert LT, Parks BO, editor. GIS and environmental modelling: progress and research issues. GIS World. 1996. p. 175–180.
Niehoff D, Fritsch U, Bronstert A. Land-use impacts on storm-runoff generation: scenarios of land-use change and simulation of hydrological response in a meso-scale catchment in SW-Germany. J Hydrol. 2002;267(1–2):80–93.
Noble JC, Greene RSB, Muller WJ. Herbage production following rainfall redistribution in a semi-arid mulga (Acacia Aneura) woodland in Western New South Wales. Rangeland J. 1998;20(2):206–25.
Odland JD. Spatial Autocorrelation. California: Sage Publications Inc.; 1988. p. 6–9.
Paruelo JM, Pütz S, Weber G, Bertiller M, Golluscio RA, Aguiar MR, Wiegand T. Long-term dynamics of a semiarid grass steppe under stochastic climate and different grazing regimes: a simulation analysis. J Arid Environ. 2008;72(12):2211–31.
Reid KD, Wilcox BP, Breshears DD, MacDonald L. Runoff and erosion in a pinon-juniper woodland: influence of vegetation patches. Soil Sci Soc Am J. 1999;63(6):1869–79.
Roth CH, Prosser I, Post D, Gross J, Webb M, 2003. Reducing sediment and nutrient export from grazed land in the Burdekin Catchment for sustainable beef production. Final report to Meat and Livestock Australia, p. 133–143.
Savary S, Rousseau AN, Quilbé R. Assessing the effects of historical land cover changes on runoff and low flows using remote sensing and hydrological modeling. J Hydrol Eng. 2009;14(6):575–87.
Scanlan JC, Pressland AJ, Myles DJ. Run-off and soil movement on mid-slopes in north-east Queensland [Australia] grazed woodlands. Rangeland J. 1996a;18(1):33–46.
Scanlan JC, Pressland AJ, Myles DJ. Grazing modifies woody and herbaceous components of North Queensland Woodlands. Rangeland J. 1996b;18(1):47–57.
Schaldach R, Alcamo J. Coupled simulation of regional land use change and soil carbon sequestration: a case study for the state of Hesse in Germany. Environ Model Softw. 2006; 21(10):1430–46.
Schurr FM, Bossdorf O, Milton SJ, Schumacher J. Spatial pattern formation in semi-arid shrubland: a priori predicted versus observed pattern characteristics. Plant Ecol. 2004; 173(2):271–82.

Sogn TA, Abrahamsen G. Simulating effects of S and N deposition on soil water chemistry by the nutrient cycling model NuCM. Ecol Model. 1997;99(2–3):101–11.
Sun G, Zuo CQ, Liu SY, Liu ML, McNulty SG, Vose M. Watershed evapotranspiration increased due to changes in vegetation composition and structure under a subtropical climate. J Am Water Resour Assoc. 2008;44(5):1164–75.
Syvitski JPM, Milliman JD. Geology, geography, and human's battle for dominance over the delivery of fluvial sediment to the coastal ocean. J Geol. 2007;115(1):1–19.
Tobler WR. A computer movie simulating urban growth in the Detroit region. Econ Geogr. 1970;46(2):234–40.
Voinov A, Costanza R, Wainger L, Boumans R, Villa F, Maxwell T, Voinov H. Patuxent landscape model: integrated ecological economic modeling of a watershed. Environ Model Softw. 1999;14(5):473–91.
Wang H, Liu G, Wang Q. Effect of vegetation as whole in watershed on protecting soil from erosion and dissection of its landscape structure. J Soil Water Conserv. 2000;14(5):73–77, 97. (in Chinese).
Wang G, Qian J, Cheng G. Current situation and prospect of the ecological hydrology. Adv Earth Sci. 2001;16(3):314–23 (in Chinese).
Wang S, Zhang Z, Sun G, Zhang M, Yu X. Effects of land use change on hydrological dynamics at watershed scale in the Loess Plateau—A case study in the Luergou watershed, Gansu Province. J Beijing For Univ. 2006;28(1):48–54 (in Chinese).
Wang L, Xu Y, Fu B, Lu Y. Landscape pattern and eco-hydrological process. Adv Earth Sci. 2009a;24(11):1238–46 (in Chinese).
Wang Y, Fu B, Lü Y, Song J, Luan Y. Local-scale spatial variability of soil organic carbon and its stock in the hilly area of the Loess Plateau China. Quatern Res. 2009b;73(1):70–6.
Wei W, Chen LD, Fu BJ, Huang ZL, Wu DP, Gui LD. The effect of land uses and rainfall regimes on runoff and soil erosion in the semi-arid loess hilly area, China. J Hydrol. 2007;335(3–4):247–58.
Wilcox BP, Breshears DD, Allen CD. Ecohydrology of a resource-conserving semiarid woodland: effects of scale and disturbance. Ecol Monogr. 2003;73(2):223–39.
Wolfram S. Cellular automata as models of complexity. Nature. 1984;311(5985):419–24.
Wu FL. Calibration of stochastic cellular automata: the application to rural urban land conversions. Int J Geog Inf Sci. 2002;16(8):795–818.
Wu J. Landscape ecology: pattern, process, scale and hierarchy. 2nd ed. Beijing: Higher Education Press; 2007. p. 1–23 (in Chinese).
Wu J, David JL. A spatially explicit hierarchical approach to modeling complex ecological systems: theory and applications. Ecol Model. 2002;153(1/2):7–26.
Wu J, Hobbs R. Key issues and research priorities in landscape ecology: an idiosyncratic synthesis. Landscape Ecol. 2002;17(4):355–65.
Xu X, Gao Q, Liu YH, Wang JA, Zhang Y. Coupling a land use model and an ecosystem model for a crop-pasture zone. Ecol Model. 2009;220(19):2503–11.
Xu Y, Fu B, Lü Y. Research on landscape pattern and ecological processes based on landscape models. Acta Ecologica Sinica. 2010;30(1):212–20 (in Chinese).
Zhang J, Chu Z, Ge Y, Zhou X, Jiang H, Chang J, Peng C, Zheng J, Jiang B, Zhu J, Yu S. TRIPLEX model testing and application for predicting forest growth and biomass production in the subtropical forest zone of China's Zhejiang Province. Ecol Model. 2008;219(3–4): 264–75.
Zhang Z, Wang S, Sun G. Runoff and sediment yield response to vegetation change at multiple scales: a review. Acta Ecologica Sinica. 2006;26(7):2356–64 (in Chinese).
Zonneveld IS. The land unit-a fundamental concept in landscape ecology, and its applications. Landscape Ecol. 1989;3(2):67–86.
Zornoza R, Mataix-Solera J, Guerrero C, Arcenegui V, Mayoral AM, Morales J, Mataix-Beneyto J. Soil properties under natural forest in the Alicante province of Spain. Geoderma. 2007;142(3–4):334–41.

Chapter 2
Strengthening Landscape Ecology's Contribution to a Sustainable Environment

Clive A. McAlpine, Leonie M. Seabrook, Tiffany H. Morrison and Jonathan R. Rhodes

Abstract The need to avert unacceptable and irreversible environmental change is one of the most urgent challenges facing society. Landscape ecology has the capacity to help address these challenges by providing spatially-explicit solutions to landscape sustainability problems. However, despite a large body of research, the real impact of landscape ecology on sustainable landscape management and planning is still limited. This essay identifies four key areas where landscape ecology can strengthen its contribution to achieving a sustainable environment. These are: recognising the growing complexity of landscape sustainability problems; adopting a formal problem-solving framework to landscape sustainability problems; helping to bridge the implementation gap between science and practice; and developing stronger links between landscape ecology and restoration ecology.

Keywords Communities · Complexity · Decision analysis · Economic constraints · Landscape sustainability problems · Risk · Uncertainty · Institutional design

2.1 Introduction

One of the most urgent challenges facing society is to avert unacceptable and irreversible environmental change arising from unsustainable land use and climate change (Rockström et al. 2009; Wijkman and Rockstrom 2012; Wiens 2012). Managers of human-modified landscapes face a large number of interrelated environmental problems stemming from a long history of cumulative land use and land cover changes, including land degradation, loss of habitat and biodiversity decline (Lindenmayer et al. 2008). Twenty-five percent of Earth's land resources

C. A. McAlpine (✉) · L. M. Seabrook · T. H. Morrison · J. R. Rhodes
School of Geography, Planning and Environmental Management,
The University of Queensland, Brisbane, QLD 4072, Australia
e-mail: c.mcalpine@uq.edu.au

B. Fu and K. B. Jones (eds.), *Landscape Ecology for Sustainable Environment and Culture*, DOI: 10.1007/978-94-007-6530-6_2,

are highly degraded, 8 % are moderately degraded, 36 % are stable or slightly degraded, and only 10 % are ranked as "improving" (Nachtergaele et al. 2011). An increasing proportion of the world's biodiversity is listed as threatened (Pimm et al. 2006; IUCN Red List). Addressing these challenges requires the development of sustainable landscape management and planning strategies that help solve current and anticipated environmental problems in a proactive, comprehensive and cost-effective manner (Wu 2006; McAlpine et al. 2010).

In recent decades, landscape ecology has made considerable progress in understanding the linkages between landscape structure, function and change, particularly for managing and designing landscapes for conservation (Lindenmayer and Hobbs 2007). Landscape sustainability is becoming an increasingly important focus for landscape ecology (Musacchio 2011; Wu 2012). It is a difficult concept to define, but can be considered in terms of a landscape's adaptive capabilities to cope with uncertainties rather than the maintenance of a landscape in a static state (Wu 2012). Ecologically, landscape sustainability requires avoiding irreversible change through careful management and continual ecological improvement (Fischer et al. 2007; Mac Nally 2007). However, ultimately, the sustainability paradigm is a human centred concept about meeting the needs of people, now and in the future (Wiens 2012).

Landscape ecology has the potential to contribute significantly to landscape sustainability and there are some signs of integrated, solution-driven research (Wu 2006), but its real impact on sustainable landscape management and planning is still limited (Naveh 2007). Similar to much environmental research, there is a lack of policy uptake, and a lack of implementation of research outcomes from landscape ecology. This stems partly from the division of knowledge into narrow, specialist fields throughout the 20th century, and partly from the institutional and social constraints to achieving landscape sustainability goals in many regions of the world, especially those where development is poorly planned or unplanned.

There are three ways through which the contribution of landscape ecology to landscape sustainability can be strengthened: (1) by better integrating bio-ecological and humanistic perspectives in landscape ecology; (2) by adopting socio-ecological thinking that focuses on the multiple functions of landscapes and the multiple actors involved in their construction; and (3) by involvement in and adoption of principles of adaptive management to deal with the complex and uncertain responses of landscapes to changing conditions (McAlpine et al. 2010). In essence, landscape ecology needs to increase involvement in and knowledge exchange between the bio-sciences that are the main focus of landscape ecology and the human-oriented decision sciences that are the main focus of land planning (Vos et al. 2007; Termorshuizen and Opdam 2009). An example of this is the pattern-process-design paradigm proposed by Nassauer and Opdam (2008) that actively links landscape science and landscape planning to achieve vitally important environmental and social outcomes. This new paradigm aims to improve the impact of landscape science in society and enhance the saliency and legitimacy of landscape ecological knowledge in addressing landscape sustainability problems (Nassauer and Opdam 2008).

This essay identifies four key areas where landscape ecology can strengthen its contribution to achieving a sustainable environment. These are: recognising the growing complexity of landscape sustainability problems; adopting a formal problem-solving framework to landscape sustainability problems; bridging the implementation gap between science and practice; and developing stronger links between landscape ecology and restoration ecology. We conclude with key principles through which landscape ecologists can contribute towards improved outcomes for landscape sustainability.

2.2 Recognise the Growing Complexity of Landscape Sustainability Problems

Complexity is at the core of landscape sustainability problems for two reasons.

First, incorporating complexity into sustainable landscape management is an essential step in attaining sustainable landscapes that are resistant and robust to future human and environmental disturbances (Parrott and Meyer 2012). There is a growing recognition that landscapes are complex social–ecological systems comprising a dynamic mosaic of land uses, and that the future of a landscape can be conceived in terms of ensembles of likely future system states, given particular management scenarios and external drivers (Parrott and Meyer 2012). The dynamics of landscapes are becoming inherently unpredictable as a result of nonlinear relationships, feedbacks between emergent patterns and processes and the components that created them, and in many cases constantly changing external drivers or boundary conditions (Ryan et al. 2007, 2010). The unpredictable impact of climate change is an example of changing boundary conditions (Solomon et al. 2009). However, recognising and managing for the potential of multiple metastable states in landscapes presents a major challenge. An important part of this challenge is buffering landscapes against the strong interaction pressures of global environmental change, both land use and climate. Another part lies in what is deemed as 'desirable' in terms of landscape components and what functions we wish to maintain or restore (Ryan et al. 2007, 2010). Much more work is necessary to fully incorporate the science of complexity into sustainable landscape management, planning and policy.

Second, landscape sustainability problems are becoming increasingly complicated, even complex, and difficult to predict (see McAlpine et al. 2010 for a typology of landscape sustainability problems). "Complicated" contexts to landscape sustainability problems have no clear cause-effect relationship and high uncertainty with no single right answer. Here, a problem-solving process requires an evaluation of multiple options, each with recognisable economic and social trade-offs. It requires an inter-disciplinary or trans-disciplinary approach where the input from planners and stakeholders is essential.

"Complex" contexts to landscape sustainability problems are synonymous with systems thinking and trans-disciplinary research, and involve problems that are inherently "wicked" with no clear solution or resolution (Brown et al. 2010). They are unpredictable with emergent behaviour, no clear cause-and-effect relationships, no right answers, high uncertainty, and they are riddled with ambiguity, dilemmas and hard choices (Snowden and Boone 2007). Complex problems often result from over exploitation of natural resources and governance failure (Jentoft and Chuenpagdee 2009), but increasingly include novel drivers arising from climate change, and their cumulative impact on landscape pattern and processes. Inherently, decision-making approaches to resolving complex problems are participatory, trans-disciplinary, based on multiple models and support tools, and focused on the governance system.

2.3 Towards an Improved Problem-Solving Framework

We suggest that landscape ecology will be better able to provide the scientific basis for the design and planning of sustainable landscapes by becoming a more applied problem-solving science (sensu McAlpine et al. 2010), which searches for real, integrated solutions to landscape sustainability problems (c.f. Costanza 2009). The framework we propose explicitly draws on the decision and planning sciences and this allows a formal integration of the different problem-solving components (see Fig. 2.1). Earlier problem solving frameworks have been criticised for being: (a) elitist (highly expert orientated with limited public participation and recognition of non-expert knowledge); (b) overly dependent on the capacity of the steering group/technology/facilitator; and (c) politically simplistic in assuming that it is possible to produce an enduring consensus which will be easy to implement (Staes et al. 2008). We believe that the modification of the problem-solving framework by landscape ecologists and planning scientists will facilitate greater disciplinary integration because it requires explicit acknowledgement of the multiple facets of each problem and the involvement of multiple stakeholders.

The framework has seven stages, which are: (1) identify and contextualise the problem; (2) set agreed objectives and management actions; (3) conduct data analysis and integration; (4) understand risks and uncertainties; (5) conduct objective and participatory decision analysis; (6) apply landscape management and planning actions; and (7) implement monitoring and adaptive management programs. While the stages take place sequentially, the implementation process can be iterative, involving modification of early stages (e.g., management objectives) based on feedback from subsequent stages (e.g., data analysis and integration) of the problem-solving process. Furthermore, stakeholder participation via an independent facilitator occurs throughout the process, for the purposes of both legitimacy (stakeholder buy-in) and efficiency (stakeholder knowledge) (Friedmann 1987).

An important component of an applied problem-solving approach is to bring the social and institutional dimensions of the problem to the fore. This is done by

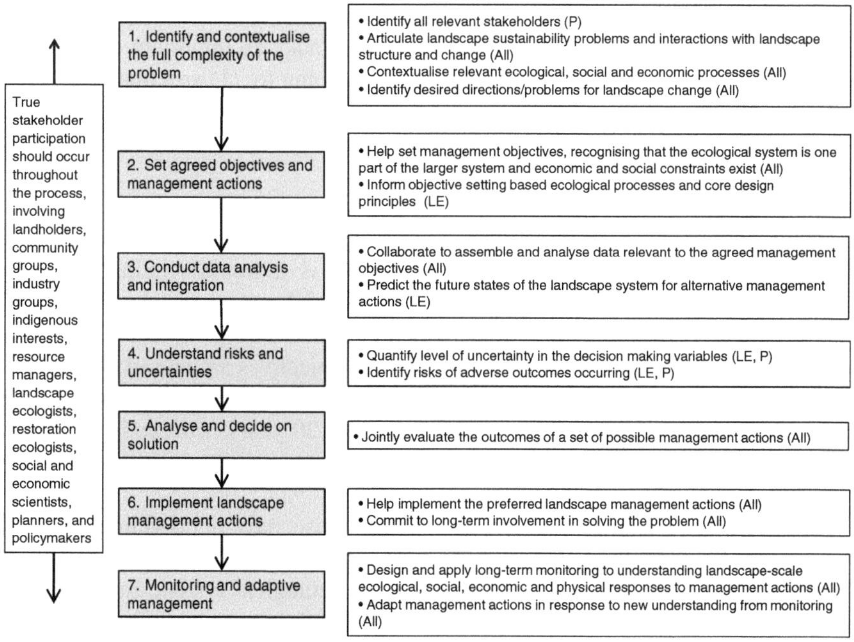

Fig. 2.1 A conceptual framework for strengthening landscape ecology's contribution to a sustainable environment. The framework is divided into seven stages of the problem-solving process. The key contributions of landscape ecologists (*LE*), planners (*P*) and all stakeholders (All) to each stage are highlighted (Modified from McAlpine et al. 2010)

incorporating stakeholder participation and institutional design considerations throughout the analytic process (Morrison 2006). There is evidence that explicitly considering the socio-institutional context compels landscape scientists and managers to think about design and implementation. This evidence comes from collaborative environmental planning and common property resource management, where it has resulted in established principles for considering stakeholder participation and institutional design factors in implementation research (Sabatier 1986; Ostrom 1990; Lane and McDonald 2005; Heikkila et al. 2011; Schmidt and Morrison 2012).

The seven stages of the framework are outlined below.

Stage 1: *Identify and conceptualise the problem.* The first stage is to identify and conceptualise the problem. Precisely defining the problem is often the most difficult but important step in the problem-solving process (Possingham and Nicholson 2007), partly because problem identification is a complex social construct, involving the aspirations of, and constraints on, the various stakeholder groups which, by necessity, brings together the social, economic and ecological sciences (Bartuska 1999). In most cases, the 'problem' will actually comprise a set of related problems, which represent the perspectives and interests of different stakeholders in a given landscape. Stakeholder participation starts at this early stage (Friedmann 1987). The involvement of these different parties during the problem identification process allows recognition of the wider issues (Crosby et al. 2000; Vos et al. 2007). Landscape ecologists can make an important contribution to this stage of the problem-solving process by helping to: (1) articulate landscape sustainability problems and their interactions with landscape structure and change; (2) identify and draft desired directions and visions for landscape change; and (3) anticipate future problems.

Stage 2: *Set agreed objectives and management actions.* The second stage in the problem-solving process is to set agreed landscape sustainability objectives and possible management actions. Objective setting is a transdisciplinary activity where landscape ecologists need to work with social and economic researchers, and stakeholders, to ensure the objectives are quantifiable, able to be prioritised and achievable. This is critical as poorly defined objectives are less likely to result in effective or measurable outcomes (Knight et al. 2006). Objectives need to be set within a medium-long time horizon, normally 15–20 years, with endpoints sufficiently ambitious so as to inspire innovative solutions to sustainability problems (Fischer et al. 2007).

Although this is an area where landscape ecologists have not traditionally had a strong involvement, they in fact have an important supportive role in setting objectives for the sustainable functioning of the ecological system. In working to ensure that objectives are achieved in a cost-effective, timely and socially acceptable manner, landscape ecologists need to recognise that the ecological system, while vital for landscape sustainability, is one part of a larger system that includes social, economic and institutional components. By adopting this approach, landscape ecologists will need to engage with other disciplines and a diverse range of stakeholders in transdisciplinary research.

Stage 3: *Re-conceptualize the problem and conduct data analysis and integration.* The aim of this stage is to re-conceptualise the problem and predict the likely outcomes and uncertainties of the set of possible management actions within the specified time frame in terms of the management objectives. Landscape ecologists are well equipped to make an important contribution to this stage of the problem-solving process by conceptualising processes operating at different scales relevant to the problem, providing landscape-scale data, models and analyses relevant to the specified management objectives and actions. Critical considerations include: (1) what are the relevant system components, processes, dynamics and scales/boundaries of the studied system; (2) what is the current level of understanding of these processes and what data (empirical and expert) and models are available to represent and predict these processes; (3) how can process models be better integrated with socio-economic models; and (4) what is the level of uncertainty of these data and model predictions. These analyses need to go beyond ecological assessments and inventories associated with landscape planning (Golley and Bellot 1999; Steinitz et al. 2003) by providing a more detailed understanding and prediction of landscape processes, functions and dynamics, and by including human activities as components of the landscape.

Landscape ecologists need to work with other discipline areas to design the methodology and assemble and analyse data relevant to these processes. This requires teams of researchers from multiple disciplines and relevant members of the community to be involved in the analysis of the possible socio-economic interactions and constraints among the agreed set of objectives and management actions (Tress et al. 2005a, b, 2007; Fry et al. 2007). This necessitates ongoing dialogue amongst the research team, stakeholder groups and the wider community. However, it needs to be acknowledged here that true socio-economic and ecological data integration may not be possible or necessary (Brown and Pullar 2011).

Stage 4: *Understand risks and uncertainties.* Risks and uncertainties are inherent in landscape sustainability problems and need to be acknowledged and incorporated into any decision making framework. Risk is the chance that an adverse outcome occurs (Burgman et al. 1993), while uncertainty arises from an imperfect understanding of a system (Regan et al. 2002). An important role for landscape ecologists is to integrate the quantification of uncertainty into decision making processes, while acknowledging that definitions of risk and uncertainty vary across stakeholders and experts (Stirling 1998; Wynne 1992).

Landscape ecologists have focussed considerable effort on developing methods for understanding and quantifying risk and uncertainty in landscape patterns and processes (See Sheppard Chap. 7, this volume; also Mooij and DeAngelis 2003; Shao and Wu 2008), but specific links to the implications for decision making and landscape planning are often not made. Consequently, landscape ecologists need to more fully integrate methods for quantifying uncertainty into the problem-solving processes. Much of the expertise for quantitatively dealing with uncertainty in decision making lies in fields of mathematics and the decision sciences (e.g., Stewart 2005; Ben-Haim 2006), and for qualitatively dealing with uncertainty lies in the planning and policy sciences (e.g., Renn et al. 1993; Innes 1996; Smith and Wales 1999). Therefore, this will require considerable collaboration among landscape ecologists and researchers in other fields.

Stage 5: *Decision analysis.* Decision analysis is a process by which the outcome of a set of management actions are evaluated in terms of explicit pre-defined management objectives and constraints (Clemen 1996). Its key strength is that it provides a formal process for bringing together the components described above in a coherent fashion. Decision analysis requires the formal linking of specific management objectives, management actions, system understanding and uncertainty within a coherent framework. Importantly, by formally linking these factors, decision analysis promotes the integration of the different components of a decision problem. Therefore, by embracing decision analysis within a formal problem-solving framework, landscape ecologists are better positioned to move towards a much more integrated science. Specific attention to the governance of the decision process also guards against vested interests, enhances ownership of the problem and ensures the incorporation of a broad range of socio-economic and environmental values (Van Driesche and Lane 2002; Rayner and Howlett 2009).

Stage 6: *Implementing landscape management and planning actions.* This stage of the framework involves implementing the preferred landscape management actions identified and evaluated in the above stages. Traditionally, landscape ecologists have had little involvement with this stage of the problem-solving process. Once the understanding of the information gleaned from the previous stages has been communicated to the stakeholders, the implementation process normally becomes the realm of the non-scientific participants. However, landscape ecologists should contribute to the design of the final plan where parts of the landscape could be viewed as long-term landscape experiments. By engaging in this stage of the problem-solving process, landscape ecologists continue to interact with policy makers, resource managers and the public, thereby helping to ensure the relevance and results of the work (Spies et al. 2002). The longer landscape ecologists can stay involved in the plans during and after their implementation, the more likely it is that the plans will take on a more adaptive form and evolve toward practices that can approach long-term sustainability (McAlpine et al. 2007).

Stage 7: Monitoring and adaptive management. This final stage involves the design and implementation of an adaptive management and monitoring protocol (Holling 1978; Walters 1986). Linking monitoring and adaptive management allows plans to be adjusted over time.

The aim of monitoring and adaptive management is to learn about landscape processes by monitoring the consequences of management actions. This then feeds back into future decision-making processes. Long-term and adaptive monitoring is required to understand adequately landscape-scale ecological, social and physical systems and their responses to management actions. Its specific application is context-dependent and will vary with the problem to be resolved (Lindenmayer and Likens 2009). Monitoring programs developed through placed-based collaborative partnerships between scientists, landscape managers and policy makers can help lead to the resolution of important environmental problems, the identification of new problems (Lindenmayer and Likens 2009) and acceptance of inevitable change (e.g., climate change, Spies et al. 2010). Monitoring of management actions must include impacts on social, physical and ecological systems and not just one of these components (Redman et al. 2004).

2.4 Address the Implementation Gap

Despite some successes in areas concerned with sustainability issues, there is a growing appreciation of our inability to tackle major sustainability issues such as biodiversity loss and climate change. The gap between research and implementation is a fundamental problem in all ecological and environmental sciences, and calls for improved integration and implementation are widespread and diverse (e.g., Bammer 2005; Knight et al. 2008). Bammer (2005) identifies three pillars for an evidence-based approach to improving integration and implementation: (a) systems thinking and complexity science, which orient us to looking at the whole and its relationship to the parts of an issue; (b) participatory methods, which recognize that all the stakeholders have a contribution to make in understanding and, often, decision making about an issue; and (c) knowledge management, exchange, and implementation, which includes a better understanding of how decisions are made, and how actions are and can be influenced by scientific evidence.

We argue that landscape ecology also suffers from a lack of effective implementation (see Naveh 2007; Wu 2012). We agree with Nassauer and Opdam (2008) that landscape design can create collaboration between scientists and practitioners and improve the impact of landscape science, and that landscape planning is the appropriate process for implementation. However, while the principles of landscape design are mostly well established (see Lindenmayer and Hobbs 2007), the enactment these principles remain a major challenge (Haila 2007). Hard choices and decisions need to be confronted, as making do with impoverished, low-productivity parts of landscapes will probably doom many landscape designs to failure (Mac Nally 2007). These choices are as much a political and social challenge as they are a scientific challenge. Most dysfunctional landscapes have resulted from poor institutional arrangements and landscape governance, and unsustainable societal values. Ultimately, solutions require reforming governance arrangements at all levels and transforming societal values (Fischer et al. 2012; Swaffield 2012).

There are several core practical approaches where landscape ecology and landscape ecologists can help bridge the implementation gap (see Bammer (2005) for a detailed discussion). One essential approach is through engagement and direct interaction among researchers and stakeholders, including policy makers (van Kerkhoff and Lebel 2006; Gibbons et al. 2008). Knowledge exchange between different research disciplines, planners, land managers and other stakeholders builds trust, establishes lines of communication, and allows the identification of common goals. Clear communication of the beneficial outcomes of landscape sustainability is essential, as the costs of implementing actions are often high and may take many decades before the benefits are realised. For instance, a meta-analysis of restoration projects found that only those which are seen to be of direct benefit to people are likely to be funded and supported in the long run (Aronson et al. 2010). However, active participation in implementation requires the involvement of all stakeholders, including landscape ecologists, which can be difficult and time consuming (Lang et al. 2012).

2.5 Strengthen Links Between Landscape Ecology and Restoration Ecology

The ultimate issue in managing landscape sustainability is to protect what works since it is difficult and expensive to replace or repair (Ehrlich 2007). However, given the current degraded state of many of the world's landscapes, ecosystem restoration is one of the most proactive approaches for reversing degradation and biodiversity loss (Hobbs et al. 2006; Tongway and Ludwig 2010). Ecological restoration is the process of assisting the recovery of an ecosystem that has been degraded, damaged, or destroyed, and is emerging as an important environmental discipline because it can provide potential solutions to real-world landscape sustainability problems. As a discipline, restoration ecology has experienced rapid growth in recent years. It assumes that many degrading stressors are temporary and that it is possible to change a damaged ecosystem to a state that is within an acceptable limit relative to a less-disturbed ecosystem (Young 2000; Palmer et al. 2006; Hobbs et al. 2009). Restoration activities include reinstating historical assemblages of plants and animals, and enhancing ecosystem functions/services such as retaining water, energy, or nutrients (Falk et al. 2006; Tongway and Ludwig 2010).

According to Bell et al. (1997), restoration and landscape ecology have an "unexplored mutualistic relationship that could enhance research and application of both disciplines". A landscape approach recognises that the whole is more than the sum of the parts, and may assist in addressing spatial and temporal prioritisation issues related to practical constraints in restoration actions (Possingham and Nicholson 2007). Restoration can also benefit landscape ecology by providing information derived from restoration ecology projects to test basic questions, especially those linked to landscape structure and function (Bell et al. 1997).

Systematic landscape restoration or landscape reconstruction is designed to reverse the adverse effects of habitat loss, fragmentation and degradation. It is critical for achieving cost-effective environmental outcomes, which balance the many conflicts of interest and competing demands for land with the need to restore landscapes for the protection of biodiversity (Crossman et al. 2007). Landscape reconstruction involves integrating a portfolio of passive and active restoration actions within a spatial framework to achieve landscape synergies between patches often with differing restoration objectives, with the ultimate ecological goal being to restore the structure, function and native biodiversity (both plant and animals) of degraded landscapes (Vesk and Mac Nally 2006). Similar to the problem-solving framework outlined above, systematic landscape restoration involves clearly defining the problem, recognising the complex makeup of landscapes and the socio-economic interests of the inhabitants, and establishing a multi-disciplinary team to work together with stakeholders to find solutions to restoration problems (Crossman et al. 2010). Ultimately, landscape reconstruction is dependent on successful restoration actions at the site-scale. The reconstruction of whole landscapes is a prerequisite for recovering threatened or declining animal populations, but the design may be species-dependent. This requires consideration of context attributes such as remnant landscape patches and riparian areas when deciding where to restore vegetation because proximity to these elements can affect restoration outcomes at the site-scale (Grimbacher and Catterall 2007).

2.6 Key Principles

In summary, if landscape ecology is to strengthen its contribution to the urgent problems hindering landscape sustainability, there are certain key principles that researchers should consider.

1. Recognise that the cause of most sustainability problems are global and national, but the solutions are regional and local.
2. Recognise the complexity of landscapes and landscape sustainability problems, arising as they do from ecological, social, economic, and institutional drivers, and incorporate complexity into the design and restoration process so that landscapes are resistant and robust to future human and environmental disturbances.
3. Protect what currently works since it is difficult and expensive to replace or repair.
4. Set goals that are sufficiently ambitious so as to inspire innovative solutions to sustainability problems.
5. Stay involved in the management plans during and after their implementation, as this increases the likelihood that the plans will take on a more adaptive form and evolve toward practices that can approach long-term sustainability.
6. Recognize and encourage the active participation of all the stakeholders in understanding and decision making to develop solutions to landscape sustainability problems.
7. Actively engage in systematic landscape restoration/reconstruction to reverse the adverse effects of habitat loss, fragmentation and degradation.

Acknowledgments Leonie Seabrook is funded through the National Environmental Research Program, Australian Department of Sustainability, Environment, Water, Population and Communities. Clive McAlpine is supported by the Australian Research Council FT100100338.

References

Aronson J, Blignaut JN, Milton SJ, Le Maitre D, Esler KJ, Limouzin A, Fontaine C, de Wit MP, Mugido W, Prinsloo P, van der Elst L, Lederer N. Are socioeconomic benefits of restoration adequately quantified? A meta-analysis of recent papers (2000–2008) in Restoration Ecology and 12 other scientific journals. Restor Ecol. 2010;18:143–54.

Bammer G. Integration and implementation sciences: building a new specialization. Ecol Soc. 2005;10:6.

Bartuska AM. Cross-boundary issues to manage for healthy forest ecosystems. In: Klopatek JM, Gardner RH, editors. Landscape ecological analysis: issues and applications. New York: Springer; 1999.

Bell SS, Fonseca MS, Motten LB. Linking restoration and landscape ecology. Restor Ecol. 1997;5:318–23.

Ben-Haim Y. Estimating an uncertain probability density. In: Lawry J, Miranda E, Bugarin A, Li S, Gil MA, Grzegorzewski P, Hyrniewicz O, editors. Soft methods for integrated uncertainty modelling. Heidelberg: Springer; 2006.

Brown VA, Harris JA, Russell JY. Tackling wicked problems through the transdisciplinary imagination. London: Earthscan; 2010.

Brown G, Pullar D. An evaluation of the use of points versus polygons in public participation geographic information systems (PPGIS) using quasi-experimental design and Monte Carlo simulation. Int J of Geogr Inf Sci. 2011;26:231–46.

Burgman MA, Ferson S, Akcakaya HR. Risk assessment in conservation biology. London: Chapman & Hall; 1993.
Clemen RT. Making hard decisions: an introduction to decision analysis. Pacific Grove: Duxbury Press; 1996.
Costanza R. The search for real, integrative solutions. Solutions. 2009;1:1.
Crosby MP, Abu-Hilal A, Al-Homoud A, Erez J, Ortal R. Interactions among scientists, managers and the public in defining research priorities and management strategies for marine and coastal resources: is the Red Sea marine peace park a new paradigm? Water Air Soil Poll. 2000;123:581–94.
Crossman ND, Bryan BA, Ostendorf B, Collins S. Systematic landscape restoration in the rural-urban fringe: meeting conservation planning and policy goals. Biodivers Conserv. 2007;16:3781–802.
Crossman ND, Connon JD, Bryan BA, Summers DM, Ginnivan J. Reconfiguring an irrigation landscape to improve provision of ecosystem services. Ecol Econ. 2010;69:1031–42
Ehrlich PR. Landscapes in Peril. Managing and designing landscapes for conservation: moving from perspectives to principles. Carlton: Blackwell Publishing; 2007.
Falk DA, Palmer MA, Zedler JB. Foundations of restoration ecology. Washington: Island Press; 2006.
Fischer J, Manning AD, Steffen W, Rose DB, Daniell K, Felton A, Garnett S, Gilna B, Heinsohn R, Lindenmayer DB, MacDonald B, Mills F, Newell B, Reid J, Robin L, Sherren K, Wade A. Mind the sustainability gap. Trends Ecol Evol. 2007;22:621–4.
Fischer J, Dyball R, Fazey I, Gross C, Dovers S, Ehrlich PR, Burlle RJ, Christensen C, Borden RJ. Human behaviour and sustainability. Front Ecol Environ. 2012;10:153–60.
Friedmann J. Planning in the public domain: from knowledge to action. Princeton: Princeton University Press; 1987.
Fry G, Tress B, Tress G. Integrative landscape research: facts and challenges. In: Wu J, Hobbs RJ, editors. Key topics and landscape ecology. New York: Cambridge University Press; 2007.
Gibbons P, Lindenmayer DB, Fischer J, Manning AD, Weinberg A, Seddon J, Ryan P, Barrett G. The future of scatters trees in agricultural landscapes. Conserv Biol. 2008;22:1309–19.
Golley FB, Bellot J. Rural planning from an environmental systems perspective. New York: Springer; 1999.
Grimbacher PS, Catterall CP. How much do site age, habitat structure and spatial isolation influence the restoration of rainforest beetle species assemblages? Biol Conserv. 2007;135:107–18.
Haila Y. Enacting landscape design: from specific cases to general principles. In: Lindenmayer DB, Hobbs RJ, editors. Managing and designing landscapes for conservation: moving from perspectives to principles. Carlton: Blackwell Publishing; 2007.
Heikkila T, Schlager E, Davis MW. The role of cross-scale institutional linkages in common pool resource management: assessing interstate river compacts. Policy Stud J. 2011;39:121–45.
Hobbs RJ, Arico S, Aronson J, Baron JS, Bridgewater P, Cramer VA, Epstein PR, Ewel JJ, Klink CA, Lugo AE, Norton D, Ojima D, Richardson DM, Sanderson EW, Valladares F, Vil M, Zamora R, Zobel M, et al. Novel ecosystems: theoretical and management aspects of the new ecological world order. Glob Ecol Biogeog. 2006;15:1–7.
Hobbs RJ, Higgs E, Harris JA. Novel ecosystems: implications for conservation and restoration. Trends Ecol Evol. 2009;24:599–605.
Holling CS. Adaptive environmental assessment and management. New York: Wiley; 1978.
Innes JE. Planning through consensus building: a new view of the comprehensive planning ideal. J Am Plan Assoc. 1996;62:460–72.
Jentoft S, Chuenpagdee R. Fisheries and coastal governance as a wicked problem. Mar Policy. 2009;33:553–60.
Knight AT, Driver A, Cowling RM, Maze K, Desmet PG, Lowbard AT, Rouget M, Botha MA, Boshoff AF. Designing systematic conservation assessments that promote effective implementation: best practice from South Africa. Conserv Biol. 2006;20:739–50.

Knight AT, Cowling RM, Rouget M, Balmford A, Lombard AT, Campbell BM. Knowing but not doing: selecting priority conservation areas and the research-implementation gap. Conserv Biol. 2008;22:610–7.
Lane MB, McDonald G. Community-based environmental planning: operational dilemmas, planning principles and possible remedies. J Env Plan Manage. 2005;48:709–31.
Lang DJ, Wiek A, Bergmann M, Stauffacher M, Martens P, Moll P, Swilling M, Thomas CJ. Transdisciplinary research in sustainability science: practice, principles, and challenges. Sustain Sci. 2012;7:25–43.
Lindenmayer DB, Hobbs RJ. Managing and designing landscapes for conservation: moving from perspectives to principles. Carlton: Blackwell Publishing; 2007.
Lindenmayer D, Hobbs RJ, Montague-Drake R, Alexandra J, Bennett A, Burgman M, Cale P, Calhoun A, Cramer V, Cullen P, Driscoll D, Fahrig L, Fischer J, Franklin J, Haila Y, Hunter M, Gibbons P, Lake S, Luck G, MacGregor C, McIntyre S, Mac Nally R, Manning A, Miller J, Mooney H, Noss R, Possingham H, Saunders D, Schmiegelow F, Scott M, Simberloff D, Sisk T, Tabor G, Walker B, Wiens J, Woinarski J and Zavaleta E. A checklist for ecological management of landscapes for conservation. Ecol Lett 2008;11:78–91.
Lindenmayer DB, Likens GE. Adaptive monitoring: a new paradigm for long-term research and monitoring. Trends Ecol Evol. 2009;24:482–6.
Mac Nally R. Remnant geometry, landscape morphology, and principles and procedures for landscape design. In: Lindenmayer DB, Hobbs RJ, editors. Managing and designing landscapes for conservation: moving from perspectives to principles. Carlton: Blackwell Publishing; 2007.
McAlpine CA, Spies TA, Norman P, Peterson A. Conserving forest biodiversity across multiple land ownerships: lessons from the Northwest forest plan and the Southeast Queensland regional forests agreement (Australia). Biol Conserv. 2007;134:580–92.
McAlpine CA, Seabrook LM, Rhodes JR, Maron M, Smith C, Bowen ME, Butler SA, Powell O, Ryan JG, Fyfe CT, Adams-Hosking C, Smith A, Robertson O, Howes A, Cattarino L. Can a problem-solving approach strengthen landscape ecology's contribution to sustainable landscape planning? Landscape Ecol. 2010;25:1155–68.
Mooij WM, DeAngelis DL. Uncertainty in spatially explicit animal dispersal models. Ecol Appl. 2003;13:794–805.
Morrison TH. Pursuing rural sustainability at the regional level: key lessons from the literature on institutions, integration and the environment. J Plan Lit. 2006;21:143–52.
Musacchio LR. The grand challenge to operationalize landscape sustainability and the design-in-science paradigm. Landscape Ecol. 2011;26:1–5.
Nachtergaele F, Biancalani R, Petri M et al. Land degradation: SOLAW background thematic report 3. Rome: Food and Agriculture Organization of the United Nations; 2011.
Nassauer JI, Opdam P. Design in science: extending the landscape ecology paradigm. Landscape Ecol. 2008;23:633–44.
Naveh Z. Landscape ecology and sustainability. Landscape Ecol. 2007;22:1437–40.
Ostrom E. Governing the commons: the evolution of institutions for collective action. Cambridge: Cambridge University Press; 1990.
Palmer MA, Falk DA, Zedler JB. Ecological theory and restoration ecology. In: Falk DA, Palmer MA, Zedler JB, editors. Foundations of restoration ecology. Washington: Island Press; 2006.
Parrot L, Meyer WS. Future landscapes: managing within complexity. Front Ecol Environ. 2012;10:382–9.
Pimm S, Raven P, Peterson A, Sekercioglu CH, Ehrlich PR. Human impacts on the rates of recent, present, and future bird extinctions. Proc Natl Acad Sci USA. 2006;103:10941–6.
Possingham HP, Nicholson E. Principles of landscape design that emerge from a formal problem-solving approach. In: Hobbs RJ, Lindenmayer DB, editors. Managing and designing landscapes for conservation: moving from perspectives to principles. Oxford: Blackwell Publishing; 2007.
Rayner J, Howlett M. Implementing integrated land management in Western Canada: policy reform and the resilience of clientelism. J Nat Resour Policy Res. 2009;1:321–34.

Redman CL, Grove JM, Kuby LH. Integrating social science into the long-term ecological research (LTER) network: social dimensions of ecological change and ecological dimensions of social change. Ecosystems. 2004;7:161–71.
Regan HM, Colyvan M, Burgman MA. A taxonomy and treatment of uncertainty for ecology and conservation biology. Ecol Appl. 2002;12:618–28.
Renn O, Webler T, Rakel H, Dienel P, Johnson B. Public participation in decision making: a three-step procedure. Policy Sci. 1993;26:189–214.
Rockström J, Steffen W, Noone K, Persson A, Chapin FS III, Lambin EF, Lenton TM, Scheffer M, Folke C, Schellnhuber HJ, Nykvist B, de Wit CA, Hughes T, van der Leeuw S, Rodhe H, Sörlin S, Snyder PK, Costanza R, Svedin U, Falkenmark M, Karlberg L, Corell RW, Fabry VJ, Hansen J, Walker B, Liverman D, Richardson K, Crutzen P, Foley JA. A safe operating space for humanity. Nature. 2009;461:472–5.
Ryan JG, Ludwig JA, McAlpine CA. Complex adaptive landscapes (CAL): a conceptual framework of multi-functional, non-linear ecohydrological feedback systems. Ecol Complex. 2007;4:113–27.
Ryan JG, McAlpine CA, Ludwig JA. Integrated vegetation designs for enhancing water retention and recycling in agroecosystems. Landscape Ecol. 2010;25:1277–88.
Sabatier P. Top-down and bottom-up approaches to implementation research: a critical analysis and suggested synthesis. J Pub Policy. 1986;6:21–48.
Schmidt P, Morrison TH. Watershed management in an urban setting: process, scale and administration. Land Use Policy. 2012;29:45–52.
Shao G, Wu J. On the accuracy of landscape pattern analysis using remote sensing data. Landscape Ecol. 2008;23:1572–9761.
Smith G, Wales C. The theory and practice of citizens' Juries. Policy Polit. 1999;27:295–308.
Snowden DJ, Boone ME. A leader's framework for decision making. Harvard Bus Rev. 2007;85: 68–76.
Solomon S, Plattner G, Knutti R, Friedlingstein P, et al. Irreversible climate change due to carbon dioxide emissions. Proc Natl Acad Sci. 2009;106:1704–1709.
Spies TA, Reeves GH, Burnett KM, McComb WC, Johnson KN, Grant G, Ohmann JL, Garman SL, Bettinger P. Assessing the ecological consequences of forest policies in a multi-ownership province in Oregon. In: Liu J, Taylor WW, editors. Integrating landscape ecology into natural resource management. Cambridge: Cambridge University Press; 2002.
Spies TA, Giesen TW, Swanson FJ, Franklin JF, Lach D, Johnson KN. Climate change adaptation strategies for federal forests of the Pacific Northwest, USA: ecological policy, and socio-economic perspectives. Landscape Ecol. 2010;25:1185–99.
Staes J, Meire P, et al. Participation aspects in the realisation of the Nete river basin management plan: methodology and application. In: Meire P, Coenen M, Lombardo C, Robba M, Sacile R, editors. Integrated water management. Netherlands: Springer; 2008.
Steinitz C, Rojo HMA, Bassett S, Flaxman M, Goode T, Maddock T III, Mouat D, Peiser R, Shearer A. Alternative futures for changing landscapes: the upper San Pedro river basin in Arizona and Sonora. Washington: Island Press; 2003.
Stewart TJ. Goal programming and cognitive biases in decision making. J Oper Res Soc. 2005;56:1166–75.
Stirling A. Risk at a turning point? J Risk Res. 1998;1:97–109.
Swaffield S. Empowering landscape ecology-connecting science to governance through design values. Landscape Ecol. 2012;1–9. doi: 10.1007/s10980-012-9765-9
Termorshuizen J, Opdam P. Landscape services as a bridge between landscape ecology and sustainable development. Landscape Ecol. 2009;24:1037–52.
Tongway DJ, Ludwig JA. Restoring disturbed landscapes: putting principles into practice. Washington: Island Press; 2010.
Tress B, Tress G, Fry G. Defining concepts and process of knowledge production in integrative research. In: Tress B, Tress G, Fry G, Opdam P, editors. From landscape research to landscape planning: aspects of integration, education and application. Heidelberg: Springer; 2005a.

Tress G, Tress B, Fry G. Clarifying integrative research concepts in landscape ecology. Landscape Ecol. 2005b;20:479–93.
Tress G, Tress B, Fry G. Analysis of the barriers to integration in landscape research projects. Land Use Policy. 2007;24:374–85.
Van Driesche J, Lane M. Conservation through conversation: collaborative planning for reuse of a former military property in Sauk County, Wisconsin, USA. Plan Theory Pract. 2002;3: 133–53.
van Kerkhoff L, Lebel L. Linking knowledge and action for sustainable development. Annu Rev Env Resour. 2006;31:445–77.
Vesk P and Mac Nally R. The clock is ticking - revegetation and habitat for birds and arboreal mammals in rural landscapes of southern Australia. Agric Ecosyst & Environ. 2006;112:356–66
Vos C, Opdam P, Steingröver E, Reijnen R. Transferring ecological knowledge to landscape planning: a design method for robust corridors. In: Wu J, Hobbs R, editors. Key topics in landscape ecology. Cambridge: Cambridge University Press; 2007.
Walters C. Adaptive management of renewable resources. Caldwell: Blackburn Press; 1986.
Wiens JA. Is landscape sustainability a useful concept in a changing world? Landscape Ecol. 2012;1–6. doi: 10.1007/s10980-012-9801-9
Wijkman A, Rockstrom J. Bankrupting nature: denying our planetary boundaries. London: Routledge; 2012.
Wu JG. Landscape ecology, cross-disciplinarity, and sustainability science. Landscape Ecol. 2006;21:1–4.
Wu J. A landscape approach for sustainability science. Sustainability science: the emerging paradigm and the urban environment. New York: Springer; 2012.
Wynne B. Uncertainty and environmental learning: reconceiving science and policy in the preventive paradigm. Glob Environ Change. 1992;2:111–27.
Young TP. Restoration ecology and conservation biology. Biol Conserv. 2000;92:73–83.

Chapter 3
Urban Landscape Ecology: Past, Present, and Future

Jianguo Wu, Chunyang He, Ganlin Huang and Deyong Yu

Abstract Cities are home to more than half of the world population. Cities have been the centers of economic and social developments, as well as sources of many major environmental problems. Cities are created and maintained by the most intense form of human-nature interactions. Cities are spatially extended, complex adaptive systems—which we call landscapes. The future of humanity will increasingly rely on cities, and the future of landscape ecology will inevitably be more urban. To meet the grand challenge of our time—sustainability—cities must be made sustainable and, to this end, landscape ecology has much to offer. In this chapter, we discuss the intellectual roots and recent development of urban landscape ecology and propose a framework for helping move it forward. This framework integrates perspectives and approaches from landscape ecology, urban ecology, sustainability science, and resilience theory.

Keywords Urban landscape ecology · Urban ecology · Urbanization · Urban ecosystem services · Urban sustainability · Research framework

J. Wu (✉)
School of Life Sciences and Global Institute of Sustainability,
Arizona State University, PO Box 874501Tempe AZ 85287, USA
e-mail: Jingle.Wu@asu.edu
URL: http://LEML.asu.edu/Jingle/

J. Wu · C. He · G. Huang · D. Yu
Center for Human-Environment System Sustainability (CHESS),
State Key Laboratory of Earth Surface Processes and Resource Ecology (ESPRE),
Beijing Normal University, Beijing 100875, China

B. Fu and K. B. Jones (eds.), *Landscape Ecology for Sustainable Environment and Culture*, DOI: 10.1007/978-94-007-6530-6_3,

3.1 Introduction

Urbanization has been a dominant driving force for global environmental changes and socioeconomic transformations across the world since the Industrial Revolution between 1750 and 1850 (Grimm et al. 2008; Wu 2008a, 2010b). This is especially true during the past several decades, with the rapid development of new cities and expansion of old ones in both developed and developing countries. For example, Beijing, one of the oldest cities in the world, and Shenzhen, one of the fastest growing young cities in China, both experienced rapid urban expansion during the past several decades (He et al. 2006, 2008, 2011; Yu et al. 2009); Phoenix—home to the Central Arizona-Phoenix Long-Term Ecological Research Project (CAPLTER)—is a relatively young, but the fastest growing, city in the US (Wu et al. 2011a, b; Grimm et al. 2012); Baltimore—home to the Baltimore Long-Term Ecological Research Project (Baltimore LTER)—is an old port city which also has gone through a profound landscape transformation since 1914 (Zhou et al. 2011) (Fig. 3.1).

As of 2008, more than 50 % of global human population live in urban areas, and the number of urban residents is currently increasing by 1 million every week (Anonymous 2010). According to the projections by the United Nations, 80 % of the global population will be urban by 2050. Even after the world population stabilizes around 2050, the urban population will continue to grow, and almost all future

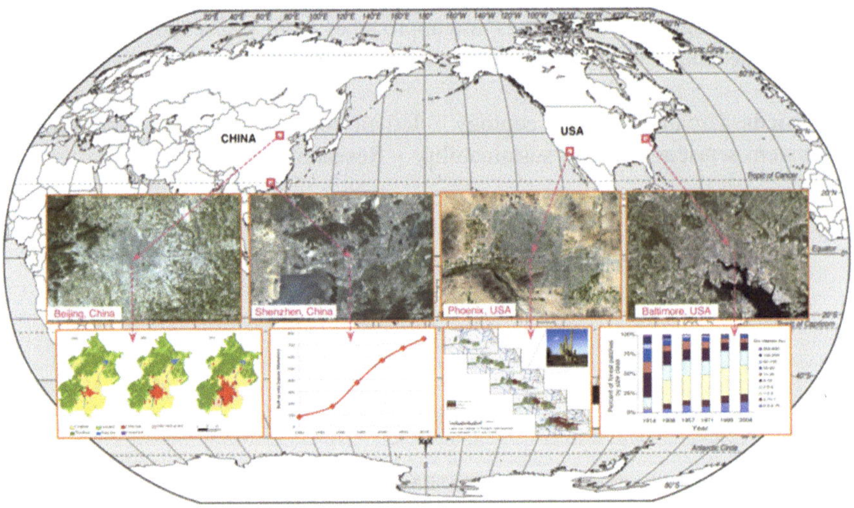

Fig. 3.1 Examples of urbanization at the landscape scale—four of the best studied cities in China and the United States: Beijing, China (39°55′ N, 116°23′ E), Shenzhen, Guangdong Province, China (22°39′ N, 114°13′ E), Phoenix, Arizona. USA (33°27′ N, 112°04′ W), and Baltimore, Maryland, USA (39°17′, 76°37′ W). The background map was obtained from http://eduplace.com/ and the *inset* images of the four cities were from Google Earth (http://www.google.com/earth/index.html). The *insets* showing urbanization patterns for the four cities are from several published sources (He et al. 2006, 2008, 2011; Yu et al. 2009; Wu et al. 2011a, b; Zhou et al. 2011; Grimm et al. 2012)

population increases will take place in urban areas (mostly in developing countries of Asia and Africa). It is certain, therefore, that our future will be increasingly urban.

Urbanization has been both a boon and a bane (Wu 2008a, 2010b). Cities have been the engines of economic growth and centers of innovation and sociocultural development. Cities usually have higher use efficiencies of energy and materials, as well as better access to education, jobs, health care, and social services than rural areas. In addition, by concentrating human populations, urbanization should be able to, at least in principle, save land for other species or nature conservation. However, cities are also places of severe environmental problems, growing socioeconomic inequality, and political and social instabilities. Although the physically urbanized land covers merely about 3 % of the earth's land surface, the "ecological footprints" of cities are disproportionally large—often hundreds of times their physical sizes (Luck et al. 2001; Jenerette and Potere 2010). Urban areas account for about 78 % of carbon emissions, 60 % of residential water use, and 76 % of the wood used for industrial purposes (Grimm et al. 2008; Wu 2008a, 2010b). As a result, urbanization has profoundly affected biodiversity, ecosystem processes, ecosystem services, climate, and environmental quality on scales ranging from the local city to the entire globe.

Until quite recently, however, ecologists have focused primarily on "natural" ecosystems, and treated cities largely as "trashed ecosystems" unworthy of study (Collins et al. 2000). This does not mean that urban ecology is really "new." In fact, the field known as "urban ecology" had already existed before the terms "ecosystem" and "landscape ecology" were coined. Nevertheless, it is during the past two decades that urban ecology has developed into a highly interdisciplinary field of study, increasingly embraced by ecologists, geographers, and social scientists. These recent and unprecedented developments in urban ecology have had much to do with the rise of landscape ecology in general and urban landscape studies in particular, resulting in a dynamic and exciting research field—urban landscape ecology. Today, studies that focus on the spatiotemporal patterns, biophysical and socioeconomic drivers, and ecological and environmental impacts of urbanization are mushrooming around the world (e.g., Fig. 3.1).

The main goal of this chapter is to provide a perspective on the scope, objectives, and recent developments of urban landscape ecology. This is not intended to be a comprehensive review of the literature on urban ecology. Rather, it is a perspective on the past, present, and future of urban landscape ecology based on our research experiences with cities in China and USA (Fig. 3.1).

3.2 Landscape Ecology and the Rising Urban Theme

Apparently, urban landscape ecology is part of landscape ecology, and thus it makes sense to discuss the former within the context of the latter. Landscape ecology is the science of studying and improving the relationship between spatial pattern and ecological (and socioeconomic) processes on multiple scales (Wu and

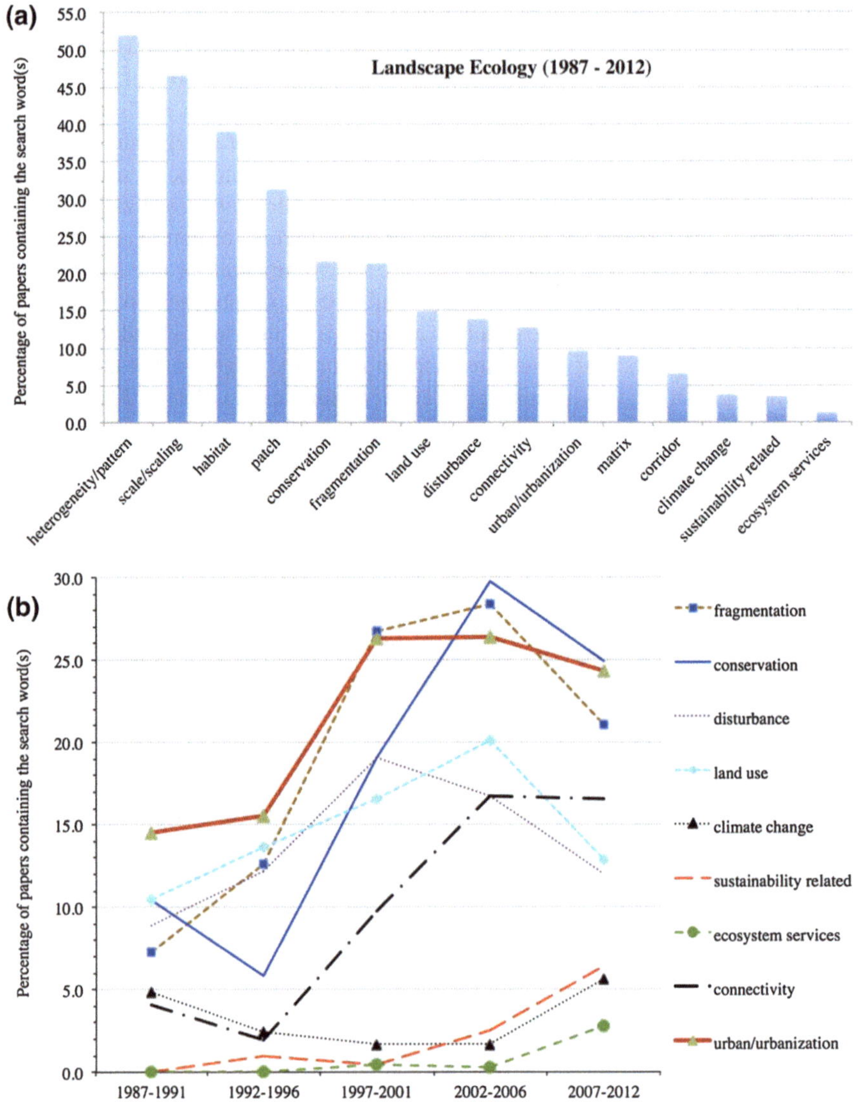

Fig. 3.2 Percentage of published papers in the journal *Landscape Ecology* between 1987 and 2012 which contain important landscape ecological terms in their titles, keywords, and abstracts: **a** ranking according to the relative frequency of occurrence, and **b** changes in relative frequency of occurrence over time (calculated by dividing the 1987–2012 period into 5 segments)

Hobbs 2007). Although landscape ecology has long considered humans and their activities as part of the landscape, its most salient feature that distinguishes itself from other ecological fields (e.g., population, community, and ecosystem ecology) is its explicit emphasis on spatial heterogeneity or pattern (Wu 2013a). This emphasis on heterogeneity should not be interpreted as stressing "structure" only

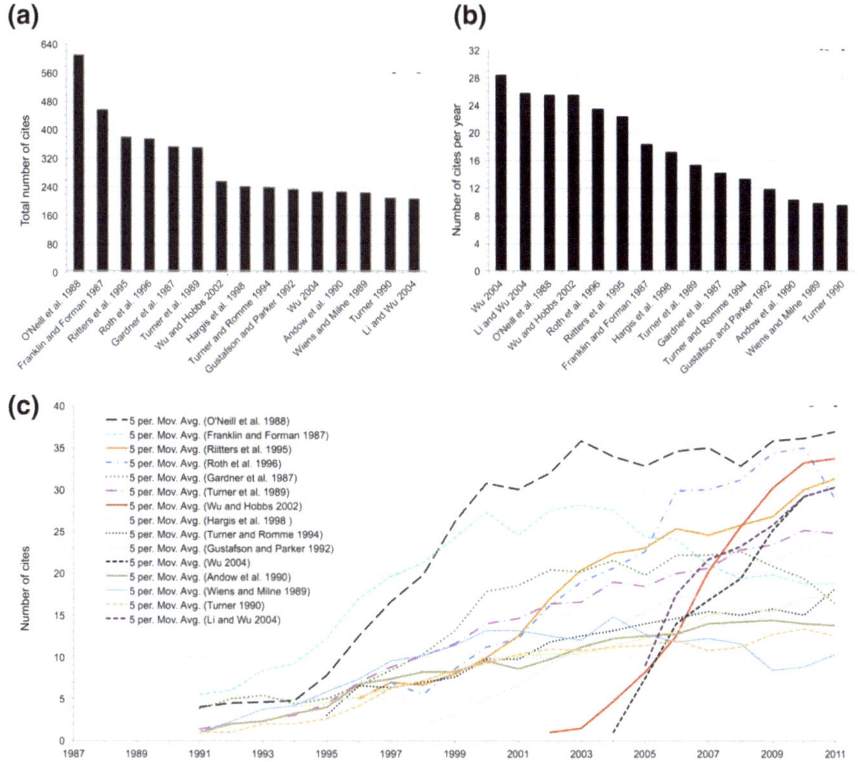

Fig. 3.3 The top 15 most-cited papers published in *Landscape Ecology* (data from the ISI web of science, http://apps.webofknowledge.com/; accessed on December 5, 2012): **a** ranking by the total number of citations, **b** ranking by the number of citations per year, and **c** temporal changes in the number of citations (5-year moving average values used here to smooth out annual fluctuations)

or deemphasizing "function." A background assumption in landscape ecology is that landscape structural patterns are related to ecological processes and ecosystem functions. In other words, the ultimate goal of analyzing spatial patterns is to get to the underlying processes or functions—pattern analysis is a "means" not an "end." Also, both landscape structural and functional attributes have "spatial patterns" which are important for ecological understanding and management.

A review of all the publications in the field's flagship journal—*Landscape Ecology*—since its establishment in 1987 confirms that landscape ecology is a spatially explicit interdisciplinary science (Figs. 3.2 and 3.3). First, the most commonly used terms in landscape ecology are those that are directly related to spatial heterogeneity or spatial pattern, including heterogeneity, pattern, fragmentation, disturbance, and connectivity (Fig. 3.2a). The frequent use of words like habitat, conservation, fragmentation, and connectivity reflects the predominance of biodiversity conservation, as a research topic, in landscape ecological studies. Another trend in the frequency of word occurrence is that urbanization,

Table 3.1 The top 15 most cited papers that were published in landscape ecology based on data from web of science (accessed on December 5, 2012)

Order	Author (year)	Article title	Vol (issue)	Total cites	Cites/ year
1	O'Neill et al. (1988)	Indices of landscape pattern	1(3)	609	25.4
2	Franklin and Forman (1987)	Creating landscape patterns by forest cutting: ecological consequences and principles	1(1)	456	18.2
3	Riitters et al. (1995)	A factor-analysis of landscape pattern and structure metrics	10(1)	378	22.2
4	Roth et al. (1996)	Landscape influences on stream biotic integrity assessed at multiple spatial scales	11(3)	374	23.4
5	Gardner et al. (1987)	Neutral models for the analysis of broad-scale landscape pattern	1(1)	352	14.1
6	Turner et al. (1989)	Effects of changing spatial scale on the analysis of landscape pattern	3(3–4)	349	15.2
7	Wu and Hobbs (2002)	Key issues and research priorities in landscape ecology: an idiosyncratic synthesis	17(4)	254	25.4
8	Hargis et al. (1998)	The behavior of landscape metrics commonly used in the study of habitat fragmentation	13(3)	240	17.1
9	Turner and Romme (1994)	Landscape dynamics in crown fire ecosystems	9(1)	237	13.2
10	Gustafson and Parker (1992)	Relationships between landcover proportion and indexes of landscape spatial pattern	7(2)	233	11.7
11	Wu (2004)	Effects of changing scale on landscape pattern analysis: scaling relations	19(2)	226	28.3
12	Andow et al. (1990)	Spread of invading organisms	4(2–3)	225	10.2
13	Wiens and Milne (1989)	Scaling of 'landscapes' in landscape ecology, or, landscape ecology from a beetle's perspective	3(2)	223	9.7
14	Turner (1990)	Spatial and temporal analysis of landscape patterns	4(1)	208	9.5
15	Li and Wu (2004)	Use and misuse of landscape indices	19(4)	205	25.6

climate change, ecosystems services, and sustainability have become increasingly dominant in landscape ecology during the past two decades (Fig. 3.2b). In addition, the top 15 most-cited papers published in *Landscape Ecology* since 1987 have been overwhelmingly dominated by topics of spatial pattern analysis and scale-related issues (Table 3.1; Fig. 3.3a, b). Again, this is indicative of the field's paramount emphasis on spatial heterogeneity and scale.

Particularly relevant to this chapter is that the number of publications on urbanization or urban landscapes in the journal has increased rapidly during the past two decades. This is not surprising because of several reasons. First, urban landscapes exhibit the most conspicuously heterogeneous patterns among all landscapes, and thus are ideal objects for applying and testing landscape metrics and spatial statistical methods. From a more dynamic perspective, urbanization is fundamentally a spatial process, and its understanding relies on spatially explicit methods that characterize landscape ecological studies. Second, urbanization and its ecological impacts have gained unprecedented impetus in research during the past 20 years as we have entered a new urbanization era. The urban landscape (the city and its surrounding areas or a metropolitan region) has emerged as a primary scale for urban studies. In fact, one may argue that a landscape approach is not only appropriate in theory but also imperative in practice for urban ecology and urban sustainability. Given the increasingly urban nature of our landscapes and the increasingly urban future of humanity, urban sustainability is becoming "an inevitable goal of landscape research" (Wu 2010b).

3.3 From Urban Ecology to Urban Landscape Ecology

To discuss the present and future of urban landscape ecology, it is helpful to recall important milestones in the history of urban ecology. This is because urban landscape ecology may be viewed as a product of the integration between landscape ecology and urban ecology. Several recent reviews on the history of urban ecology can be found elsewhere (Pickett et al. 2001, 2011; Wu 2008a, b, 2013b; McDonnell 2011). To illustrate how urban landscape ecology is related to urban ecology, here we provide a synopsis of the evolution of different perspectives and approaches in urban ecological research since its early years (Fig. 3.4).

The earliest version of "urban ecology" was developed in the 1920s, as part of human ecology, by the Chicago school of sociology, championed by Robert E. Park and Ernest W. Burgess (Park et al. 1925). In other words, urban ecology was born in a "social science family," as a sociological approach that uses ecological concepts (e.g., competition, ecological niches, and succession) and natural selection theory as organic analogies to study the social life and societal structures in the city. The key idea of this urban ecology approach is that competition for land and resources in an urban area leads to the continuous structuring of the city space into ecological niches (e.g., zones) through "invasion-succession" cycles (to put it blatantly, the poor and immigrants come in and the rich and "original" move out). Spatial and social differentiations occur consequently, and different social groups occupy different zones (or niches). This idea is epitomized in the concentric zone theory (Park et al. 1925). The Chicago school urban ecology was quite influential for a few decades, but largely disregarded by the 1950s as criticisms mounted of its neglecting the roles cultural and social factors (e.g., race and ethnicity) as well as planning and industrialization. This sociological tradition of urban ecology is still alive today as one may often find a chapter or a section on urban ecology in most sociology textbooks (but rarely in classic ecology texts). In fact, one may argue that understanding the relationship between spatial and social structures in the city

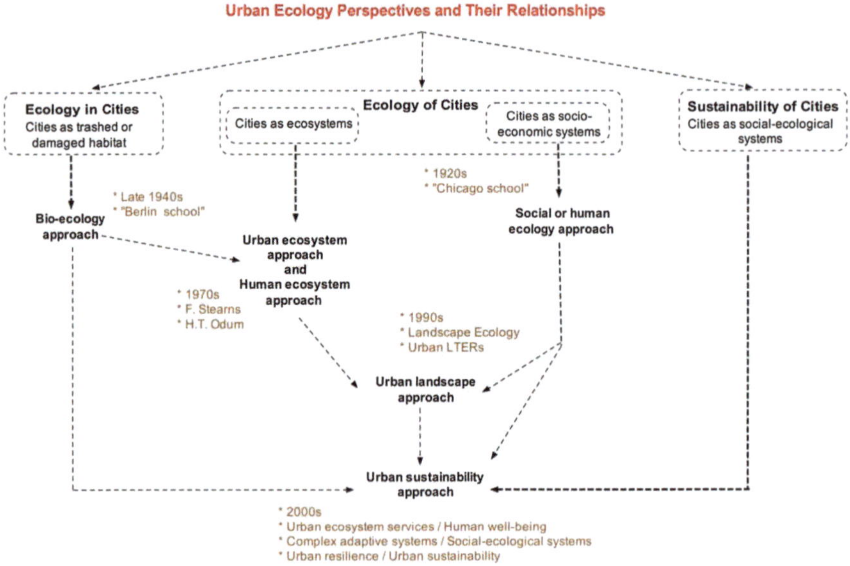

Fig. 3.4 Different perspectives in urban ecology and the rising prominence of the urban landscape ecology approach to the studies of cities and human-dominant areas (modified from Wu 2008a, b, 2013a, b)

is a key component in urban landscape ecology, particularly when urban sustainability is considered as its ultimate goal (Fig. 3.4).

In the late 1940s, European ecologists, most noticeably the "Berlin school," began to study remnant plant and animal species in cities—a bio-ecological approach or the "ecology in cities" approach (Grimm et al. 2000; Pickett et al. 2001; Wu 2008a). Excellent reviews of these studies are found in Sukopp (1990, 2002). In the 1970s forest ecologists (e.g., Forest Stearns) and ecosystem ecologists (e.g., the Odum brothers) advocated ecosystem-based approaches to studying the structure and function of cities (Stearns and Montag 1974; Odum 1983). H. T. Odum's emergy-based urban approach is still being used by some (Huang and Chen 2009; Lee et al. 2013). Not until the early 1990s did urban ecology start to move into the mainstream of ecology. A seminal paper during this time period was McDonnell and Pickett (1990) that introduced the well-established gradient analysis method in plant community ecology and vegetation science to the study of urban ecosystems—the urban–rural gradient approach.

During the 1980s, landscape ecology was developing swiftly in North America and beyond, and many of the landscape studies dealt with land use and land cover change including urbanization. With the rapidly increasing availability of remote sensing data, GIS, and spatial pattern analysis methods (e.g., landscape metrics), the number of studies on the spatiotemporal patterns and socioeconomic drivers of urbanization began to soar (many such "patterns and drivers studies" continue to be done by physical geographers, remote sensing scientists, and the like). The launching of the two Long-Term Ecological Research projects on urban ecology (Urban LTERs) by the US National Science Foundation in 1997 played an instrumental role in promoting the integration between human ecosystem-based functional approaches and pattern-oriented landscape approaches (Pickett et al. 1997; Grimm et al. 2000; Jenerette and Wu 2001; Luck et al. 2001; Luck and Wu 2002; Wu and David 2002; Wu et al. 2003; Jenerette et al. 2006; Buyantuyev and Wu 2009, 2012). An urban landscape ecology that couples spatiotemporal patterns with ecological processes began to take form in the early 2000s.

Since the publication of the Millennium Ecosystem Assessment in 2005, ecosystem services (and their relationship with human well-being) have increasingly become mainstream in ecology. This trend has been accompanied by the rapid development of sustainability science that focuses on the dynamic relationship between society and nature (Kates et al. 2001; Wu 2006). Consequently, a nascent but robust research direction in urban landscape ecology now is focused on urban sustainability (Fig. 3.4). This emerging urban sustainability approach integrates the various urban ecology perspectives, and its scientific core develops around the structure, function, and services of the urban landscape, frequently invoking hierarchy theory, complex adaptive systems theory, and resilience theory (Wu and David 2002; Alberti 2008; Wu 2010b; Ahern 2013; Wu and Wu 2013).

3.4 A Framework for Urban Landscape Ecology

So, how should urban landscape ecology be defined? Simply put, urban landscape ecology is landscape ecology of urban areas. More specifically, it is the science of studying and improving the relationship between urban landscape pattern and ecological processes for achieving urban sustainability. While urban sustainability may be defined in a number of ways, here we define it as an adaptive process of maintaining and improving ecosystem services and human well-being in the urban landscape (Wu 2010a, 2013b). As such, urban landscape ecology consists of three interactive major components: quantifying the spatiotemporal patterns and understanding the drivers and mechanisms of urbanization ("patterns/drivers studies"), assessing the ecological and environmental impacts of urbanization ("impacts studies"), and understanding and improving urban sustainability (Figs. 3.5 and 3.6).

The first component is to characterize the spatiotemporal patterns and driving processes of the urban landscape. This involves mapping and quantifying urban morphological attributes and landscape patterns over time, identifying key socioeconomic and environmental drivers, and understanding urban pattern-process relationships on multiple scales ranging from the parcel to the metropolitan region. Both landscape ecologist and geographers have done a great deal in this front (Jenerette and Wu 2001; Luck and Wu 2002; Batty 2005; Schneider and Woodcock 2008; Wu et al. 2011a, b). Recent years have seen a rapid increase in the number of this sort of studies. For these studies to be really relevant to societal

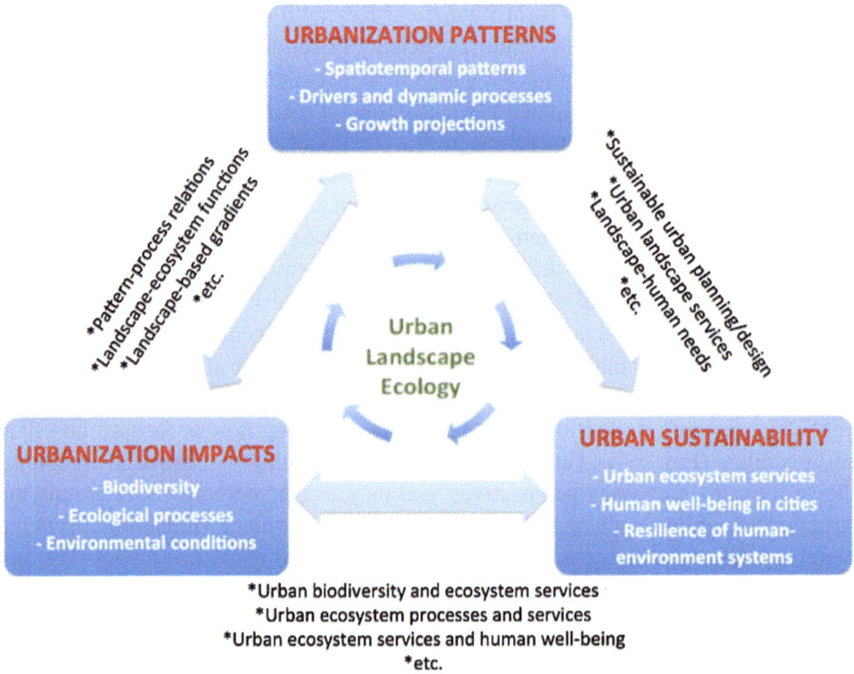

Fig. 3.5 The scope of urban landscape ecology: three key components and their relationship

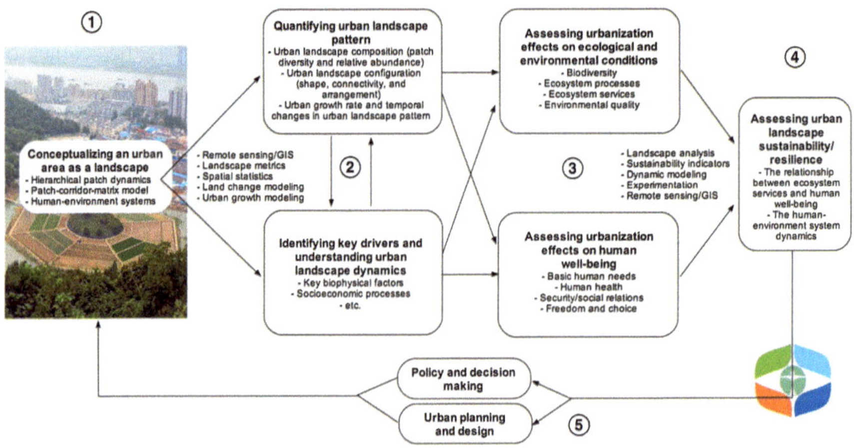

Fig. 3.6 A framework for urban landscape ecology that integrates "patterns/drivers studies" with "impacts studies," and promotes urban sustainability as its ultimate goal

needs and policy making, they must be integrated with the other two components. An example demonstrating two major methods on these topics—landscape pattern metrics and landscape gradient analysis—is illustrated in Figs. 3.7 and 3.8, based

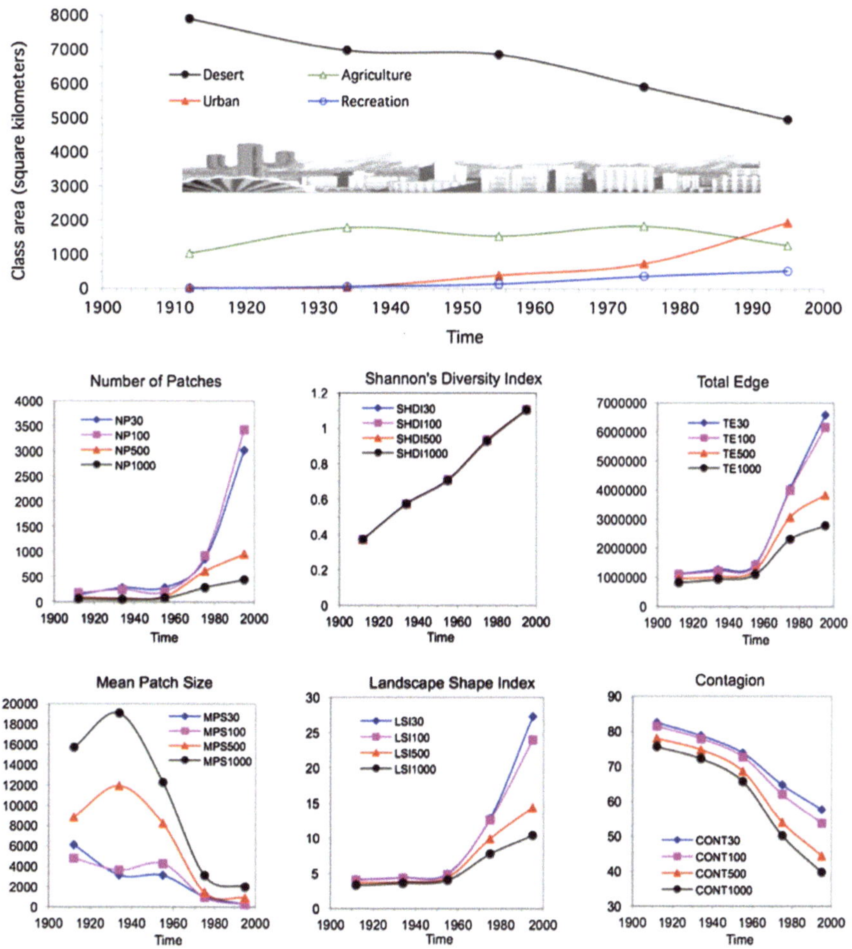

Fig. 3.7 Quantification of the spatiotemporal patterns of urbanization in the Phoenix metropolitan region, Arizona, USA, using landscape pattern metrics (modified from Wu 2004; Wu et al. 2011a, b). A large number of landscape metrics have been used to characterize urbanization patterns, and seven of them are shown here

on the Central Arizona-Phoenix Long-Term Ecological Research project (CAPLTER). Among many other studies using these methods are those at Baltimore, Beijing, and Shenzhen (Fig. 3.1).

The second component is focused on "impacts studies" that investigate how urbanization affects biodiversity, population and community processes, ecosystem functions, and ecosystem services. Most studies on cities that have been carried out by bio-ecologists and environmental scientists belong to this category, and several recent books have reviewed these studies (Carreiro et al. 2008; McDonnell et al. 2009; Niemela 2011). It is well documented that urbanization may decrease native species richness but increase the number of exotic species; increase landscape-level

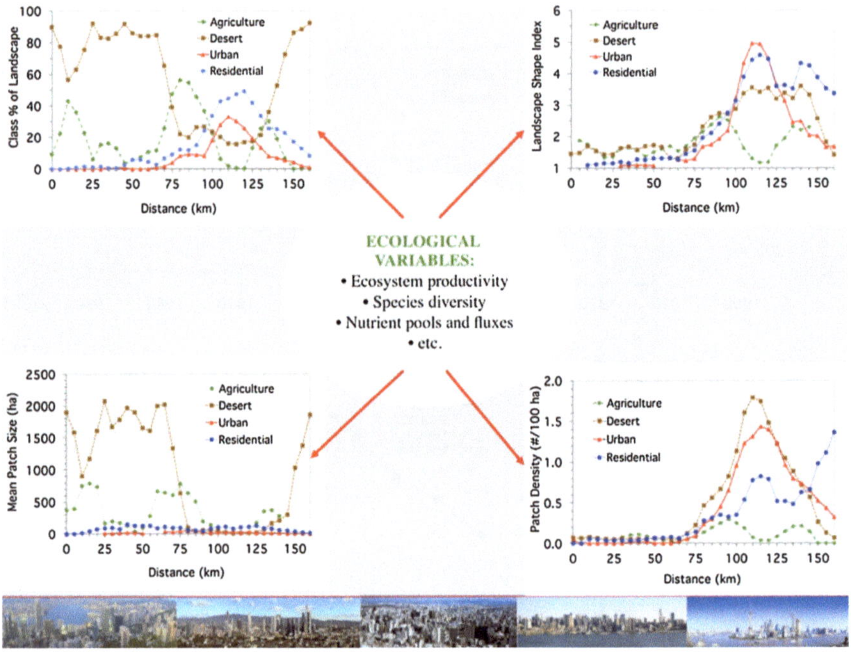

Fig. 3.8 Landscape pattern gradient analysis as used in the quantification of the spatial pattern of the Phoenix metropolitan region, USA (modified from Luck and Wu 2002)

ecosystem primary production due to irrigation but reduce environmental quality; and alter soil properties and biogeochemical and hydrological cycles (Pickett et al. 2001, 2011; Wu 2013b). Also, urban heat islands—pronounced increases in air and surface temperatures (especially nighttime) over non-vegetated impervious surfaces due to enhanced longwave radiation—and their effects on air quality and human health have long been studied (Oke 1982; Jenerette et al. 2007; Buyantuyev and Wu 2010; Jenerette et al. 2011; Li et al. 2011). While understanding the various effects of urbanization is important and necessary, the "impacts studies" need to address how these effects can be eliminated, mitigated, or adapted through urban design and planning actions. This requires the integration among the three components (Fig. 3.5).

The third component of urban landscape ecology focuses on the sustainability of urban areas—urban sustainability. Rigorous research in urban sustainability is still nascent, and a cohesive framework is yet to be developed. However, several core research questions are emerging, including the kinds, amounts, and spatial interactions of urban ecosystem services, human well-being (measured as the degrees of satisfying the basic, psychological, and spiritual needs of humans as influenced by landscape structural and functional attributes), and the resilience of coupled human-environment systems in the urban landscape (Wu 2010b, 2013b). To address these questions, it is imperative to integrate the three components (Fig. 3.5). These new developments in urban landscape sustainability differ from previous studies focused on urban sustainability indicators in terms of both key research questions and methodologies. For example, the urban landscape ecology approach to urban sustainability increasingly emphasizes ecosystem services and their relationship with human well-being, with spatially explicit methods that consider both ecosystem properties and landscape structural attributes (Ahern 2013; Wu 2013b). From a broader perspective, this serge of interest in urban sustainability by landscape ecology is part of the recent movement towards a "landscape sustainability science" (Wu 2006, 2012, 2013a; Musacchio 2009, 2011).

Complementary to the three-component framework is a more detailed 5-step strategy that outlines the major steps for urban landscape ecological studies (Fig. 3.6). To follow this strategy, the first step is to conceptualize an urban area as a spatially heterogeneous human-environment system (i.e., a landscape). This can be done based on, for example, the patch-corridor-matrix model (Forman 1995) or the hierarchical patch dynamics paradigm (Wu and Loucks 1995; Wu and David 2002). Then, in the second step the spatiotemporal patterns, including the kinds, amounts, diversity, connectivity, and spatial configuration of the urban landscape and their temporal changes, can be quantified, and key biophysical and socio-economic drivers identified. These patterns-and-drivers studies can be, and have frequently been, done with a combination of methods—remote sensing, GIS, landscape metrics, spatial statistics, simulation modeling, and, to a much lesser extent, experiments (mainly longitudinal). The third step is to link the spatio-temporal patterns of urbanization to ecological and environmental variables of interest so that the impacts of urbanization can be assessed. The impacts studies need to go beyond environmental quality, biodiversity, and ecosystem functions and services to include variables that are directly related to human well-being (e.g., those of human survival, security, and psychological needs). These impacts studies can be done using a number of statistical and modeling methods, including those in step one. The fourth step is assess the sustainability and resilience of both ecosystem services and human well-being in the urban landscape. The tradeoffs and synergies among ecosystem services and between ecosystem services and human well-being in the urban landscape need to be understood, and scenarios for sustaining natural capital and flows as well as human well-being need to be sought. These scenarios have to be investigated in concert with landscape planning and design because they involve intentional alterations of landscape composition and configuration. In addition to the methods mentioned above, sustainability indicators may play an important role in accomplishing these goals.

3.5 Concluding Remarks

The world has become increasingly urban, and the ecology of landscapes needs to reflect this reality in its science. Indeed, this has been happening in the past few decades, and studies of urban areas now are prominent in landscape ecology.

A research area that can be called urban landscape ecology is identifiable, which is part of landscape ecology and also related to urban ecology (as well as urban geography and urban sociology). The existing studies on this topic, however, do not yet form a cohesive framework or have a unified goal. In this chapter, we have reviewed the intellectual roots of urban landscape ecology, and proposed a framework to help move the field forward.

Landscapes and regions represent arguably the most operational scales for sustainability research and practice (Forman 1990, 2008; Wu 2006). To meet the challenge of urban sustainability, cities need to be studied as spatially extended, complex adaptive systems with interdisciplinary approaches integrating ecological, economic, social, and design/planning sciences (Wu 2013b). This seems to be the main theme of urban landscape ecology or the future direction it is moving towards. Landscape ecology needs to be more "urban;" urban ecology needs to be more landscape-realistic; both need to focus more on sustainability.

References

Ahern J. Urban landscape sustainability and resilience: the promise and challenges of integrating ecology with urban planning and design. Landscape Ecol. 2013. doi:10.1007/s10980-012-9799-z.

Alberti M. Advances in urban ecology: integrating humans and ecological processes in urban ecosystems. New York: Springer; 2008.

Andow DA, Kareiva PM, Levin SA, Okubo A. Spread of invading organisms. Landscape Ecol. 1990;4:177–188.

Anonymous. The urban equation. Nature. 2010;467(7318):899–901.

Batty M. Agents, cells, and cities: new representational models for simulating multiscale urban dynamics. Environ Plann A. 2005;37:1373–94.

Buyantuyev A, Wu J. Urbanization alters spatiotemporal patterns of ecosystem primary production: a case study of the Phoenix metropolitan region, USA. J Arid Environ. 2009;73(4–5):512–20.

Buyantuyev A, Wu JG. Urban heat islands and landscape heterogeneity: linking spatiotemporal variations in surface temperatures to land-cover and socioeconomic patterns. Landscape Ecol. 2010;25(1):17–33.

Buyantuyev A, Wu JG. Urbanization diversifies land surface phenology in arid environments: interactions among vegetation, climatic variation, and land use pattern in the Phoenix metropolitan region, USA. Landscape Urban Plann. 2012;105:149–59.

Carreiro MM, Song YC, Wu JG, editors. Ecology, planning and management of urban forests: international perspectives. New York: Springer; 2008.

Collins JP, Kinzig A, Grimm NB, Fagan WF, Hope D, Wu J, Borer ET. A new urban ecology. Am Sci. 2000;88(5):416–25.

Forman RTT. Ecologically sustainable landscapes: the role of spatial configuration. In: Zonneveld IS, Forman RTT, editors. Changing landscapes: an ecological perspective. New York: Springer; 1990. p. 261–78.

Forman RTT. Land mosaics: the ecology of landscapes and regions. Cambridge: Cambridge University Press; 1995.

Forman RTT. The urban region: natural systems in our place, our nourishment, our home range, our future. Landscape Ecol. 2008;23(3):251–3.

Franklin JF, Forman RTT. Creating Landscape Patterns by Forest Cutting: Ecological Consequences and Principles. Landscape Ecol. 1987;1:5–18.
Gardner RH, Milne BT, Turner MG, O'Neill RV. Neutral models for the analysis of broad-scale landscape pattern. Landscape Ecol. 1987;1:19–28.
Grimm N, Grove JM, Pickett STA, Redman CL. Integrated approaches to long-term studies of urban ecological systems. Bioscience. 2000;50(7):571–84.
Grimm NB, Redman CL, Boone CG, Childers DL, Harlan SL, Turner II BL. Viewing the urban socio-ecological system through a sustainability lens: lessons and prospects from the central Arizona–Phoenix LTER programme. In: Singh SJ, et al., editors. Long term socio-ecological research: human-environment interactions 2. New York: Springer; 2012.
Grimm NB, Faeth SH, Golubiewski NE, Redman CL, Wu J, Bai X, Briggs JM. Global change and the ecology of cities. Science. 2008;319(5864):756–60.
Gustafson EJ, Parker GR. Relationships between Landcover Proportion and Indexes of Landscape Spatial Pattern. Landscape Ecol. 1992;7:101–110.
Hargis CD, Bissonette JA, David JL. The behavior of landscape metrics commonly used in the study of habitat fragmentation. Landscape Ecol. 1998;13:167–186.
He C, Okada N, Zhang Q, Shi P, Zhang J. Modeling urban expansion scenarios by coupling cellular automata model and system dynamic model in Beijing, China. Appl Geogr. 2006;26:323–45.
He C, Okada N, Zhang Q, Shi P, Zhang J. Modeling dynamic urban expansion processes incorporating a potential model with cellular automata. Landscape Urban Plann. 2008;86:79–91.
He C, Tian J, Shi P, Hu D. Simulation of the spatial stress due to urban expansion on the wetlands in Beijing, China using a GIS-based assessment model. Landscape Urban Plann. 2011;101:269–77.
Huang SL, Chen CW. Urbanization and socioeconomic metabolism in Taipei: an emergy synthesis. J Ind Ecol. 2009;13(1):75–93.
Jenerette GD, Wu J. Analysis and simulation of land use change in the central Arizona—Phoenix region. Landscape Ecol. 2001;16(7):611–26.
Jenerette GD, Potere D. Global analysis and simulation of land-use change associated with urbanization. Landscape Ecol. 2010;25(5):657–70.
Jenerette GD, Wu J, Grimm NB, Hope D. Points, patches, and regions: scaling soil biogeochemical patterns in an urbanized arid ecosystem. Glob Change Biol. 2006;12:1532–44.
Jenerette GD, Harlan SL, Stefanov WL, Martin CA. Ecosystem services and urban heat riskscape moderation: water, green spaces, and social inequality in Phoenix, USA. Ecol Appl. 2011;21(7):2637–51.
Jenerette GD, Harlan SL, Brazel A, Jones N, Larsen L, Stefanov WL. Regional relationships between surface temperature, vegetation, and human settlement in a rapidly urbanizing ecosystem. Landscape Ecol. 2007;22(3):353–65.
Kates RW, Clark WC, Corell R, Hall JM, Jaeger CC, Lowe I, McCarthy JJ, Schellnhuber HJ, Bolin B, Dickson NM, Faucheux S, Gallopin GC, Grubler A, Huntley B, Jager J, Jodha NS, Kasperson RE, Mabogunje A, Matson P, Mooney H, Moore B III, O'Riordan T, Svedin U. Sustainability science. Science. 2001;292:641–2.
Lee YC, Yeh CT, Huang SL. Energy hierarchy and landscape sustainability. Landscape Ecol. 2013;. doi:10.1007/s10980-012-9706-7.
Li HB, Wu JG. Use and misuse of landscape indices. Landscape Ecol. 2004;19:389–399.
Li JX, Song CH, Cao L, Zhu FG, Meng XL, Wu JG. Impacts of landscape structure on surface urban heat islands: a case study of Shanghai China. Remote Sens Environ. 2011;115(12):3249–63.
Luck M, Wu JG. A gradient analysis of urban landscape pattern: a case study from the Phoenix metropolitan region, Arizona, USA. Landscape Ecol. 2002;17(4):327–39.
Luck M, Jenerette GD, Wu J, Grimm NB. The urban funnel model and the spatially heterogeneous ecological footprint. Ecosystems. 2001;4:782–96.

McDonnell MJ. The history of urban ecology: a ecologlist's perspective. In: Niemelä J, Breuste JH, Guntenspergen G, McIntyre NE, Elmqvist T, James P, editors. Urban ecology: patterns, processes, and applications. New York: Oxford University Press; 2011. p. 5–13.
McDonnell MJ, Pickett STA. Ecosystem structure and function along urban-rural gradients: an unexploited opportunity for ecology. Ecology. 1990;71:1232–7.
McDonnell MJ, Hahs AK, Breuste JH, editors. Ecology of cities and towns: a comparative approach. Cambridge: Cambridge University Press; 2009.
Musacchio LR. The scientific basis for the design of landscape sustainability: a conceptual framework for translational landscape research and practice of designed landscapes and the six Es of landscape sustainability. Landscape Ecol. 2009;24(8):993–1013.
Musacchio LR. The grand challenge to operationalize landscape sustainability and the design-in-science paradigm. Landscape Ecol. 2011;26(1):1–5.
Niemela J, editor. Urban ecology: patterns, processes, and applications. Oxford: Oxford University Press; 2011.
Odum HT. Systems ecology: an introduction. New York: Wiley; 1983.
O'Neill RV, Krummel JR, Gardner RH, Sugihara G, Jackson B, DeAngelis DL, Milne BT, Turner MG, Zygmunt B, Christensen SW, Dale VH, Graham RL. Indices of landscape pattern. Landscape Ecol. 1988;1:153–162.
Oke TR. The energetic basis of the urban heat island. Q J Roy Meteorol Soc. 1982;108:1–24.
Park RE, Burgess EW, McKenzie R. The city. Chicago: University of Chicago Press; 1925.
Pickett STA, Burch WR, Dalton SE, Foresman TW, Grove JM, Rowntree R. A conceptual framework for the study of human ecosystems in urban areas. Urban Ecosyst. 1997;1:185–99.
Pickett STA, Cadenasso ML, Grove JM, Nilon CH, Pouyat RV, Zipperer WC, Costanza R. Urban ecological systems: linking terrestrial ecological, physical, and socioeconomic components of metropolitan areas. Ann Rev Ecol Evol Syst. 2001;32:127–57.
Pickett STA, Cadenasso ML, Grove JM, Boone CG, Groffman PM, Irwin E, Kaushal SS, Marshall V, McGrath BP, Nilon CH, Pouyat RV, Szlavecz K, Troy A, Warren P. Urban ecological systems: scientific foundations and a decade of progress. J Environ Manage. 2011;92(3):331–62.
Riitters KH, Oneill RV, Hunsaker CT, Wickham JD, Yankee DH, Timmins SP, Jones KB, Jackson BL. A factor analysis of landscape pattern and structure metrics. Landscape Ecol. 1995;10:23–39.
Roth NE, Allan JD, Erickson DL. Landscape influences on stream biotic integrity assessed at multiple spatial scales. Landscape Ecol. 1996;11:141–156.
Schneider A, Woodcock CE. Compact, dispersed, fragmented, extensive? A comparison of urban growth in twenty-five global cities using remotely sensed data, pattern metrics and census information. Urban Stud. 2008;45(3):659–92.
Stearns F, Montag T, editors. The urban ecosystem: a holistic approach. Stroudsburg: Dowden, Hutchinson & Ross, Inc; 1974.
Sukopp H. Urban ecology and its application in Europe. In: Sukopp H, Hejny S, Kowarik I, editors. Urban ecology: plants and plant communities in urban environments. The Hague: SPB Academic Publishing bv; 1990. p. 2–22.
Sukopp H. On the early history of urban ecology in Europe. Preslia. 2002;74:373–93.
Turner MG, O'Neill RV, Gardner RH, Milne BT. Effects of changing spatial scale on the analysis of landscape pattern. Landscape Ecol. 1989;3:153–162.
Turner MG. Spatial and temporal analysis of landscape patterns. Landscape Ecol. 1990;4:21–30.
Turner MG, Romme WH. Landscape dynamics in crown fire ecosystems. Landscape Ecol. 1994;9:59–77.
Wiens JA, Milne BT. Scaling of 'landscape' in landscape ecology, or, landscape ecology from a beetle's perspective. Landscape Ecol. 1989;3:87–96.
Wu J, Hobbs R. Key issues and research priorities in landscape ecology: An idiosyncratic synthesis. Landscape Ecol. 2002;17:355–365.
Wu J. Effects of changing scale on landscape pattern analysis: scaling relations. Landscape Ecol. 2004;19:125–138.

Wu JG. Landscape ecology, cross-disciplinarity, and sustainability science. Landscape Ecol. 2006;21(1):1–4.

Wu JG. Making the case for landscape ecology: an effective approach to urban sustainability. Landscape J. 2008a;27(1):41–50.

Wu JG. Toward a landscape ecology of cities: beyond buildings, trees, and urban forests. In: Carreiro MM, Song YC, Wu JG, editors. Ecology, planning and management of urban forests: international perspectives. New York: Springer; 2008b. p. 10–28.

Wu JG. Landscape of culture and culture of landscape: does landscape ecology need culture? Landscape Ecol. 2010a;25(8):1147–50.

Wu JG. Urban sustainability: an inevitable goal of landscape research. Landscape Ecol. 2010b;25(1):1–4.

Wu JG. A landscape approach for sustainability science. In: Weinstein MP, Turner RE, editors. Sustainability science: the emerging paradigm and the urban environment. New York: Springer; 2012. p. 59–77.

Wu JG. Key concepts and research topics in landscape ecology revisited: 30 years after the Allerton Park workshop. Landscape Ecol. 2013a. doi:10.1007/s10980-012-9836-y.

Wu JG. The state-of-the-science in urban ecology and sustainability: a landscape perspective. Landscape Urban Plann. 2013b. (In review).

Wu JG, Loucks OL. From balance-of-nature to hierarchical patch dynamics: a paradigm shift in ecology. Q Rev Biol. 1995;70:439–66.

Wu JG, David JL. A spatially explicit hierarchical approach to modeling complex ecological systems: theory and applications. Ecol Model. 2002;153(1–2):7–26.

Wu JG, Hobbs RJ, editors. Key topics in landscape ecology. Cambridge: Cambridge University Press; 2007.

Wu JG, Wu T. Ecological resilience as a foundation for urban design and sustainability. In: Pickett STA, Cadenasso ML, McGrath BP, editors. Resilience in urban ecology and design: linking theory and practice for sustainable cities. New York: Springer; 2013.

Wu JG, Jenerette GD, David JL. Linking land use change with ecosystem processes: a hierarchical patch dynamics model. In: Guhathakurta S, editor. Integrated land use and environmental models. Berlin: Springer; 2003. p. 99–119.

Wu JG, Jenerette GD, Buyantuyev A, Redman C. Quantifying spatiotemporal patterns of urbanization: the case of the two fastest growing metropolitan regions in the United States. Ecol Complex. 2011a;8:1–8.

Wu JG, Buyantuyev A, Jenerette GD, Litteral J, Neil K, Shen W. Quantifying spatiotemporal patterns and ecological effects of urbanization: a multiscale landscape approach. In: Richte M, Weiland U, editors. Applied urban ecology: a global framework. Oxford: Blackwell; 2011b. p. 35–53.

Yu DY, Shao HB, Shi PJ, Pan YZ, Zhu WQ. How does the conversion of land cover to urban use affect net primary productivity? A case study in Shenzhen city, China. Agric For Meteorol. 2009;149:2054–60.

Zhou W, Huang G, Pickett STA, Cadenasso ML. 90 years of forest cover change in an urbanizing watershed: spatial and temporal dynamics. Landscape Ecol. 2011;26(5):645–59.

Chapter 4
Biocultural Diversity for Sustainable Ecological, Cultural and Sacred Landscapes: The Biocultural Landscape Approach

Gloria Pungetti

Abstract The interactions between natural, cultural and spiritual components of landscape, and the values that people associate with them, are key elements to landscape research based on both ecological and cultural principles. The understanding of these interactions and values is an essential prerequisite for the healthy and sustainable management of our planet. Such understanding can be achieved through research on biocultural diversity, meant as the diversity of life on earth in both nature and culture. This paper explores recent developments in ecological landscape research to support biocultural diversity, and offers a new approach based on biocultural landscape.

Keywords Biocultural landscape research · Biocultural diversity · Ecological, cultural and sacred landscapes · Landscape values and heritage · Traditional ecological knowledge · Right to landscape

If we take care of the earth, the earth will take care of us.
Native Hopi saying

4.1 Introduction: Holistic Landscape Approach

To advance studies on the natural and cultural diversity of landscape, a holistic approach is required. Different aspects of landscape, i.e. natural, cultural, analytical, political and interventional, should be addressed in all their dimensions (Fig. 4.1). These have been comprehensively illustrated in previous research based on ecological landscape design and planning (Pungetti 1999), and landscape

G. Pungetti (✉)
Cambridge Centre for Landscape and People (CCLP) and Darwin College, University of Cambridge, Cambridge, UK
e-mail: cclp@hermes.cam.ac.uk

B. Fu and K. B. Jones (eds.), *Landscape Ecology for Sustainable Environment and Culture*, DOI: 10.1007/978-94-007-6530-6_4,

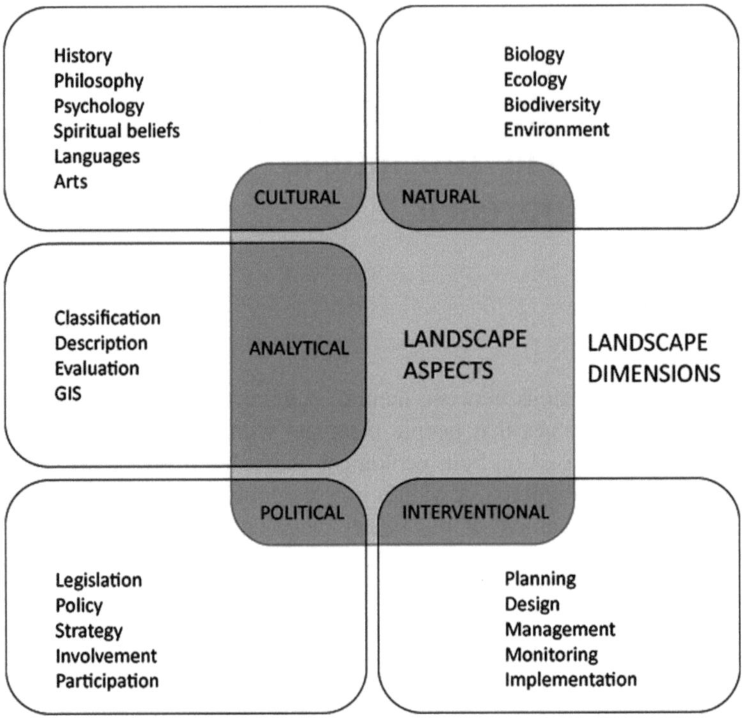

Fig. 4.1 Holistic landscape approach: landscape aspects (*gray*) and landscape dimensions (*white*)

dimensions have also been considered within the structure of the 10th Council of Europe meeting of workshops for the implementation of the European Landscape Convention.

Past research (Naveh 2011; Grove and Rackham 2003; Makhzoumi and Pungetti 1999, 2003) has indicated that the ecological landscape approach could serve as a vehicle for bringing together natural and cultural landscape research. Indeed, landscape ecology studies on landscape mosaic (e.g. Forman and Wilson 1996) and on landscape resources (e.g. Farina 2006) have proposed several directions for studying and preserving our common landscape by the humans who are shaping it (Fig. 4.2).

Natural and cultural landscapes form a total ecosystem entity where all are united and responsible (Naveh 2011). It was in this light that the 8th World Congress of IALE addressed the importance of landscape ecology for a sustainable environment and culture. Landscape ecology, furthermore, can be applied in understanding cultural landscapes and maintaining indigenous knowledge. This topic was addressed at the Symposium ‘Landscape ecological perspectives on biocultural diversity and sacred landscape’ of the Congress, and it is discussed in Sect. 4.6.

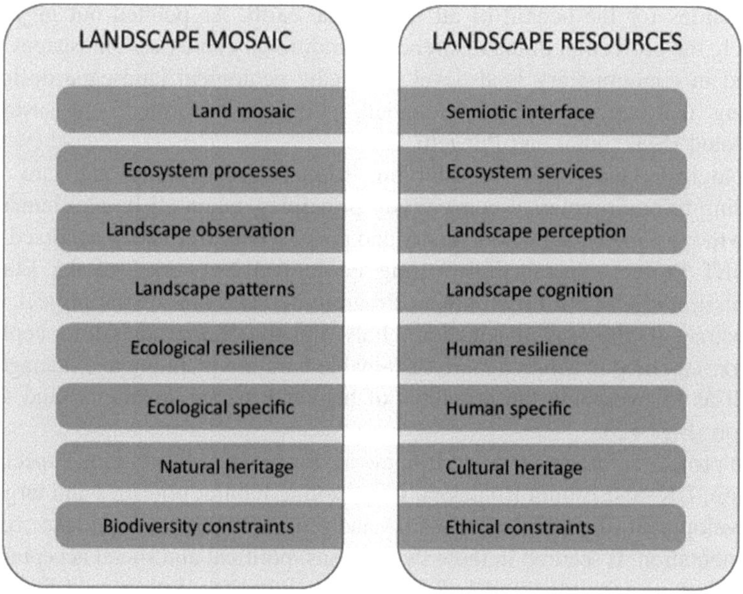

Fig. 4.2 The ecological landscape model background: *mosaic* versus *resources*

The above holistic approach in landscape research constitutes the foundation for the new approach based on biocultural landscape proposed here. Holistic landscape research is further illustrated with the ecological, cultural and sacred heritage of our landscapes.

4.2 Ecological Landscape and Heritage

The natural and the cultural components of landscape are the two pillars in holistic landscape research. Natural landscape is generally understood as a landscape unaffected by human activity. Yet this notion can rarely be applied today, particularly in places built on ancient civilisations as they have been under the influence of people for a long period of time; even a decision to 'keep areas wild' reflects human choice and intervention.

Nevertheless, ecological landscape is a landscape with ecological character and components. In the holistic approach it is perceived as inclusive of abiotic, biotic and human factors (Pungetti 1991) and it reflects ecological principles to its design and planning.

Ecological landscape design and planning, in turn, support the conservation and development of landscape considering the ecological patterns, processes and functions of a site. They therefore sustain the preservation and creation of landscapes and habitats that enhance their interconnected human and natural local communities for the benefit of all life on the earth. As pointed out in previous research, the protection of the resilience of traditional ecological landscapes can be assisted in contemporary land development by ecological landscape design and planning that are ecologically founded, culturally informed and sustainably anticipated (Makzoumi and Pungetti 1999).

Ecological landscapes, in addition, support biodiversity and are intact, according to environmental connectivity principles, when all living elements are able to move and change. Biodiversity and connectivity have been endorsed by the ECONET Project 'Sustainability using Ecological Networks' of the European Commission (EC) Life Environment Programme. The aim of the project was to demonstrate in the United Kingdom, Italy and the Netherlands how ecological networks can help to achieve more sustainable land use planning and management, as well as to overcome the problems of habitat loss, fragmentation and species isolation (Fig. 4.3).

The project defined a strategic framework for nature conservation especially in Cheshire, UK, and Emilia Romagna, Italy, setting habitat priorities and targets for the development of ecological networks and providing practical guidance for their implementation. It secured in these two regions, political and social acceptance for the concept of expanded and linked areas for wildlife. With over 1,500 people involved in the UK, Italy and the Netherlands, the project raised awareness on the concept of ecological networks and supported its integration into farming, forestry and land regeneration, including the restoration of mineral workings and landfill sites.

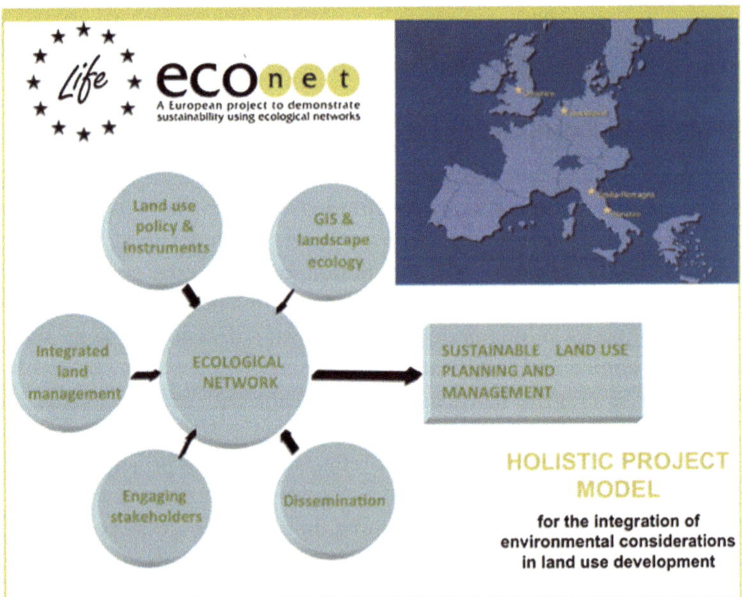

Fig. 4.3 Life ECONET holistic model for ecological networks and sustainability

Moreover, the ECONET Project incorporated the ecological network idea into regional and local policies, for example the Regional Planning Guidance and Community Strategy in Cheshire, and the Provincial Territorial Plan of the Provinces of Bologna and Modena, that was further developed into the Emilia Romagna Regional Territorial Strategy. The project thus far has been a milestone for the application of ecological networks throughout Europe (Pungetti 2009a) and received widespread recognition including a showcase at the World Summit in Johannesburg in 2002 and a WWF Golden Panda in 2004.

The establishment of ecological networks in Europe has required some of the most advanced applications of the principles of landscape ecology to land use planning (Bennett 1991; Jongman et al. 2011). Developments in this field, combining theoretical concepts of landscape ecology with the practice of landscape planning and management, were illustrated in the ECONET Research (Jongman and Pungetti 2004, 2011).

In addition to biological and physical considerations important to biodiversity protection and restoration, cultural and aesthetic issues are equally important to illustrate how sympathetic land use policies can be implemented. Examples were analysed for large scale areas such as Estonia (Remm et al. 2004), as well as regional areas such as Milan (Massa et al. 2004), and it was demonstrated that networks and greenways have relevance not only to landscape and biodiversity conservation, but also to the planning process.

Both the ECONET Project and ECONET Research, conversely, supported a multifunctional landscape with (a) a particular view to socio-ecological heritage; (b) an integrative approach and a holistic model, and (c) a co-occurrence principle where landscape serves multiple demands.

4.3 Cultural Landscape and Heritage

'A cultural landscape is fashioned from a natural landscape by a culture group. Culture is the agent, the natural area is the medium. The cultural landscape is the result' (Sauer 1963, 343). Besides, cultural landscape can be seen as the 'combined works of nature and of man' (UNESCO 2008, 14). Moreover, with the assertion that landscape is 'an area, as perceived by people, whose character is the result of the action and interaction of natural and/or human factors' (Council of Europe 2000, 3) the European Landscape Convention bridges both cultural and natural aspects of landscape, expanding to its social dimension. With this in mind, the EUCALAND Project 'European Culture expressed in Agricultural Landscapes' of the EC Culture Programme fostered cooperation between the United Nations Educational, Scientific and Cultural Organization (UNESCO) and the Council of Europe in the area of agricultural landscape.

The vision of the project was for Europeans to recognise agricultural landscapes as a significant part of their cultural heritage. These landscapes therefore have been investigated in the light of the meanings that people have for them, and then described from past to present, in view of a common debate and classification. The final goal was to reach general recommendations on alternative ways for dealing with future agricultural landscapes, addressed not only to scientists and planners, but also to policy makers and especially to the people of Europe (Pungetti and Kruse 2010).

The EUCALAND Project brought together 40 institutions from 27 countries with an interdisciplinary and intercultural vision for long-term cooperation on European agricultural landscapes (Fig. 4.4). Among these, experts from 13 countries formed 6 interlinked and coordinated multi-disciplinary teams. Each team undertook research and drew conclusions on European agricultural landscape issues such as their description, history, classification, policy, planning and dissemination.

Agricultural landscapes are not just perceived in terms of farming and natural features (Amend et al. 2008), but also as a common heritage carrying social and cultural values (Pungetti 2009b). The characteristic components of the European agricultural landscapes have been identified, highlighting the cultural, social and psychological benefits for the well-being of citizens, as well as their future trends. The development of many of the characteristic features of these landscapes, in addition, shows the historical passage of time. Their history, accordingly, has been outlined with the similarities and differences between the countries involved.

A first classification of agricultural landscapes, debated throughout Europe, finally reached a consensus. The analysis moved on from the various existing

Fig. 4.4 The EUCALAND Project on European agricultural landscape and cultural heritage

European landscape classifications towards a new focus on agricultural landscape types, viewed as the product of history, underlining general aspects and similarities between them at a European level.

Agricultural landscapes have also been examined under the framework of international heritage policies, with the European Landscape Convention playing a central role. Moreover, different European view points, as well as economic, social and ecological trends, have been incorporated when considering the heritage of these landscapes. From this, recommendations were drawn up as guidelines for politicians, scientists and planners, aimed at making the wider population in Europe more aware of their cultural heritage and hence better able to plan for their future landscape.

4.4 Biocultural Diversity

As illustrated above, nature and culture are the pillars for holistic landscape research. Ecological landscapes support wilderness within their geological, morphological and ecological settings. However geology, topology and habitats are just parts of what constitutes an ecological landscape, where man today continuously intervenes. Yet nature and culture are mutually intertwined to create together a particular character to the landscape, which goes beyond its underlying natural and physical features. The human imprint is clearly marked in the cultural landscape.

As Naveh (2011) pointed out, cultural landscape is the tangible meeting point between nature and mind, and the conservation of its cultural assets should be an integral part of holistic and dynamic landscape management. This kind of landscape is more than a puzzle of mosaics in repeated patterns of ecosystems, as it retains a multidimensional organised landscape complexity.

Cultural landscape moreover can convey not just cultural but also spiritual relationships with nature, and often reflects traditional techniques of sustainable land use. It supports biological diversity in many regions of the world, but above all supports cultural diversity. The protection of traditional cultural landscape, with its tangible and intangible values, is therefore helpful in maintaining biocultural diversity (Pungetti 2012a).

Biocultural diversity, specifically, comprises the diversity of life manifested in biology and ecology, as well as in cultures, languages and spiritual beliefs (Fig. 4.5). These are interrelated within a 'complex socio-ecological adaptive system' (Maffi and Woodley 2010).

The CO@ST Initiative 'Natural and Cultural Heritage of Coasts and Islands' at CCLP, under the United Nations Environment Programme (UNEP) and the International Union for Conservation of Nature (IUCN), supports biocultural diversity with research on the conservation of natural and cultural heritage in islands and coastal areas. This clearly requires a holistic interdisciplinary approach able to provide prevention of, and responses to, natural and human hazards.

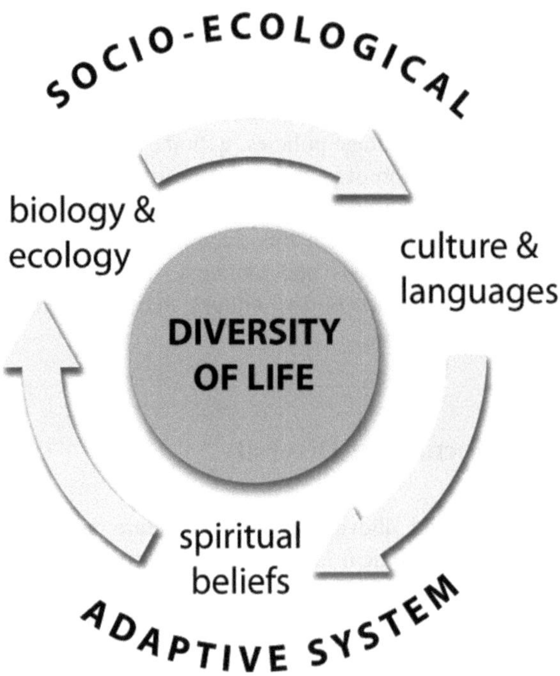

Fig. 4.5 The socio-ecological adaptive system, linking the different manifestations of life

CO@ST embraces this concept with innovative methods, tools and measurements, and offers a novel approach to the analysis, mapping, planning and management of world coastlines and islands. The main goal is to preserve their natural and cultural values, including their landscapes, seascapes, beauty, ecosystems and services.

The ESLAND Project 'European Culture expressed in Island Landscapes' of the EC Culture Programme is looking further with research on European island landscapes, considering their historical and cultural heritage, character and identity, and providing scenarios and tools for their future development. Islands clearly highlight the richness of global landscape diversity and are ideal for the application of natural and cultural approaches in landscape research (see Vogiatzakis et al. 2008).

In particular, the conceptual framework of the ESLAND Project supports consideration of cultural heritage in European island landscapes of different size, including the unique identity and values people attach to them. The concept is applied carrying out an interdisciplinary approach for an intercultural discussion, leading to an agreed methodology on the study of European island landscape evolution, classification, identity and scenario. The ultimate goal is to contribute to the implementation of European and international policies, providing tools for a type of landscape development that should be more culturally and sustainable oriented than at present (Pungetti 2012b).

Starting with a core of experts in island landscape research coming from eight organisations in six European countries, the ESLAND partnership has expanded to cover all European countries with islands of any size, and a few leading global

Fig. 4.6 The ESLAND Project on European island landscape and cultural heritage, with its partnership

organisations, for a total partnership of 33 organisations in 24 countries (Fig. 4.6). The core group specifically considers the history of European island landscapes fundamental not only for the understanding of present day landscape features, but also for preserving their heritage and for conveying traditional practices to modern developers. It has moreover recognised that landscape character assessment of European islands should include both natural and cultural aspects in order to contribute to heritage knowledge, as indicated above.

Cultural heritage and identity are indeed precious values that need to be preserved. They are unique to each place and in islands they embed a strong character. The ESLAND Research accordingly promotes an interdisciplinary approach to identify the key island landscape values in order to raise awareness about European cultural heritage and identity, and to support more interaction between local communities—the stewards of their landscapes. It also seeks to produce innovative scenarios for planning European island landscapes, visualising their future developments and utilising a participatory approach to preserve, as well as to develop, these landscapes with a more culturally-oriented perspective.

Such a perspective has been clearly underlined by Naveh (2011, 6), who has left us with a call: ‘we have to realize that we are dealing not only with a need for a sustainable environment, but with a sustainable world of our Total Human

Ecosystem, by which we humans, together with all other organisms, are integrated with our total environment into an irreducible whole'. He has advocated a change in the way of thinking, with a true synthesis between nature and culture indicated in interdisciplinary landscape science which serves as a bridge between nature and mind, between bio-ecological processes and cognitive-perceptual dimensions, and between tangible and intangible values.

Tangible values in traditional landscape ecology have been extensively studied, analysed, designed and managed. They include physical elements, species and natural resources. The need to preserve our natural heritage, endangered species and non-renewable natural resources has led governments to put strategies in place for biodiversity and nature conservation, based also on landscape ecology principles (see IUCN 2008 and Convention on Biological Diversity 2010).

Intangible values in holistic landscape ecology, instead, need further exploration. They include cultural elements, language diversity, traditional knowledge, spiritual practices and cultural heritage. Intangible cultural heritage is recreated by communities and groups in response to their environment, their interaction with nature and their history (see UNESCO 2002, 2007). Besides, cultural and spiritual heritage, language diversity and traditional knowledge are increasingly valued in international programmes related to biodiversity, nature conservation and landscape ecology (see Pungetti et al. 2012).

4.5 Sacred Landscape and Heritage

The above intangible values, and in particular the spiritual dimension of ecology, has been emphasised by scientists, as well as numerous religious organisations, indigenous peoples and local communities as illustrated in literature (Næss 1989; Berkes 1999; Tucker and Grim 2000; Dudley et al. 2005; Pungetti 2008). The common effort is to prevent destruction of sacred landscapes and places, thus healing social wounds and aiding ecological and environmental struggles.

Long ago indigenous people marked their territories, leaving traces visible today in archaeological sites, and shaping a land for which they claim spiritual responsibility—their sacred landscape. If, on the one hand, several of these landscapes are today protected, on the other hand many shrines and pilgrim trails, crucial for indigenous religions, are left outside the protected zones and are under threat, if not destroyed, by human impact such as mineral and water extraction, quarrying, aggressive tourism and ski resorts among others.

One of these threatened landscapes is the San Francisco Peaks. To the Hopi people they represent the spirit of the land, what we call 'genius loci'. The Hopi, the legend tells, promised to their genius loci, the bringer of rain, to be good stewards of the earth (Beggs and McLeod 2003). Evidence of this spiritual stewardship is visible on the petroglyphs and shrines of sacred places, still used by the native people who, through this use, maintain their claim to these sacred landscapes.

A sacred landscape is as an area with spiritual significance to peoples and communities. It can contain sacred species, which have strong spiritual values for the community they refer to. It can also include sacred natural sites, which blend natural and spiritual values (Pungetti 2012a). The 3S Initiative 'Sacred Species and Sites' has been set up by CCLP under IUCN and World Wildlife Fund (WWF) to support research in the above field. The main goal is to improve recognition of the spiritual values of species and sites connected with landscapes, cultures and traditions.

A wide spectrum of experts from academia and international conservation organisations gathered in Cambridge in 2007 to demonstrate ways in which sacred species and sites contribute to landscape ecology and conservation biology. Together with these scholars, spiritual leaders from around the world helped to provide new insights into biocultural diversity conservation and sacred landscape.

Since then, key conceptual topics have been connected to over 50 case studies worldwide, and described by the authors in the 3S Research, highlighting issues from fundamental theory to practical applications. Sacred landscapes, sites, plants and animals from around the world have been examined to demonstrate the links between traditional spiritual beliefs, practices and nature conservation. The research group has proposed further topics for the biocultural agenda, providing guidelines for future research and practice.

The 3S Research promoted the integration of spiritual values of ecosystems and landscapes into policy, planning and management. In demonstrating ways in which sacred species and sites can contribute to landscape ecology and conservation biology, the 3S had special significance for advancing studies in these fields (Pungetti et al. 2012).

The authors concluded that for species, being considered sacred is often a protector from threats. Sacredness alone, however, is not enough to preserve a sacred species. No one set of values will be sufficient on its own to achieve ideal conservation outcomes. In the contemporary world, therefore, an integrated holistic approach is required for sustainable biocultural conservation and to preserve heritage of sacred landscapes.

4.6 Traditional Ecological Knowledge and Landscape

Over time, the elders of indigenous people have passed their knowledge on to the younger generations via legends, myths, oral stories, drawings and texts, often engraved on rocks and recorded on natural materials. Due to the lack of written texts, these cultural expressions constitute what the natives call their 'history book on the land', and are in the hands of those designated to maintaining their traditions. Such knowledge acquired and preserved through generations in an indigenous or local society, consisting of experience in working to secure subsistence from nature with an ecological basis, is recognised as 'Traditional Ecological Knowledge' (TEK) or indigenous peoples' ecological knowledge.

TEK, together with cultural and sacred landscapes, can contribute to modern techniques of sustainable land use and can maintain and enhance natural, cultural and spiritual values of land and their communities (see Berkes 1999). These local communities and indigenous peoples were taught to care for land and life, thus the land where they lived, and where native people still live, is their holy land. However over the centuries they have struggled to protect their sacred places: sacred grounds where their ancestors rest, sanctuaries for medicinal plants, and landscapes of unusual natural power and cultural significance.

Devils Tower, in the Black Hills of Wyoming, USA, is an example. The story tells that if a dying man, poor in body and spirit, went into the Black Hills, he would emerge from there restored and in excellent health. These hills were hence considered by the local indigenous people as being the centre of life, and all around them the places and landscapes were considered sacred and kept 'in light of reverence' (Beggs and McLeod 2003). However human activities could be devastating, as in the case of a mining operation using large quantities of water, with the effect of drastically reducing the production of holy springs in the area (*ibid.*). In 1906, President Roosevelt declared Devils Tower as the first United States National Monument. Today it is also listed under the protected landscapes and seascapes of IUCN (Category V).

Sacred groves are another expression of TEK. They are one type of Japan's Satoyama cultural landscape and play an important role in the preservation of habitats and their biodiversity. These small-scale woods are isolated from larger forests, and surrounded by Shinto shrines and other places of Shinto and Buddhist worship all over Japan. Vegetation in sacred groves has been protected for centuries as a subject of worship, and can therefore often provide vital hints on the original plants and on the relationship between nature and humans in a particular region. Sacred groves, moreover, can serve as stepping stones to link the vegetation of larger forests, or can provide green space for recreation not only in rural but also in peri-urban and urban areas. Furthermore, they host traditional Shinto ceremonies and festivals, and are highly valued as cultural properties. As illustrated at the Symposium 'Landscape ecological perspectives on biocultural diversity and sacred landscape' of the IALE 8th World Congress, today the preservation of sacred groves is struggling due to the impact of development around them. Research is currently being carried out on their vegetation and the relationships with local people from a landscape ecology perspective, with the aim to suggest strategies to conserve Japan's sacred groves (Fukamachi and Rackham 2012).

East Asian landscape character, besides, is often associated with the concept of *Feng-shui*, which is a combination of the terms *feng* (wind) and *shui* (water). The purpose of *Feng-shui* is not limited to traditional land use, but extends to landscape ecology, contemporary land use management and cultural landscape policies. The notion of *Chisanchisu*, nevertheless, is common in countries like Korea, meaning that the happiness of people can be secured through the virtuous management of mountains and water sources (Hong and Kim 2011). This philosophy has been influential in East Asia for centuries, and is now reflected in the national land use policies of several countries. However, the excessive development of industries and land brought by modernisation has resulted in rapid changes in the traditional land use practices, and analogously the concept of *Feng-shui* is gradually vanishing. It is indeed difficult today to find those characteristic cultural landscapes surrounded by breathtaking lakes and mountains, which were once painted on silk.

As a contribution to TEX, the conservation of agricultural heritage systems and practices worldwide has been pursued through the global Initiative ‘Conservation and adaptive management of Globally Important Agricultural Heritage Systems’ (GIAHS) initiated by FAO in 2002. GIAHS aims to establish the basis for the international recognition, dynamic conservation and sustainable management of such systems, agricultural biodiversity and their associated biodiversity knowledge systems, food and livelihood security, landscapes and cultures. Worldwide agricultural systems, with their natural resources and landscapes, have been shaped and conserved by generations of farmers using locally adapted management practices. These systems, built on local knowledge and experience, express their landscape and cultural evolution through the diversity of ecosystems and knowledge, as well as the close relationship of man with nature. As pointed out at the previously mentioned Symposium of the IALE 8th World Congress, agricultural heritage systems have resulted not only in outstanding landscape maintenance and adaptation of resilient ecosystems, together with globally significant agricultural biodiversity systems based on indigenous knowledge, but above all in the sustained provision of multiple goods and services, food and livelihood security, and quality of life. Parallel to this, the need to investigate further the concept of landscape services has been addressed by previous research (Termorshuizen and Opdam 2009).

China, which was one of the first pilot countries of GIAHS, is rich in agricultural history and heritage systems. Therefore three pilot systems were selected by FAO: Qingtian Traditional Rice-fish System in Zhejiang province, Hani Rice Terraces System in Yunnan province, and Wannian Rice Culture System in Jiangxi province. These systems were studied for their agro-biodiversity characteristics and multi-values, searching for dynamic conservation approaches to be practiced in loco. The experiences and lessons could certainly be applied in modern agricultural practices, and to other agricultural heritage systems.

Another example of TEK is the lemon gardens in the Sorrentino-Amalfitana Peninsula, Italy. A candidate to be listed on the GIAHS systems, this agricultural landscape gives character to the entire peninsula. Lemon pergolas, chestnut windbreaks, wall terraces and narrow footpaths have been built and preserved over centuries to guarantee the conservation of the local lemon varieties (*Citrus limonum*). These varieties were exchanged for gold on Mediterranean ships in the sixteenth century for their healing properties against scurvy. Due to their economic value, local communities had to find alternative ways of cultivation on a land with particularly steep terrain and environmental constraints. This resulted in the construction of stone terraces on very steep slopes. In this way the local communities have succeeded in protecting their territory and have contributed to preserving the soil from hydrogeological instability. What is more, they have shaped the land to create an outstanding coastal landscape of incredible natural and cultural value.

4.7 The Right to Landscape

Because of the strong interaction of landscape values with great diversity, i.e. natural, cultural, social and spiritual, there are tensions and conflicts between the parties involved. Native Americans for example, claim the right to worship and protect the genius loci and the earth in their sacred places, whilst other citizens claim the right to mine and extract natural resources there, or to climb mountains that are considered sacred for the natives. Although this happens on public land, indigenous people, who have lived in those landscapes from time immemorial, claim to keep their land in balance by worshipping. In fearing careless destruction by society, they feel deprived of religious freedom and their right to landscape.

Scientists are therefore asking for landscape rights to be respected. The RtL Initiative 'The Right to Landscape' under the International Federation of Landscape Architects (IFLA) and Amnesty International has been set up at CCLP with this goal, proposing a novel approach for an international multidisciplinary academic discourse associated with landscape and human rights. RtL is based on the premise that landscape is full of meanings, and comprises an underpinning component for ensuring the well-being and dignity of people (Egoz et al. 2011). The aim is to collectively define the concept of 'The Right to Landscape' and provide a body of knowledge that supports human rights.

The 60th anniversary of the Declaration of Human Rights, celebrated in December 2008, called for a reflection on ethical human dilemmas and for a critical examination of future ways of dealing with human rights in the context of crises such as climate change, economic recession and civil wars. In order to do so, it will be necessary to design holistic frameworks that capitalise on the connections of different resources at different levels, between natural and cultural settings (Fig. 4.7).

Landscape is therefore proposed here as an umbrella concept, which allows its multiple tangible elements to unite with its intangible values, and in turn to generate alternative scenarios for constructing new approaches to land use and human well-being. By expanding on the concept of human rights in the context of landscape as a container of both tangible and intangible values, it has been possible to produce a discourse that includes different contexts.

This new discourse on landscape and human rights served as a platform to inspire a diversity of ideas and conceptual interpretations highlighted at the RtL Conference in Cambridge in December 2008. The worldwide case studies discussed, interdisciplinary in the theoretical situation of their authors, broke fresh ground for an emerging critical dialogue on the convergence of landscape and human rights. The results of the RtL Research (Egoz et al. 2011) have shown landscape as 'a concept indispensable to the probing of human nature and human well-being, drawing on and cross-fertilizing such diverse fields as the study of nature, history, anthropology, psychology, politics, and law' (Tuan 2011, 310).

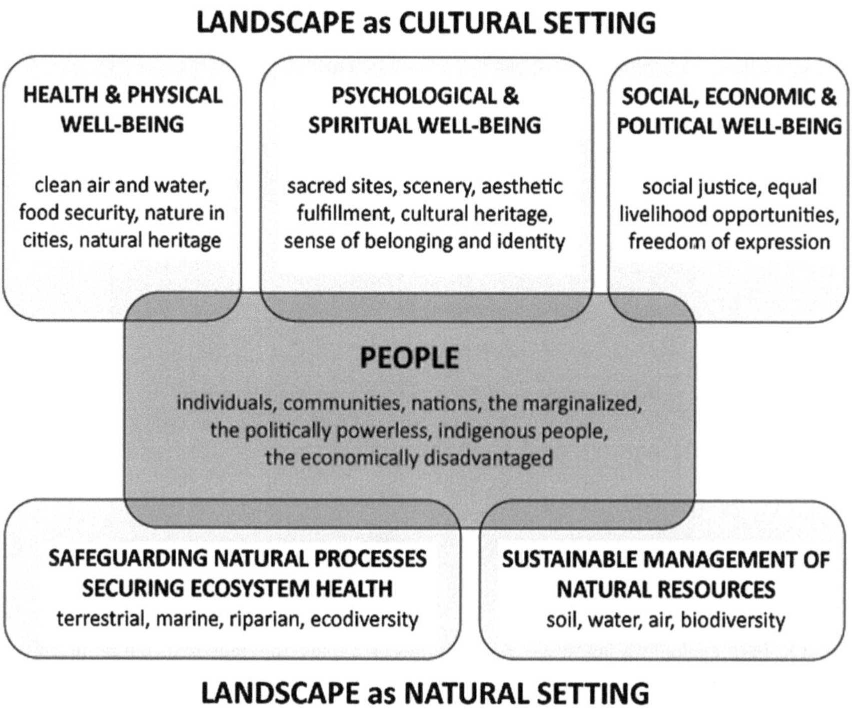

Fig. 4.7 The right to natural and cultural landscape resources, after Egoz et al. (2011)

4.8 Biocultural Landscape Approach

Ecological, cultural and sacred landscapes support sustainable resource management, ecosystem services and well-being. They sustain biocultural diversity and can be developed by applying a new research model based on biocultural landscape.

Although several attempts have been made to move from a landscape mosaic to a landscape resource approach in ecological landscape research (see Fig. 4.2), in the past natural heritage was the main focus, while interpretation of the cultural heritage of landscape was limited. As nature conservation was kept separate from other kinds of conservation, past landscape models proposed, more often than not, management policies aimed at minimising changes. These models were initially designed using a top-down approach, where communities were peripheral to the landscape process.

The model proposed here, instead, is inclusive; it bridges natural and cultural heritage with an expansive interpretation of culture comprehensive of its manifestations, i.e. food, music, agriculture, fishing, forest, spirituality, art, language and poetry. By integrating different kinds of conservation, this model intends to be

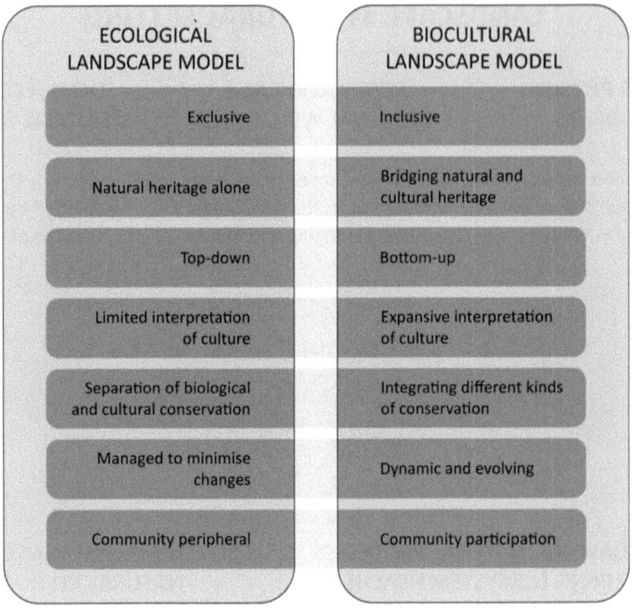

Fig. 4.8 Past ecological landscape research model versus the proposed future biocultural landscape model

dynamic and evolving, accepting and managing change. It uses a bottom-up approach, where communities are involved and participate in landscape development (Fig. 4.8).

The new model provides the framework for a biocultural landscape approach in landscape research and landscape ecology. Biocultural landscapes are understood as areas with biotic and cultural elements, connected to each other by historical/ecological interaction on the territory (Pungetti 2012a). They form holistic systems which include diverse elements such as land tenure, land use patterns, production systems, cultural identity, spiritual dimension and genius loci among others.

Indeed nature and culture are interlinked in the biocultural landscape framework (Brown et al. 2005) which puts people at the core of the process (see Fig. 4.7). Biological and cultural diversity are here coupled: if cultural diversity is vanishing, biological diversity is threatened and vice versa. Biocultural landscapes, conversely, are created by people who have special TEK. They are precisely a transmission of TEK in the everyday life of 'very normal people' with their culture, education, identity, values, traditions, life-style and land use practices.

Seen as sustainable interactions between people and their environment, cultural landscapes and sacred sites offer possibilities to elaborate approaches to sustain biocultural diversity on the appropriate spatial and temporal scales (Schaaf and Lee 2006). Because of their critical ecological, but also cultural, historical and institutional dimensions, cultural and sacred landscapes can thus be viewed as sustainability units, well adapted for the exploration of the mutually beneficial interactions between biological and cultural diversity which are vital for enhancing environmental integrity and human well-being.

Recognising the inextricable link between biological and cultural diversity, UNESCO and the Secretariat of the Convention on Biological Diversity (SCBD) joined forces to understand and address the interaction between biological and cultural diversity, and the common challenges posed by contemporary processing affecting the current diversity trends. The result was the SCBD-UNESCO Joint Programme on Biological and Cultural Diversity, which aimed at deepening global awareness of the inter-linkages between cultural and biological diversity. State Parties and other relevant stakeholders were invited to support the implementation of the joint programme. In addition to linking grassroots and community initiatives with local, regional, national and global policy processes, one of the key goals is to advance knowledge on the ways in which cultures have shaped and continue to shape biodiversity in a sustainable way, so as to be able to identify and implement management and policy approaches to sustain our planet's diversity (Persic and Martin 2008).

Building on the above, the biocultural landscape model can be implemented at different scales—global, national and local—each with its own objectives, as shown in Fig. 4.9.

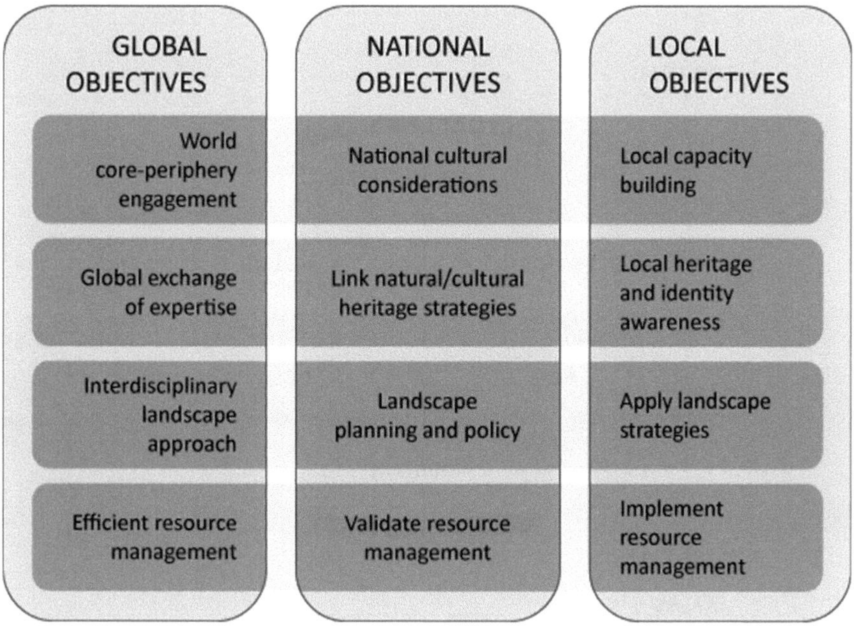

Fig. 4.9 Different scales of implementation of the biocultural landscape model with their objectives

The new model, furthermore, offers a fresh approach to biocultural conservation based on biocultural landscape research, considering:

- a paradigm shift in nature conservation and conservation biology
- integration of biological and cultural diversity
- ecosystem and landscape services
- biocultural landscape and sacred landscape ecology
- traditional knowledge and spiritual beliefs
- environmental ethics and human rights
- know-how, empowerment and participation of local and indigenous people
- sustainable management of natural and cultural resources
- preservation of natural, rural and cultural heritage (Fig. 4.10).

In this context, a group of scientists presented their research at the already mentioned Symposium 'Landscape ecological perspectives on biocultural diversity and sacred landscape' of the 8th World Congress of IALE, with the aim of offering a perspective from the point of view of landscape ecology on biocultural diversity conservation and sacred landscape. The Symposium, coordinated by the CCLP core group, was initiated by the results of the 3S Initiatives on Sacred Species and Sites and enriched by studies on biocultural landscape from all the participants, thus expanding the concept to landscape ecology.

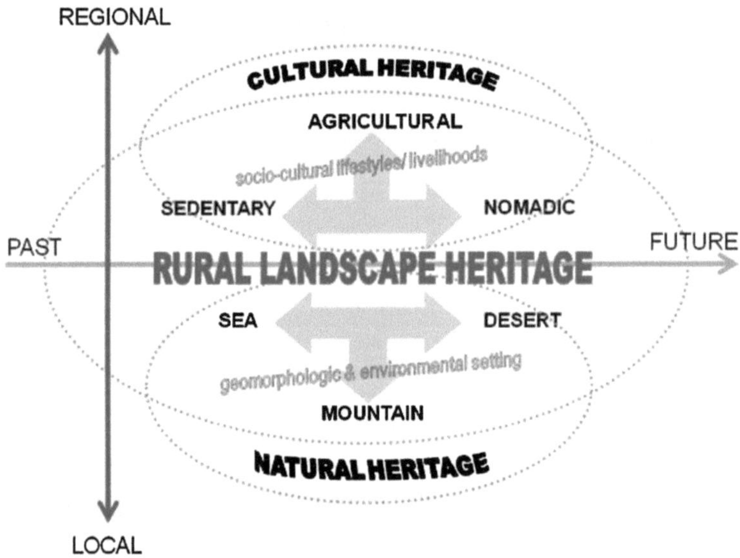

Fig. 4.10 Preservation of natural, rural and cultural heritage through time and space, after Makhzoumi (2013)

From here, the IALE Biocultural Landscape Working Group has been set up to support research on biocultural landscapes from a worldwide landscape ecology perspective. The ultimate goal is to demonstrate ways in which biocultural landscape and landscape ecology can contribute to the conservation of biodiversity and cultures.

Research findings of the working group have demonstrated so far that a sense of sacredness helps in safeguarding natural and cultural heritage of sites and species, although it sometimes puts the latter in danger. However, research has just started to systematically study biocultural landscape; qualitative analyses and assessment of intangible values require further attention. International, intercultural and interreligious understanding, together with respect, dialogue and cooperation between communities, are needed on the ground, to couple research on biocultural landscape and diversity with a multidisciplinary and multifunctional approach.

4.9 Conclusions

In acknowledging the current degradation and rapid transformation of European landscapes, the 10th Council of Europe meeting of workshops for the implementation of the European Landscape Convention has underlined the urgent need to contribute to preserving the quality of our landscapes. To reach this goal, the multifunctional value of these landscapes should be taken into account, and those implementing the Convention should take steps to secure their intrinsic quality and multiple values. This paper has built on this precondition to correlate it with the emerging landscape ecology perspectives on the link between ecological and cultural diversity.

As outlined before (Council of Europe 2000) European landscape enriches the quality of life of people and plays a key role in the ecological, cultural, social, and we add spiritual, realms. It therefore constitutes a resource from which it is possible to develop a more sustainable future with the support of responsible organisations, public authorities and citizens, able to define the framework in which the cultural landscape function, quality and value can be secured.

The framework proposed here is the biocultural landscape approach. Biocultural diversity, it is argued, needs to be considered in landscape ecology studies in order to promote respect for landscape and nature, and ultimately to integrate the spiritual and cultural values of land and local communities into ecological landscape design and planning, nature conservation and sustainable development. Conversely, biocultural landscapes become pillars for interdisciplinary landscape science, which serves as a connection between nature and culture, and between their tangible and intangible values.

A biocultural vision implies equally that heritage conservation, whether natural or cultural, should consider the human context and local communities, and should expect to achieve optimal results only when taking an approach that employs and integrates the cultural outlook in nature conservation.

Concluding, per se or in relation to biocultural diversity, biocultural landscape offers new synergies and knowledge, useful for advancing sustainable landscape conservation and development, and for providing an integrated perspective for a truly holistic approach in landscape ecology. The protection, conservation and restoration of these landscapes are indispensable for a sustainable planet, and for passing their intrinsic intangible values and heritage on to future generations.

Acknowledgments The author would like to thank her inspired maestri and friends at CCLP, Oliver Rackham, Jala Makhzoumi, Jay Appleton and Dick Grove, together with Rob Jongman, Toby McLeod and the IALE 8th World Congress Board. This paper is in memory of Francesco Di Castri and Zev Naveh.

References

Amend T, Brown J, Kothari A, Philips A, Stolton S. Protected landscapes and agrobiodiversity values. Gland: IUCN; 2008. p. 45–58.

Beggs M, McLeod C. The sacred land reader. Berkley: Earth Island Institute; 2003.

Bennett G, editor. Towards a European ecological network. Brussels: Institute for European Environmental Policy; 1991.

Berkes F. Sacred ecology: traditional ecological knowledge and resource management. London: Taylor & Francis; 1999.

Brown J, Mitchell N, Beresford M, editors. The protected landscape approach: linking nature, culture, and community. Gland: IUCN; 2005.

Convention on biological diversity. In-depth review of the implementation of the programme of work on protected areas. UNEP/CBD/SBSTTA/14/L.5. Nairobi: CBD; 2010.

Council of Europe. European landscape convention. CET NO. 176. Florence: Council of Europe; 2000.

Dudley N, Higgins-Zogib L, Mansourian S. Beyond belief: linking faiths and protected areas to support biodiversity conservation. Gland: WWF, Equilibrium and ARC; 2005.

Egoz S, Makhzoumi J, Pungetti G, editors. The right to landscape: contesting landscape and human rights. London: Ashgate; 2011.

Farina A. Principles and methods in landscape ecology: towards a science of the landscape. Dordrecht: Springer; 2006.

Forman RTT, Wilson EO. Land mosaics: the ecology of landscapes and regions. Cambridge: Cambridge University Press; 1996.

Fukamachi K, Rackham O. Sacred groves in Japanese satoyama landscapes: a case study and prospects for conservation. In Pungetti G, Oviedo G, Hooke D, eds. Sacred species and sites: advances in biocultural conservation. Cambridge: Cambridge University Press; 2012. p. 419–424.

Grove AT, Rackham O. The nature of Mediterranean Europe: an ecological history. New Haven: Yale University Press; 2003.

Hong SK, Kim JE. Changing biocultural diversity of sacred landscapes in Southwestern Korea. In: Jones B, Fu B, editors. Landscape ecology for sustainable environment and culture. Proceedings of the 8th world congress of the international association for landscape ecology. Beijing: IALE; 2011.

IUCN. IUCN and UNESCO publish guidelines on sacred natural sites in protected areas. WCC, 07 Oct 2008. Barcelona: IUCN Press Release; 2008.

Jongman RHG, Bouwma IM, Griffioen A, Jones-Walters L, Van Doorn AMM. The Pan European Ecological Network – PEEN. Landscape Ecol. 2011; 26(3):311–326

Jongman RHG, Pungetti G, editors. Ecological networks and greenways: concept, design, implementation. Cambridge: Cambridge University Press; 2004.

Jongman RHG, Pungetti G, editors. Ecological networks and greenways: concept, design, implementation (Chinese edition). Cambridge and Beijing: Cambridge University Press and China Architecture & Building Press; 2011.

Maffi L, Woodley E. Biocultural diversity conservation: a global sourcebook. London: Earthscan; 2010.

Makhzoumi J. Is rural heritage relevant in an urbanizing Mashreq? Exploring the discourse of landscape heritage in Lebanon. In: Maffi I, Daher R, editors. Practices of patrimonialization in the Arab World: positioning the material past in contemporary societies. London: I.B. Tauris, (in progress); 2013.

Makhzoumi J, Pungetti G. Ecological landscape design and planning: the Mediterranean context. London: Spon-Routledge; 1999.

Makhzoumi J, Pungetti G. Ecological Landscape design and planning: The Mediterranean context (Chinese edition). Taiwan: Liu-Ho Publisher; 2003.

Massa R, Bani L, Baietto M, Bottoni L, Padoa-Schioppa E. An ecological network for the Milano region based on focal species. In: Jongman RHG, Pungetti G, editors. Ecological networks and greenways: concept, design, implementation. Cambridge: Cambridge University Press; 2004. p. 188–99.

Næss SH. Ecology, community and lifestyle. Cambridge: Cambridge University Press; 1989.

Naveh Z. Landscape ecology for sustainable environment and culture. In: Jones B, Fu B (eds). Landscape ecology for sustainable environment and culture. Proceedings of the 8th world congress of the international association for landscape ecology. Beijing: IALE; 2011.

Persic A, Martin G, editors. Links between biological and cultural diversity: concepts, methods and experiences. Paris: UNESCO; 2008.

Pungetti G. Water environment landscape: a comparison between Dutch and Italian planning. Bologna: Pitagora; 1991.

Pungetti G. A holistic approach to landscape research in the Mediterranean. In: Makhzoumi J, Pungetti G. Ecological landscape design and planning: the Mediterranean context. London: Spon-Routledge; 1999.

Pungetti G. Cultural and spiritual values in landscape conservation and management. In: Teofili C, Clarino R, editors. Riconquistare il paesaggio: la convenzione europea del paesaggio e la conservazione della biodiversità in Italia. WWF: Rome; 2008. p. 360–6.

Pungetti G. From ecological networks to conservation networks. In: Catchpole R, Smithers R, Baarda P, Eycott A, editors. Ecological networks: science and practice. IALE-UK: Gartstang; 2009a. p. 129–34.

Pungetti G. Natural and cultural heritage of European agricultural landscapes in transformation. In: Breuste J, Kozová M, Finka M, editors. European landscapes in transformation: challenges for landscape ecology and management. European IALE conference 2009. Salzburg and Bratislava: University of Salzburg, Slovak University of Technology in Bratislava and Comenius University in Bratislava; 2009b. p. 135–6.

Pungetti G. Sacred species and sites: dichotomies, concepts and new directions in biocultural diversity conservation. In: Pungetti G, Oviedo G, Hooke D, editors. Sacred species and sites: advances in biocultural conservation. Cambridge: Cambridge University Press; 2012a. p. 13–27.

Pungetti, G. Islands, culture and landscape. J Marine Island Cultures 2012b;1(2):51–54.

Pungetti G, Kruse A, editors. European culture expressed in agricultural landscapes: perspectives from the Eucaland project. Rome: Palombi Editori; 2010.

Pungetti G, Oviedo G, Hooke D, editors. Sacred species and sites: advances in biocultural conservation. Cambridge: Cambridge University Press; 2012.

Remm K, Külvik M, Mander Ü, Sepp K. Design of the Pan-European ecological network: a national level attempt. In: Jongman RHG, Pungetti G, editors. Ecological networks and greenways: concept, design, implementation. Cambridge: Cambridge University Press; 2004. p. 151–70.

Sauer CO. The morphology of landscape. In: Leighly J, editor. Land and life: a selection from the writings of Carl Ortwin Sauer. Berkeley: University of California Press; 1963. p. 315–50.
Schaaf T, Lee C, editors. The role of sacred natural sites and cultural landscapes. Paris: UNESCO; 2006.
Termorshuizen JW, Opdam PFM. Landscape services as a bridge between landscape ecology and sustainable development. Landscape Ecol. 2009;24(8):1037–52.
Tuan YF. The right to landscape review. In: Egoz S, Makhzoumi J, Pungetti G, editors. The right to landscape: contesting landscape and human rights. London: Ashgate; 2011. p. 310.
Tucker ME, Grim J. Religions of the world and ecology series. Cambridge, MA: Center for the Study of World Religions, Harvard Divinity School, University of Harvard; 2000.
UNESCO. UNESCO Universal declaration on cultural diversity. Paris: UNESCO; 2002.
UNESCO. First inscriptions on UNESCO's intangible cultural heritage lists to take place in Sep 2009. Paris: UNESCO; 2007.
UNESCO. Operational guidelines for the implementation of the world heritage convention. Intergovernmental Committee for the World Cultural and Natural Heritage. Paris: UNESCO; 2008.
Vogiatzakis IN, Pungetti G, Mannion AM, editors. Mediterranean island landscapes: natural and cultural approaches. Dordrecht: Springer; 2008.

Chapter 5
Using Ecosystem Services in Community-Based Landscape Planning: Science is Not Ready to Deliver

Paul Opdam

Abstract Community-based landscape governance is considered as conditional to achieving sustainable landscape. I consider landscape governance from the point of view of adapting landscapes to create value out of ecosystem services, using the social–ecological system model as a theoretical framework. I advocate the use of the term landscape services because it can serve as a common ground between science and local communities, and between scientists from different disciplines. Six principles for sustainable landscape change are presented, which can be developed as a checklist in planning, and as requirements to scientific methods. From the current literature it is obvious that ecosystem service research does not provide the type of science that is required to support sustainable, community-based landscape planning. Research is mainly science driven, focussed on assessments at large spatial scale, and with policy users in mind. Active involvement of local stakeholders is scarce. There is a strong demand for approaches that are able to involve local governance networks and move the ecosystem services research out of the static mapping and evaluation approaches towards dynamic systems thinking. The chapter ends with a research agenda.

Keywords Landscape services · Criteria for sustainability · Social-ecological system · Green infrastructure · Adaptive governance · Knowledge application · Habitat networks · Landscape change

P. Opdam (✉)
Land Use Planning Chair Group and Nature and Society Group,
Wageningen University and ALTERRA, Droevendaalsesteeg 3 36708 Wageningen,
The Netherlands
e-mail: Paul.Opdam@wur.nl

B. Fu and K. B. Jones (eds.), *Landscape Ecology for Sustainable Environment and Culture*, DOI: 10.1007/978-94-007-6530-6_5,

5.1 Introduction: Placing Ecosystem Services in the Context of Sustainable Change of Landscapes

Since time immemorial, humans have been changing landscapes. Recent trends in using land to increase economic benefits have raised concern about the repercussions on valuable landscape services to do with the quality of human life and nature. There is a growing public demand for high-quality landscapes, especially in urbanizing areas and in regions with industrial farming (e.g. Jackson 2008; Kaplan 2007; Stephenson 2008; Tzoulas et al. 2007). In the Netherlands, for example, where the distinctions between urban and rural life are rapidly disappearing, citizens and businesses are challenging farmers by demanding that the landscape fulfil functions other than food production. If landscape development moves away from aiming to improve economic value towards gaining economic, social, and ecological values (Wu and Hobbs 2007), the adapting of landscapes enters the domain of sustainable development (Antrop 2006).

However, the uptake of the concept of sustainable development in landscape change planning has been problematic. For example, the concept of ecosystem services has failed to inspire spatial planning science (Termorshuizen and Opdam 2009). The reasons for this have not been studied extensively, but some causes have been proposed. Firstly, planning for sustainability requires knowledge to be integrated from scientific disciplines ranging from environmental to social sciences. A recent review on sustainability assessment methodologies (Singh et al. 2009) revealed that only a few methods take into account the environmental, economic, and social aspects. Secondly, the wide variety of interpretations of the sustainability notion in scientific disciplines, interest groups, and cultures (Antrop 2006; Singh et al. 2009) confuses decision makers and causes them to question the concept's credibility. Also, it allows actors to bend the implications of the concept according to their underlying interests, values, and convictions (after Borch 2007). Thirdly, the application of sustainability to on-the-ground landscape change is accompanied by a plethora of uncertainties: for example, with respect to future land use demands, climate change implications, and the impacts of land use change on landscape functions.

These characteristics are typical of an unstructured (wicked) problem (Hisschemöller and Hoppe 2001) which causes conflict and controversy and also interdependence—the latter because achieving one's objectives is conditional on others achieving theirs. Jiggins et al. (2007) and Van Bommel et al. (2009) showed how difficult it can be for stakeholders in a resource dilemma to accept such interdependence and its consequences. Unstructured problems are further marked by the multiple perspectives of stakeholder groups. Therefore, defining common visions on the need to adapt landscapes and finding the best solution to achieve them is a complex, usually unpredictable process, in which many factors (biophysical, social, economic, and political) interact. In such conditions, were uncertainty, contest and negotiation prevail over facts, landscape change planning adopts characteristics of adaptive governancewith a strong learning component embedded in local communities rather than organized as a government-led top down process. In the context of social–ecological systems, Armitage et al. (2009) described the term adaptive co-management as "a flexible system of resource management, tailored to specific places and situations, supported by and working in conjunction with various organizations at different scales". Based on this I understand community-based landscape planning as a process in which local communities develop a shared vision on their desired landscape, and collectively take decisions and organize action to adapt the physical structure of the landscape to realize their aims. In this chapter I explore how the emerging concept of ecosystem services, which typically connects the functioning of the landscapes to the multiple interests and benefits of its users, can be developed in science to become a common ground in community-based sustainable landscape planning.

To achieve this aim, I consider the multifunctional landscape as a social-ecological system (as defined by Walker et al. 2006), providing services to users and visitors (Termorshuizen and Opdam 2009), and spatially limitable by its specific patterns of physical attributes and land use (Turner 1989). I explore a new way of looking at landscape change, based on supply and demand of landscape services. I consider how in the multifunctional landscape context local actors may link landscape services to physical networks of relatively natural landscape elements, while keeping in mind principles of sustainable landscape change. In such a planning process science has a facilitating role. I will explore how the current literature on ecosystem services provides knowledge and tools to facilitate such community-based landscape planning. I close off with summarizing priority research to build a much stronger foundation for landscape planning based on ecosystem services, which will enhance the development of the sustainability concept in landscape ecology.

5.2 Landscape Services: Why a New Term is Needed

If humans thus value the landscape for its benefits, we may describe the relationship with the term *ecosystem service*. The term ecosystem service was born in the meeting zone between ecology and economy in the last decade of the 20th century, and got a boost in science and in policy by the Millennium Ecosystem Assessment (MEA) organized on behalf of the United Nations (Carpenter et al. 2009). There has been a long lasting debate on what would be an appropriate classification of types of ecosystem services (Wallace 2007; Constanza 2008). I will use here three categories that can be directly linked to stakeholder interests: (1) production services, such as the production of agricultural crops and fibres, (2) regulation services such as the purification of water by marsh vegetation and the pollination of commercial plants by wild bees, and (3) social services such as the perception of the beauty of nature and the influence on human mental health (following Hein et al. 2006, but leaving out provisioning services which are conditional to the other types rather than directly used).

The concept of ecosystem services connects ecosystems to humans. I will use the definition proposed by Fischer et al. (2009): "ecosystem services are the aspects of ecosystems utilized (actively or passively) to produce human well-being". Fisher et al. pointed out that services must be ecological phenomena, but that they do not have to be utilized directly. The original motive for launching this concept is that framing the value of biodiversity in the context of socio-economic benefits might improve the effectiveness of biodiversity conservation strategies, either as an alternative or as an extension to the intrinsic value of biodiversity. Most of the literature on ecosystem services is still framed in the context of conservation planning: large geographical scale, protected areas, biodiversity conservation and economic value of biodiversity and natural ecosystems. Therefore, although the term ecosystem services might serve as a boundary concept in the science-policy interface, this will not automatically be the case for bringing together scientists and local communities. In this section I advocate the use of the term *landscape services* as a special case of ecosystem services for use in local level landscape planning.

Sustainable landscape development is all about values in the landscape that are recognized by the human society. If, then, the landscape is framed as a value-producing system (following Termorshuizen and Opdam 2009), it appears that the unique contribution of landscape ecology to sustainable landscape development is the understanding of how the spatial pattern relates to the functioning of the landscape as a system. To make spatial pattern explicit in understanding the added value of landscape change is most critical in multifunctional, fragmented landscapes, because it is here that the provision of services heavily depends on the spatial configuration of small landscape elements, on top of the characteristics of the individual elements. An example is pest suppression in crops by parasitic and predatory insects, which need a specific structure of landscape elements to develop a viable population (Bianchi et al. 2006). The spatial position of the supply of services compared with the position of the service users is also important. For example, the effect of placing ditches or vegetation strips for flood regulation very much depends on the relative position of the elements in the catchment area. These intricate relationships between the spatial pattern of landscape elements and the provision of services is one reason for preferring the term landscape services over ecosystem services (Termorshuizen and Opdam 2009). The term emphasizes the importance of spatial pattern and the spatial relations linking landscape patches, whereas the ecosystem concept highlights the functional (vertical) relationships within ecosystem patches (a.o. O'Neill 2001).

A second reason for preferring landscape services is that the landscape is the result of the interplay between natural and human processes. While in the definition of ecosystems natural processes prevail (Fisher et al. 2009), the provisioning of landscape services very much depends on systems in which humans have a dominant role.

A third reason for preferring landscape services over ecosystem services, as proposed by Termorshuizen and Opdam (2009), is that the term better unifies scientific disciplines and therefore fosters interdisciplinary approaches which are so crucial for sustainability science. The term "landscape" is used by planners and landscape architects and is also used in social sciences, whereas "ecosystem" is not. For example, Macleod et al. (2007), in their paper on sustainable catchment management, use the word landscape to denote the management unit, but not "ecosystem". While "ecosystem" is increasingly becoming a core concept in environmental science and associated with nature, biodiversity, and environmental protection, "landscape" is a broader concept because more disciplines recognise it as a meaningful concept.

Furthermore, the term landscape services is more relevant and legitimate to local communities. Although this is not investigated, local actors may associate landscape services with the multidisciplinary character of their environment, so with the place where they live and work and for which they are responsible, whereas ecosystems are associated with areas where natural processes prevail. When the term ecosystem services is used in papers on collaborative management, the subject is often about managing large areas with a natural character, such as semi-natural farms in Arizona or a Wetland area in Sweden (Olsson et al. 2004; Schultz et al. 2007).

Based on these lines of argument, I prefer landscape services (as proposed by Termorshuizen and Opdam 2009) as a unifying concept linking the biophysical landscape to the local community, to apply in community-based sustainable landscape planning. It is essentially modified from the ecosystem services concept, but emphasizes that the service producing system is the people's local environment, the place for which they feel responsible, with distinct spatial elements that they can change to generate ecological, social, and economic value.

5.3 A Focus on Green Infrastructure

Cultural landscapes often are intensively used by humans for food production, working and living. The physical pattern of these landscapes, rural as well as urban ones, often takes the shape of a mosaic in which patches for production of food or for housing and commercial functions are intertwined by a pattern of longitudinal and small patch-shaped elements with a more natural character. These elements include waterways and their margins, roads and their margins, margins of arable land, woodlots and hedgerows, and amenity grassland. For this pattern several terms are being used that emphasize their functional connectedness: ecological networks (Opdam et al. 2006), green infrastructure (Benedict and MacMahon 2006; Horwood 2012), green–blue networks (Steingröver et al. 2010) and green veining (Grashof-Bokdam and Van Langevelde 2004). In this chapter I use "green infrastructure" to underline the fine-grained network structure composed of both terrestrial and aquatic landscape elements.

The green infrastructure, by supporting biophysical processes, provides functions that if valued by humans (Fig. 5.1) turn into services. For example, green infrastructure supports water regulation functions (Herzon and Helenius 2008),

Fig. 5.1 The structure-function-value chain linking *green infrastructure* size and shape to human value (modified after Termorshuizen and Opdam 2009). Value includes benefit, the value a service has to specific users. In assessments, the chain is applied from *left to right* (starting with: how has the pattern changed?), in landscape design it is applied in the reversed order (starting with: which are the preferred values?)

connectivity functions to link protected habitat areas (Davies and Pullin 2007), and perception of beauty and cultural history by tourists (Ode et al. 2009). Also, it supports a relatively large part of the landscape's biodiversity (Duelli and Obrist 2003). Biodiversity, the species of plants and animals occurring naturally in the landscape, may be valued for its presence as such and can then be considered as a social service. On the other hand, biodiversity may also provide production and regulating services that are valued from an economic perspective, for example a pest regulation service in agricultural crops (Steingröver et al. 2010). Many landscape services depend on the spatial pattern of green infrastructure, because the underlying biodiversity depends on it. Examples of such natural processes are flows of surface water (including organic matter) depending on the dimensions and connectivity of water bodies, and flows of individuals of wild species depending on distances and connectivity between forest patches. Because these pattern-process relations determine benefits to human users, they form the knowledge base of sustainable landscape change. Or to frame it from a planning perspective: ecology-inclusive spatial planning should aim for the management and spatial adaptation of green infrastructure networks. Opdam et al. (unpublished mscr.) have shown that in community-based landscape planning green infrastructure and the connected landscape services connected individual interests to common interests. In agricultural landscapes, the remaining part of the landscape is used for food or biomass production, and therefore focusses on a single landscape service which is not directly depending on the spatial structure of the landscape mosaic, and is of individual interest to the farmer. In urban landscapes, the remaining landscape is infrastructure and built structures, where economy and technology rules. Buildings only provide landscape services if roofs are green (which connects them to green infrastructure).

5.4 Defining Sustainable Change of Landscapes

In this section I discuss principles related to sustainability that have been suggested in the literature on land use or landscape planning. Reed et al. (2006) distinguished two approaches of sustainability indicators, which I will call the physical system approach and the social system approach. The physical system approach is typical for the environmental and economic sciences; it is often expert-led and does not take into account the variety of resource user perspectives. The social science approach is based on bottom–up, context-specific participatory approaches. Typical for the physical system approach are lists of specific, technical criteria. For example, for agricultural systems, Ghersa et al. (2002) proposed 14 farm-level criteria (including farm and field size, incidence of drought, agrochemical efficiency, tillage efficiency, crop yield variation coefficient) and 3 landscape-level criteria (e.g. frequency of field size classes), all of which are quite technical and have no relevance for non-food services. Other similar examples can be found in Sheppard and Meitner (2005) for forestry landscapes (9 criteria) and in George (1999) for environmental assessment (18 criteria). Within the landscape planning literature, Leitão and Ahern (2002) discussed a range of technical metrics for landscape pattern, emphasizing the need to link pattern and process across scales as a sustainability criterion, but without considering the social and economic aspects of sustainability. I concur with the conclusion of Robert et al. (1997) that "one-dimensional physical measures" do not account for the interdependency of the physical and societal components of sustainable landscape change. In participatory planning, long lists of criteria are off-putting rather than inviting. By contrast, the social sciences literature considers sustainability as the emergent property of human interaction, not to be captured in unambiguous generalizable criteria (Röling 2004). Sustainability becomes translated into concrete social practice by a joint effort of all relevant actors. Issues such as social learning, cooperation and equal distribution of costs and benefits are therefore core themes in social science literature on sustainability. The functioning of the physical system and its relation to long-term profits to humans is rarely considered.

An important debate between proponents of the physical and social systems approaches is whether there are tipping points in landscape change, beyond which the landscape no longer functions properly. In clarifying this debate, Farrell and Hart (1998) distinguished two competing conceptions on how social, environmental, and economic values should be balanced. The Critical Limits View follows a positivistic point of view by proposing objective generic limits of acceptable change. It assumes that the Earth's environmental carrying capacity and resource limitations impose limits to its use. Related to this view is the concept of strong sustainability, which assumes the maintenance of ecological capital (Antrop 2006; Dietz and Neumayer 2007). Alternatively, the Competing Objectives View of sustainability assumes that social, economic, and ecological goals are balanced in the context of a broad range of human needs. Related to this view is the concept of weak sustainability, which promotes the idea that the natural capital of an area can be substituted by other forms of capital (Dietz and Neumayer 2007; Antrop 2006). Farrell and Hart (1998) noted that the idea of resource limits is entirely absent from the competing objectives view.

To create a common ground for sustainable landscape planning, these two opposing approaches have to be converged. For example, with respect to the issue of thresholds, Termorshuizen and Opdam (2009) have pointed out that because relations between the landscape pattern and landscape functioning are often non-linear, the existence of thresholds is obvious. A well-known example is that species diversity increases steeply with ecosystem area until a certain threshold is reached, after which the curve levels off. Such a relationship is very relevant to decision making about whether an intervention in the landscape results in added value. In our example, if the aim is to increase the level of biodiversity, interventions in the landscape are most cost-effective below the threshold. Above the threshold, investments create less added value from the perspective of species number, but may be necessary from another point of view, for example to create conditions for the long-term persistence of a particular species that the community gives high priority. The threshold is a fact, but what it means to action depends on the local aspiration levels, trade-offs with other services etc. As Rockström et al. (2009) stated it: "although current scientific understanding underpins the analysis of the existence, location and nature of thresholds, normative judgements influence the definition and the position of planetary boundaries".

Literature contains some promising attempts to include both physical and social system components. For example, Mendoza and Prahbu (2000) proposed 6 principles to apply in forest management: (1) Policy, planning and institutional framework are conducive to sustainable forest management; (2) Ecosystem integrity must be maintained; (3) Forest management maintains or enhances fair intergenerational access to resources and economic benefit; (4) The stakeholders concerned have an acknowledged right to co-manage forest equitability and the means to do this; (5) The health of the forest actors, cultures and the forest is acceptable to all stakeholders; (6) The yield and quality of forest goods and services are sustainable. Papaik et al. (2008) reduce this set to three general principles: (1) take long-term processes in the forest system ("system inertia") into account, (2) consider local and broader scale perspectives concomitantly, and (3) empower local stakeholders. The principles are derived from two principles of democracy: intra-generational equity and intergenerational equity, and combined with scale levels, and the concept of weak and strong sustainability. I strongly agree with Potschin and Haines-Young (2012) advocating that ecosystem services have to be considered in the context of "place". An alternative set of principles was proposed by Musacchio (2009) which she has called the six e's of sustainable landscape design: Economy, Environment, Equity, Aesthetics, Ethics and Experience. The last three e's can be seen as referring to the social aspects of the human–landscape relationship, which can be expressed as social value, parallel to economic and environmental values.

Building on these attempts, and from a perspective of community-based planning for landscape services, I suggest 6 principles for sustainable landscape *change*.

1. *The landscape is a service providing system.* The landscape is a physical system resulting from the interplay between the natural and socio-economic processes: humans use landscape resources for their benefit. This principle defines the landscape concept based on the nature-human relationship, which is central in sustainability thinking. Because the landscape is spatially heterogeneous, the provision of landscape services varies across the landscape mosaic.

2. *Change results in added value*. Humans change the pattern of the landscape to gain more benefits, or prevent loss of value due to external causes. The change creates added value. This principle implies a definition of change and why change is done. Because the landscape is spatially heterogeneous, the added value of change is spatially variable across the landscape mosaic.
3. *Landscape change is community-based*. This principle characterizes the social process of landscape change planning. Individuals differ in the benefits they use, hence in their motive for change. They also live in different parts, where they may own different parts of the landscape mosaic. Landscape change creates different values in different places, perceived differently by members of the local community, and costs of change are spatially variable. Moreover, a change at one site may affect the value at other sites. Therefore, decision making about adapting landscapes is a collaborative process aiming for an even distribution of costs and benefits among stakeholders, hence with a strong responsibility in the local community (community-based environmental governance).
4. *Multi-scale perspective*. The environmental system is hierarchically structured along a spatial scale, including the local landscape spatial level. Hence, the pattern-process relationships at the landscape level interact with and are influenced by processes at higher levels of scale. Sustainability calls for taking into account implications of local level change on ecological and socio–economic processes elsewhere in the hierarchy of scale levels. Because local values depend on relationships with surrounding landscape systems, local goal setting and design should be based on the opportunities and constraints offered by these surroundings.
5. *Long term perspective*. Sustainable use means utilizing the benefits of landscape services while maintaining the potential of the landscape to provide these resources to future generations. A widely accepted basic idea of sustainability is that resources are not depleted but remain available for future generations (the well-known principle of intergenerational equity). Because biodiversity plays a key role in many landscape services and because biodiversity depends on the spatial pattern of ecosystems in the landscape, managing the spatial cohesion of the ecosystem network for long term viability of a substantial amount of species (Opdam et al. 2003) is conditional to a sustainable use of landscape resources.
6. *Resilient to fluctuations*. Services link the ecological to the human system. Both are inherently variable over time, due to internal variability and to external influences, such as weather fluctuations, climate change and water system variability, and on the social system side change of political power and evolving public views on the value of landscape. Therefore, all the decisions made about change according to the first five principles are very much subject to uncertainty. Concepts that have been proposed to deal with uncertainty in environmental governance are resilience and adaptive capacity (Walker et al. 2004). Thus, sustainable landscape management aims for ensuring that the physical landscape is resilient to resist or absorb unintended change without losing its capacity to provide desired services (Vos et al. submitted). Resilience

also requires capacity building in the social component, including social learning, multilevel governance, adaptive management and transition management (Lebel et al. 2006; Pahl-Wostl 2006; Foxon et al. 2009).

5.5 Community-Based Planning of Sustainable Landscape Change: What Science can Deliver

5.5.1 Landscape Change as Social-Ecological System Dynamics

Considered as the result of a long standing interaction between humanity and the biophysical system, the landscape can be described as a social-ecological system (SES) (Walker et al. 2004; Matthews and Selman 2006). Human users adjust it for better performance (Taylor Lovell and Johnson 2009). This interdependency is two-way (Fig. 5.2): (1) *use and valuation*: humans value the current performance of the landscape for its benefits, and (2) *intentional landscape change*: humans intervene in the biophysical system, aiming at improved benefits or at ensuring its

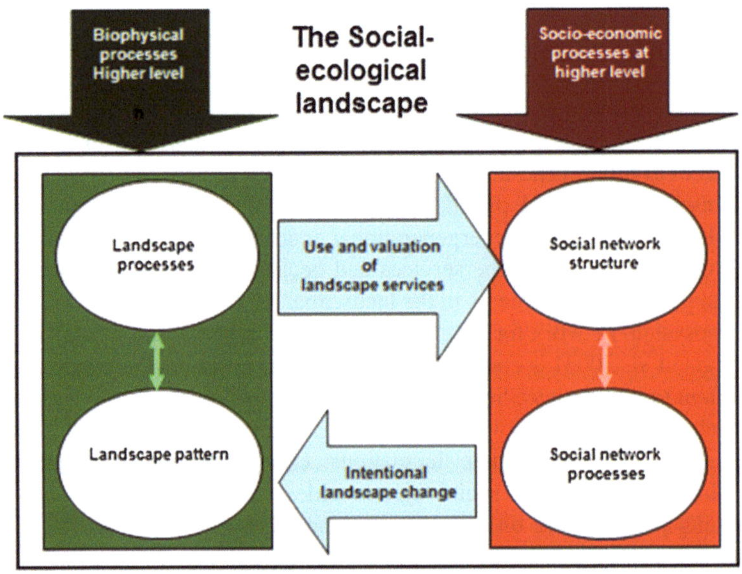

Fig. 5.2 Schematic representation of the landscape as a social-ecological system, consisting of a biophysical and social component, each including a pattern-process relationship. Ecological and social networks functionally link sites across the landscape area. Defined at the local scale level, it is being affected by biophysical, socio-economic and political processes at higher levels of scale

current performance under changing external conditions (such as climate change). The process of landscape change can be viewed in terms of these dynamics of the SES.

Landscape services may be valued from an individual or a common perspective. Crop growing allows the farmer to earn money, but the pattern of crops influences the attractiveness of the landscape to human visitors. Thus, valuation is a community process in which individual and common values are determined and negotiated. This is done in the social component of the SES, where farmers, water board representatives, citizens, entrepreneurs, visitors, as well as local governmental, NGO and pressure group representatives constitute a social network with formal or informal structures (Jansen et al. 2006). Their interactions interfere with actors at higher governance scale levels. For example, for the implementation of international or national legislation, local authorities may have regular debates in a regional setting. Also, for discussing water management, a local water management authority will take part in regional governance networks.

The outcome of a landscape services valuation process can be a motive to adapt the local landscape. Such a need may be elicited within the boundaries of the SES, because of emerging aspirations (for example entrepreneurs who see opportunities to expand their business) or evolving perceptions of value (for example, urbanizing populations attributing more value to experiencing nature). Inhabitants may want to improve the quality of the surface water running past their back yard, while farmers may be interested in the delivery of a natural pest control service. But the demand may also come from higher levels of spatial scale. Weekend visitors from the city next door may call for a restoration of historic landscape character. A response in the SES may also be elicited by changes at higher level of institutional or jurisdiction scale level (due to a change in national environmental policy) or economic scale level (e.g. a change in world market prices for crops). National health insurance companies may discover the value of green infrastructure in improving the mental health of citizens (Ward Thompson 2011). All examples refer to opportunities to create added value. But a demand for change may also be invoked by reported threats of current values, for example if predicted changes in precipitation patterns make a future increase in flood damage risk plausible. So, emerging demands may originate from various levels of scale, and the local community will be challenged to find an appropriate response, which balances their own needs with those of the wider society.

The outcome of this process can be some form of formal agreement among representatives of community groups concerning why, where and how to intervene in the physical pattern of the green infrastructure to achieve the envisioned level of landscape services. Such a plan has to take into account that the functioning of green infrastructure partially depends on higher levels of scale, the green infrastructure may continue beyond the planning boundaries, extending its total area to a more robust level able to support more species than the local planning area (see Opdam 2013 for an overview of scale sensitive landscape governance).

As for the role of landscape ecology in community-based planning, three fields of research emerge from this process cycle, which will be discussed in the following sections:

- Assessment and valuation methods, by which local communities can assess how and where the existing pattern of green infrastructure generates landscape functions that are valued as services
- Design methods, by which local communities can find cost-effective ways to adapt the landscape, based on knowledge of the response of the landscape to land use and structural change
- Monitoring methods, by which local communities are informed about the response of the landscape system, as a learning feed-back to the social system

5.5.2 Assessing Landscapes for the Provision of Landscape Services

In the community-based planning process vision building is a key step; at its basis is information from assessment and valuation. Typical questions are: what in the actual landscape is valuable to me or to my group, which are the essential underlying processes connected to those values? However, different actors have different answers to these questions, and an important part of the planning processes is that they deliberate about these diverging perceptions of the landscape. Causes of such divergence may be different ethical views on the human-nature relationship, different perceptions of system dynamics, different locations within the area and different social and economic interests. This is the context of mapping and valuation methods, which make Potschin and Haines Young (2012) to propose the term place-based assessment of ecosystem services.

Mapping methods should inform local communities about the spatially explicit relationship between biophysical patterns and value of landscape services. First and for all, this requires a level of resolution that allows local actors to recognize the pattern of the landscape where they live, the parts they own and the sites they love. However, most of the published mapping attempts have chosen national (Scolozzi et al. 2012 for Italy, Egoh et al. 2008 for South Africa) or regional (Chan et al. 2006 part of California USA; Nelson et al. 2009, part of Oregon USA; Sherrouse et al. 2011, Colorado USA, focussing at social values; Koschke et al. 2012, administrative region in Germany) levels of spatial scale. For a recent local scale mapping of ecosystem services, see Petz and Van Oudenhoven (2012). Martinez-Harms and Balvanera (2012), after reviewing papers on mapping approaches, concluded that services critical for human welfare (such as scenic beauty, cultural identity, disease regulation) are rarely addressed in mapping. These mapping approaches typically deliver spatially explicit information of where in the area the actual landscape mosaic provides which ecosystem services, where the locations are with bundles of services, and how these 'hotspots' could be valued. Such maps are often acquired by linking land cover types to generic estimates of value, often drawn from studies done elsewhere and often assuming that the conversion from biophysical assets to value is identical across the range. The resulting maps are too much generalized and with too little spatial detail to serve in community-based landscape planning. Furthermore, Eigenbrod et al. (2010) showed that maps based on such course proxies may strongly deviate from directly measured ecosystem services, and conclude that such methods are unsuitable for identifying priority areas for multiple services. At a local scale, Grêt-Regamey et al. (2008) came to a similar conclusion. More detailed approaches are often limited because of lack of detailed land use data and landscape services value estimates. An interesting approach to explore is how coarse grained maps might get more detail by bringing in stakeholder knowledge.

However, most studies thus far discussed lack public involvement. Seppelt et al. (2011) reviewed 153 publications on ecosystem services over two decades, and concluded that stakeholder involvement is in its infancy, either in identifying relevant ecosystem services, in providing ground truthing for management options or in assigning weights of importance to different services. Often, if stakeholders are involved, they provide information to the scientists to incorporate into the analytical model (for example, Vihervaara et al. 2010). However, if stakeholders would actively map landscape services themselves and be supported by scientists, they would develop much more insight into the ecological complexity of their system. A discourse-based mapping method could help to converge the variety of opinions among stakeholders on what is valuable, but such attempts are scare (Wilson and Howarth 2002). There is a need for mapping methods which give stakeholders a central role, which help them to define important services, locate sites for action to improve benefits, and organize the change process. One of the few examples is the study by Raymond et al. (2009), who developed a mapping method in interaction with a large group of decision makers that revealed place-specific differences in ecosystem services values. They founded their approach on theories of social–ecological systems and sense of place. Fagerholm et al. (2012) experimented with mapping of landscape services providing sites by the inhabitants of a local community in Zanzibar, Tanzania (a 61 km^2 area). The results show that the located sites are spatially clustered and that sites provide several services at once. A similar stakeholder-based mapping of geographic hotspots of social value with concern for multiple objectives and related management was carried out in Australia by Bryan et al. (2010).

A large unmet challenge is to understand how landscape services are distributed among different groups in the local community (Tallis and Polasky 2009). An important step in the planning process is a demand and supply analysis (Fig. 5.3). It shows how land owners, who could optimize landscapes to provide demanded benefits, are linked to interest groups who (may) have a demand for services; such groups may reside outside the area. Such an analysis could show shared interests among stakeholder groups, which may lead to coalitions in demands or even in investments in the area. On the supply side, spatial clusters of landscape elements

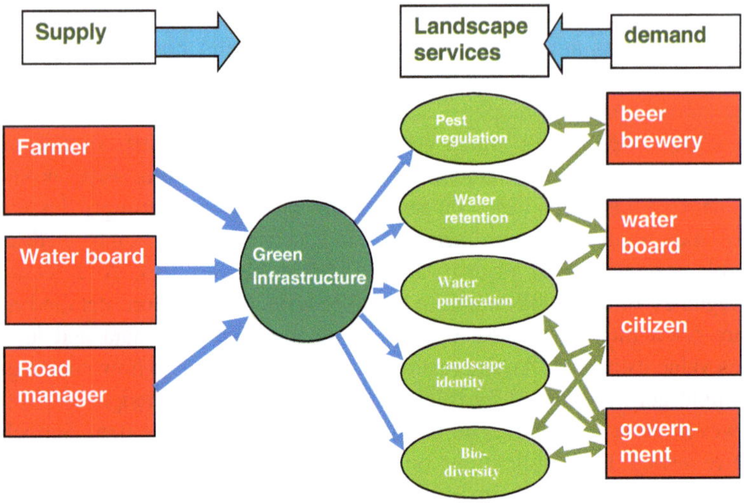

Fig. 5.3 Theoretical example of the network of relationships between land owners and land managers (who can influence the infrastructure of the landscape to produce services) and beneficiaries of landscape services. The network structure emphasizes the need for cooperation at the landscape level and for building coalitions of actors at the demand side

providing the same set of ecosystem services could be informative for land owners to build consortia of service providers. Therefore, there is a need for methods that not only map localities with a potential to provide a cluster of services, but also localities where a demand exists. Some recent studies make a step towards providing such information. Nedkov and Burkhard (2011) provided information of the spatial variation in storm water retention capacity. Their method also identified sites with a flood risk, which could be interpreted as parts of the area with a demand for a water regulation service upstream. Based on suggestions by Fisher et al. (2009), Syrbe and Walz (2012) proposed to distinguish *service providing areas* (ecosystem sites or networks) and *service benefiting areas* (for example, rural settlements, urban agglomerations, farms). They provided an example for flood regulating services in the region of Saxony. Such analysis could be the basis for a spatially explicit supply and demand analysis. Again, I would like to emphasize that such endeavours should be able to incorporate social demands from specific interests groups. An inspiring example of how this could be done was reported by Pinto-Correiaand and Carvalho-Ribeiro (2012), who combined user-based preferences of landscape patterns with land cover indicators, which may offer a road towards characterizing the land cover pattern that users prefer.

Mapping and assessment approaches have not yet included spatial interdependencies between levels of scale, and the same can be said for approaches that consider the provisioning of landscape services over long time frames. Considerable progress could be made here if in future science is able to link indicators of biological community composition to the service level and the reliability of service provision (Feld et al. 2009). Such advanced knowledge could be linked to the large landscape ecological knowledge-base which links biodiversity to landscape structure at various levels of spatial scale.

5.5.3 Intervening in the Physical Landscape

In science, the usual approach in considering landscape change is either to describe and quantify it as a process over time and analyse its drivers, or to focus on its impact on biophysical phenomena, such as biodiversity, and socio-economic processes such as the perception of change by inhabitants. Searching with Google scholar for articles with the exact term landscape change planning (August 7, 2012) rendered only 5 hits, and with "community-based landscape change" I got no hit at all. Under "environmental governance" most articles deal with social processes, with little or no explicit attention to a change in the physical system. It seems that while humans change landscapes throughout history, science has either considered it as an unwanted "impact", or as something outside scientific focus. Landscape design is mostly the domain of landscape architects, and not seen as a scientific experiment (but see Nassauer and Opdam 2008).

Therefore, in the context of planning for added value by landscape services in social–ecological systems, many unanswered questions can be posed. One challenge is to determine which actions create added value: which adaptation of the physical pattern will turn into a better service provisioning, higher service reliability or added economic or social value. The shape of the relationships between structure and function and function and value is crucial information for decision-making—will a certain investment lead to added value? Many of these relationships are supposed to be nonlinear (Eiswerth and Haney 2001; Vos et al. 2001; Barbier et al. 2008). Therefore, investments may vary between big gains in value and a decrease in value, depending on the actual landscape structure (Farber et al. 2002; DeFries et al. 2004). If the relationship shows a threshold, it depends on the actual landscape structure whether it is possible to improve a service or that it is already at its maximum (Fig. 5.4a). For example, increasing the spatial cohesion of an area initially improves the performance of a specific population, but when certain cohesion has been achieved, additional investments will not lead to further improvement of performance. If the relationship follows a bell-shaped curve with a single optimum (Fig. 5.4b), investing in structure may initially increase the function level, but further investment will lead to a loss of functioning (and thereby of value). For example planting more and more trees to increase landscape quality for recreation will eventually result in a forest which is less valued than half-open landscape. Very little work has been done in quantifying such relationships, and it is notable that recent reviews (Carpenter et al. 2009; De Groot et al. 2010; Seppelt et al. 2011) did not address the importance of knowing how measures in the landscape turn into added value.

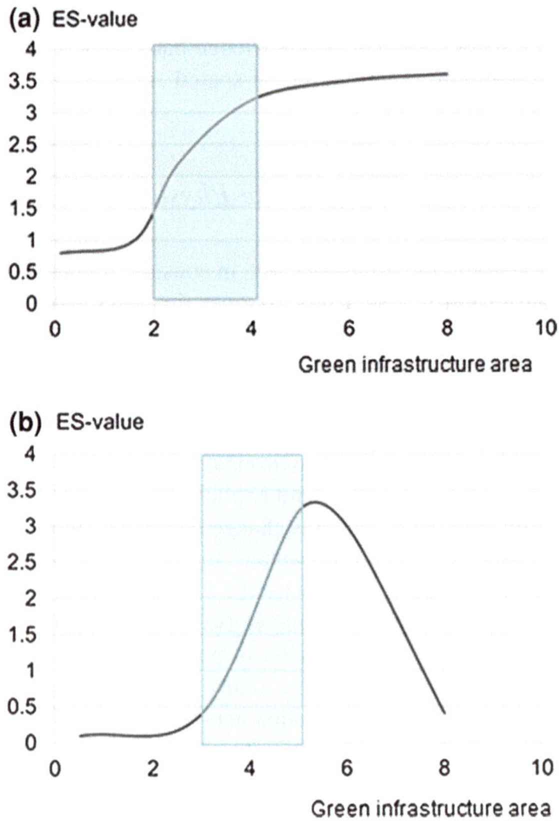

Fig. 5.4 Knowing the shape of the relationship between green infrastructure and landscape service benefits are essential for deciding about investments to gain added value. The *rectangle* indicates the range of *green* infrastructure area in which investments are most profitable. **a** Below and above a critical segment of the size range adaptation is not profitable. **b** Beyond the maximum value investments to further increase the size of the network turns into loss of value

A second challenge is to know *where* the service can be improved with the best result. As stated before, mapping and evaluation studies do not inform about where change is effective. The cost-effectiveness of measures depends on several factors. One is the spatial relationships between places where landscape services are produced and the localities where services are in demand. For example, pest control measures need to be taken at the farm and in its surroundings, flood control need to be taken upstream. Secondly, the service provision depends on the type of vegetation of the green infrastructure. For example, grassy strips and woody elements have different contributions to the perception of scenic beauty. Thirdly, if the service depends on spatial connectivity, places may be found where the green infrastructure is narrow or interrupted; measures at such sites are particularly cost-effective because a relatively small investment may merge two networks into one large. Some authors touch upon these spatial implications [e.g. Fisher et al. (2009) and Syrbe and Walz (2012) draw attention to spatial relationships between service production areas and service benefit areas], but attempts to bring science and practitioners together and develop a science-based landscape design approach are hard to find. Promising lines of research are attempts to develop GIS-tools (including 3-D visualizations) and scenario approaches in which stakeholders play an strong role as decision maker (Nassauer and Opdam 2008; Southern et al. 2011; Van Berkelet al. 2011). For example, Fürst et al. (2010) reported on an interactive online tool in which stakeholders can experiment with proposing changes in land cover types, find out effects on several land use functions, and communicate about the implications of land use change. Steingrover et al. (2010) developed a design approach by which a group of stakeholders could scan the existing green infrastructure network in a farm landscape from the perspective of the landscape services natural pest suppression in crops. Stakeholders identified places which required broader field margins, improved connectivity or an adapted vegetation management.

Sustainable change of landscapes implies a multifunctional approach. If landscapes are to be planned for several services at the same time, interdependencies between services become an issue. Services influence each other, for example regulating services influence production services (Bennett et al. 2009). Such interdependencies may inspire a hierarchical stepwise design procedure focussed on creating synergies. Stakeholder groups could go through a process of identifying bundles of landscape service provisioning. If top-sites for several services coincide, one change produces several benefits to a variety of beneficiaries simultaneously. Such a synergy of services enhances the formation of coalitions in the local community. Alternatively, trade-offs between services cause conflicts between stakeholders. The best locations to adapt the landscape might also be found there where there is the strongest demand for benefits, not only because the return on investment is advantageous, but also because of the strong support of local stakeholders.

According to sustainability principle 4 (calling for a multi-scale-level approach), designing landscape for landscape services needs to take into account the spatial hierarchy of the environmental system (Opdam 2013). Landscape elements producing landscape services often do so as part of a larger network of landscape elements, which extends across the whole planning area, and beyond its boundaries. The usual land cover-based approaches do not take into account that service provision may become better or more reliable in clusters of identical land cover cells. Similarly, the implications of local landscape change on surrounding areas has not been addressed. Seppelt et al. (2011) reported that none of the 153 studies considered the consequences of local decisions on distant ecosystems.

5.5.4 Monitoring Responses of Social-Ecological Systems

Social learning and adaptive management requires feed-back information on how the physical landscape responds to change. Data are needed for all parts of the pattern-process-value chain (Fig. 5.1): how the landscape was changed, how its performance responded to change, and how this resulted in added value, and to whom? In addition, it is interesting to know the institutional change in the SES, for example the formation of new cooperative bonds in the governance network.

Carpenter et al. (2009) paid attention to monitoring from the point of view of the Millennium Ecosystem Assessment, emphasizing assessment and large spatial scales. Systematic monitoring at the local level, as a source of information to the local adaptive management network, seems to be virtually absent. Apparently, there is lot of work that needs to be done here before monitoring results can be incorporated into a social learning process.

5.6 Perspective Sand Conclusions

I have viewed community-based landscape planning as a change process in a social-ecological system, and discussed how the concept of landscape services could be evolved as a boundary concept to enhance and structure communication and negotiation about sustainable change between different interest groups having a stake in the local landscape.

From the reviewed literature in this chapter it is obvious that the scientific state of the art is not ready to deliver adequate tools to support community-based landscape planning. Scientific efforts on ecosystem services have been primarily focussed on assessments at large spatial scale, and with policy users in mind. For application in community-based landscape planning, scientific development of theories and tools are still in their childhood. There is a strong demand for tools that are able to support local governance networks, where the users are citizens, farmers, entrepreneurs and local authorities, the object of governance is the local landscape system, and the aim is to adapt the landscape to meet the expected demands without losing its potential to provide services in the long term. That science tends to be focussed on the policy level and on assessment tools while neglecting the deliberation and implementation phase of designing solutions has been argued several time before (Opdam 2010) and is not specific of ecosystem services research. However, it seems that in ecosystem service research the need for this is not yet recognized.

A second conclusion is that ecosystem service science has not yet been able to merge with several emerging themes relevant to sustainability science: governance networks, multiple level governance and complex adaptive systems, and resilience in social-ecological systems (Folke et al. 2005; Ernstson et al. 2008). There is very little attention to multiple scale effects and implications for decision making (Opdam 2013), to creating conditions that ensure long term provision of services. A recent attempt by Burkhard et al. (2011) to link ecosystem services thinking to resilience theory may be a good start of theory building.

The Social-Ecological model is helpful to move the ecosystem services research out of the static mapping and evaluation approaches towards dynamic systems thinking. It helps understanding that the valuation of landscape services is not a static outcome of a scientific assessment, based on generic standardized estimates of value. On the contrary, values attributed to services may be dynamic, subject to social and political change, and variable over social-ecological space.

Setting aims for the delivery of services by the future landscape should be the result of place-based and context-based deliberation, negotiation and evolution, acknowledging that 'landscape service value' means something different to stakeholders with different interests and different views on the human-nature relation, and living in different places of the landscape. As Potschin and Haines Young (2012) put it: "context matters" in the relationship between ecosystem patterns and service valuation. To be able to support this kind of planning process, scientists need to understand better how interventions in the landscape structure depend on the structure of the governance network, and develop methods based on that insight. It also calls for a reinterpretation of classic landscape ecological knowledge in terms of a response of the biophysical system to change (instead of framing human intervention as undesired impacts). Therefore, by incorporating valuation and intervention into a landscape planning cycle, the SES model makes clear the essence of the first three principles of sustainable landscape change that were suggested in this chapter. Moreover, the SES concept is born out of resilience thinking and theories of adaptive governance (Dietz and Neumayer 2007) and adaptive co-management (Armitage et al. 2009), and incorporates the need for long term resource availability and scale-sensitive decision making.

A challenging approach to better understand these interactions is modelling socio-ecological systems with Agent Based Models (Schlüter et al. 2012). This approach allows the performance of experiments with different interdependencies among actors, with different sets of knowledge, and with different incentives. Such experiments can be validated with case studies and comparative analysis to feed the outcome of case studies back into theory. The SES model can be used as a theoretical framework for sustainable landscape change, and helps to formulate questions about the interactions between the social and physical component of the landscape, and on the role of scientific knowledge in shaping this interaction.

This chapter will close with suggesting research priorities, for which I use the 6 sustainability principles as a source of inspiration.

1. *Understanding landscapes as Social-Ecological Systems.* In a SES, the landscape is the physical part of the system that offers opportunities for sustainable socio–economic development. How does this view align with the many different views on landscape in literature and in society (Stephenson 2008), and how does it interfere with views on protecting cultural heritage? How can sense of place become the conceptual basis for local landscape adaptation (Nassauer 2012)? How does the SES model helps us to move from impact and threat thinking towards opportunity thinking?
2. *Mapping supply and demand of landscape services*, essentially guiding the actions of local land owners and land users to intervene in the landscape system. Maps need to show where investments in the landscape structure are profitable, and where the demanding stakeholders are located. Maps should go beyond informing about current value from services, and guide cost-effective interventions (where, how) for creating *added* value. It also calls for maps to

show details of the landscape that matters to its users and owners, such as linear elements and ponds

3. *Mechanisms to drive the supply and demand of landscape services*. This requires an analysis of the structure of the network of land owners and land users as connected by landscape services (for example as in Fig. 5.3). Mechanisms may be based on a set of market-based rules, such as financial arrangements and governmental incentives for producing services of common interest, such as agri-environmental payments (GLB). How are such mechanisms capable of organizing the desired change, do they promote coordinated action (see Schouten et al. 2013) and what is the outcome in terms of the redistribution of benefits over stakeholders?
4. *Conditions to ensure the long term delivery of landscape services*. Many services depend on assemblages of species, which are inherently fluctuating over time and, secondly, may respond slowly to physical changes in the landscape. This issue requires a reinterpretation of fragmentation and metapopulation literature in the light of the necessary species assemblages for particular services and what physical conditions are required to ensure their existence. Important topics include minimum thresholds of area and connectivity of green infrastructure for service provision, not only for the level of the service, but also for the reliability of service provisioning (Naeem 1998) under variable weather conditions and increased occurrence of extreme weather due to climate change
5. *Multiple scale approaches* how the added value of landscape change depends on ecological and hydrological processes at higher levels of spatial scale, how demands for landscape services are distributed across spatial scales, and methods to incorporate such scale interdependencies into local decision making (Opdam 2013)
6. *Developing resilience and adaptive capacity* in governing landscapes for service provisioning. This is an emerging topic in science (Carpenter and Folke 2006) about dealing with uncertainty, complexity and perturbations in social-ecological systems, but still largely theoretical. Walker and Salt (2006) have proposed a set of criteria for developing resilience which have been used by Schouten et al. (2012) to develop a framework for rural policy assessment. This attempt could be further developed as an information system to create feedback between characteristics of the physical landscape and the social network.

Acknowledgments This chapter was inspired by many discussions with colleagues and on the ground applications in research led by Alterra. I thank Eveliene Steingröver, Sabine van Rooij, Carla Grashof, Severine van Bommel, JolandeTermorshuizen, Willemien Geertsema, Florence van den Bosch and many others. The EU-financed Interreg project GIFT-T! has financially contributed to the preparation of this chapter.

References

Antrop M. Sustainable landscapes: contradiction, fiction or utopia? Landscape Urban Plan. 2006;75:187–97.

Armitage DR, Plummer R, Berkes F, et al. Adaptive co-management for social-ecological complexity. Front Ecol Environ. 2009;7:95–102.

Barbier EB, Koch EW, Silliman BR, et al. Coastal ecosystems-based management with nonlinear ecological functions and values. Science. 2008;319:321–3.

Benedict MA, McMahon ET. Green infrastructure. Linking landscapes and communities. Washington: Island Press; 2006.

Bennett EM, Peterson GD, Gordon LJ. Understanding relationships among multiple ecosystem services. Ecol Lett. 2009;12:1394–1404.

Bianchi FJJA, Booij CJH, Tscharntke T. Sustainable pest regulation in agricultural landscapes: a review onlandscape composition, biodiversity and natural pestcontrol. Proc R Soc Lond B Biol Sci. 2006;273:1715–27.

Borch K. Emerging technologies in favour of sustainable agriculture. Futures. 2007;39:1045–66.

Bryan BA, Raymond CM, Crossman ND, Macdonald DH. Targeting the management of ecosystem services based on social values: where, what and how? Landscape Urban Plan. 2010;97:11–122.

Burkhard B, Fath BD, Müller F. Adapting the adaptive cycle: hypotheses on the development of ecosystem properties and services. Ecol Model. 2011;222:2878–90.

Carpenter SR, Folke C. Ecology of transformation. Trends Ecol Evol. 2006;21:309–15.

Carpenter SR, Mooney HA, Agard J, et al. Science for managing ecosystem services; beyond the millennium ecosystem assessment. PNAS. 2009;106:1305–12.

Chan KM, Shaw MR, Cameron DR, Underwood EC, Daily GC. Conservation planning for ecosystem services. PLOS Biol. 2006;4:2138–52.

Constanza R. Ecosystem services: multiple classification systems are needed. Biol Conserv. 2008;141:350–2.

Davies ZG, Pullin AS. Are hedgerows effective corridors between fragments of woodland habitat? An evidence-based approach. Landscape Ecol. 2007;22:333–51.

DeFries RS, Foley JA, Asner GP. Land-use choices; balancing human needs and ecosystem function. Front Ecol Environ. 2004;2:249–57.

De Groot RS, Alkemade R, Braat L, Hein L, Willemen L. Challenges in integrating the concept of ecosystem services and values in landscape planning, management and decision making. Ecol Complex. 2010;7:260–72.

Dietz S, Neumayer E. Weak and strong sustainability in the SEEA: concepts and measurement. Ecol Econ. 2007;61:617–26.

Duelli P, Obrist MK. Regional biodiversity in an agricultural landscape: the contribution of seminatural habitat islands. Basic Appl Ecol. 2003;4:129–38.

Egoh B, Reyers B, Rouget M, Richardson DM, Le Maitre DC, van Jaarsveld AS. Mapping ecosystem services for planning and management. Agric Ecosyst Environ. 2008;127:135–40.

Eigenbrod F, Armsworth PR, Anderson BJ, Heinemeyer A, Gillings S, Roy DB, Thomas CD, Gaston KJ. The impact of proxy-based methods on mapping the distribution of ecosystem services. J Appl Ecol. 2010;47:377–85.

Eiswerth ME, Haney JC. Maximizing conserved biodiversity: why ecosystem indicators and thresholds matter. Ecol Econ. 2001;38:259–74.

Ernstson H, Sörlin S, Elmqvist TH. Social movements and ecosystem services-the role of social network structure in protecting and managing urban green areas in Stockholm. Ecol Soc. 2008;13(2):39. http://www.ecologyandsociety.org/vol13/iss2/art39/.

Fagerholm N, Käyhkö N, Ddumbaro F, Khamis M. Community stakeholders' knowledge in landscape assessments-mapping indicators for landscape services. Ecol Ind. 2012;18:421–33.

Farber SC, Constanza R, Wilson MA. Economic and ecological concepts for valuing ecosystem services. Ecol Econ. 2002;41:375–92.

Farrell A, Hart M. What does sustainability really mean? The search for useful indicators. Environment. 1998;40(4/9):26–31.

Feld CK, Martins da Silva P, Sousa JP, et al. Indicators of biodiversity and ecosystem services: a synthesis across ecosystems and spatial scales. Oikos. 2009;118:1862–71.

Fisher B, Turner RK, Morling P. Defining and classifying ecosystem services for decision making. Ecol Econ. 2009;68:643–53.

Folke C, Hahn Th, Olsson P, Norberg J. Adaptive governance of social ecological systems. Ann Rev Environ Resour. 2005;30:441–73.

Foxon T, Reed MS, Stringer LC. Governing long-term social-ecological change: what can the adaptive management and transition management approaches learn from each other? Environ Policy Gov. 2009;19:3–20.

Fürst C, König H, Pietzsch K, Ende H-P, Makeschin F. Pimp your landscape-a generic approach for integrating regional stakeholder needs into land use planning. Ecol Soc. 2010;15(3):34. http://www.ecologyandsociety.org/vol15/iss3/art34/.

George C. Testing for sustainable development through environmental assessment. Environ Impact Assess Rev. 1999;19:175–200.

Ghersa CM, Ferraro DO, Omacini M, Martínez-Ghersa MA, Perelman S, Satorre EH, Soriano A. Farm and landscape level variables as indicators of sustainable land-use in the Argentine Inland-Pampa. Agric Ecosyst Environ. 2002;93:279–93.

Grashof-Bokdam CJ, Van Langevelde F. Green veining: landscape determinants of biodiversity in European agricultural landscapes. Landscape Ecol. 2004;20:417–39.

Grêt-Regamey A, Walz A, Bebi P. Valuing ecosystem services forsustainable landscape planning in Alpine regions. Mountain Res Dev. 2008;28:156–65.

Hein L, Chr Van Koppen, de De Groot RS, Van Ierland EC. Spatial scales, stakeholders and the valuation of ecosystem services. Ecol Econ. 2006;57:209–28.

Herzon I, Helenius J. Agricultural ditches, their biological importance and functioning. Biol Conserv. 2008;141:1171–83.

Hisschemöller M, Hoppe R, Dunn W, Ravetz J, editors. Knowledge, power and participation in environmental policy analysis. New Brunswick: Transaction Publishers; 2001.

Horwood K. Green Infrastructure: reconciling urban green space and regional economic development: lessons learnt from experience in England's north-west region. Local Environ: Int J Justice Sustain. 2012;16:963–75.

Jackson LL. Who "designs" the agricultural landscape? Landscape J. 2008;27:23–40.

Janssen MA, Goosen H, Omzigt N. A simple mediation and negotiation support tool for water management in the Netherlands. Landscape Urban Plan. 2006;78:71–84.

Jiggins J, Van Slobbe E, Röling N. The organisation of social learning in response to perceptions of crisis in the water sector of The Netherlands. Environ Sci Policy. 2007;10:526–36.

Kaplan R. Employees' reactions to nearby nature at their workplace: the wild and the tame. Landscape Urban Plan. 2007;82:17–24.

Koschke L, Fürst C, Frank S, Makeschin F. A multi-criteria approach for an integrated land-cover-based assessment of ecosystem services provision to support landscape planning. Ecol Ind. 2012;21:54–66.

Lebel L, Anderies JM, Campbell B, Folke C, Hatfield-Dodds S, Hughes TP, Wilson J. Governance and the capacity to manage resilience in regional social-ecological systems. Ecol Soc. 2006;11(1):19. http://www.ecologyandsociety.org/vol11/iss1/art19/.

Leitão AB, Ahern J. Applying landscape ecological concepts and metrics in sustainable landscape planning. Landscape Urban Plan. 2002;59:65–93.

Macleod CJA, Scholefield D, Haygarth PM. Integration for sustainable catchment management. Sci Total Environ. 2007;373:591–602.

Martinez-Harms MJ, Balvanera P. Methods for mapping ecosystem service supply: a review. Int J Biodivers Sci, Ecosyst Serv Manage. 2012;8:17–25.

Matthews R, Selman P. Landscape as a focus for integrating human and environmental processes. J Agric Econ. 2006;57:199–212.

Mendoza GA, Prahbu R. Development of a methodology for selecting criteria and indicators of sustainable forest management: a case study on participatory assessment. Environ Manage. 2000;26:659–73.

Musacchio LR. The scientific basis fort he design of landscape sustainability: a conceptual framework for translation all and scape research and practice of designed landscapes and the six Es of landscape sustainability. Landscape Ecol. 2009;24:993–1013.

Naeem S. Species redundancy and ecosystem reliability. Conserv Biol. 1998;12:39–45.

Nassauer JI. Landscape as medium and method for synthesis in urban ecological design. Landscape Urban Plan. 2012;106:221–9.

Nassauer J, Opdam P. Design in science: extending the landscape ecology paradigm. Landscape Ecol. 2008;23:633–44.

Nedkov S, Burkhard B. Flood regulating ecosystem services-mapping supply, and demand, in the Etropole municipality, Bulgaria. Ecol Ind. 2011;21:67–79.

Nelson E, Mendoza G, Regetz J, et al. Modelling multiple ecosystem services, biodiversity conservation, commodity production, and tradeoffs at landscape scales. Front Ecol Environ. 2009;7:4–11.

Ode A, Fry G, Tveit MS, Messager P, Miller D. Indicators of perceived naturalness as drivers of landscape preference. J Environ Manage. 2009;90:375–83.

O'Neill RV. Is it time to bury the ecosystem concept? Ecology. 2001;82:3275–84. (With full military honors of course!).

Olsson P, Folke C, Hahn T. Social-ecological transformation for ecosystem management: the development of adaptive co-management of a wetland landscape in Southern Sweden. Ecol Soc. 2004;9, 4 (online).

Opdam P. Learning science from practice. Landscape Ecol. 2010;25:821–3.

Opdam P. Incorporating multiple ecological scales into the governance of landscape services. In: Padt F, Opdam P, Termeer C, Polman N, editors. Scale sensitive governance of the environment. London: Wiley Blackwell. 2013 (in press)

Opdam P, Verboom J, Pouwels R. Landscape cohesion: an index for the conservation potential of landscapes for biodiversity. Landscape Ecol. 2003;18:113–26.

Opdam P, Steingröver E, Van Rooij S. Ecological networks: a spatial concept for multi-actor planning of sustainable landscapes. Landscape Urban Plan. 2006;75:322–32.

Pahl-Wostl C. The importance of social learning in restoring the multifunctionality of rivers and flood plains. Ecol Soc. 2006;11(1):10. http://www.ecologyandsociety.org/vol11/iss1/art10/.

Papaik MJ, Sturtevant B, Messier C. Crossing scales and disciplines to achieve forest sustainability. Ecol Soc. 2008;13(1):30. http://www.ecologyandsociety.org/vol13/iss1/art30/.

Petz K, Van Oudenhoven APE. Modelling land management effect on ecosystem services: a study in The Netherlands. Int J Biodivers Sci, Ecosyst Services Manage. 2012;8:135–55.

Pinto-Correia T, Carvalho-Ribeiro S. The index of function suitability (IFS): a new tool for assessing the capacity of landscapes to provide amenity functions. Land Use Policy. 2012;29:23–34.

Potschin M, Haines-Young R. Landscapes, sustainability and the place-based analysis of ecosystem services. Landscape Ecol. 2012 (on line).

Raymond CM, Bryan BA, MacDonald DH, Cast A, Strathearn S, Grandgirard A, Kalivas T. Mapping community values for natural capital and ecosystem services. Ecol Econ. 2009;68:1301–15.

Reed MS, Fraser EDG, Dougill AJ. An adaptive learning process for developing and applying sustainability indicators with local communities. Ecol Econ. 2006;59:406–18.

Robert K-H, Daly H, Hawken P, Holmberg J. A compass for sustainable development. Int J Sustain Dev World Ecol. 1997;4:79–92.

Rockström, J, Steffen W, Noone K, et al. Planetary boundaries: exploring the safe operating space for humanity. Ecol Soc. 2009;14(2):32. http://www.ecologyandsociety.org/vol14/iss2/art32/.

Röling NG. Teaching interactive approaches to natural resource management: a key ingredient in the development of sustainability in higher education. In: Corcoran PB, Wals AEJ, editors.

Higher education and the challenge of sustainability. Dordrecht: Kluwer Academic Publishers.
Schlüter M, MCallister RRJ, Arlinghaus R, Bunnefeld N, Eisenack K, Hölker F, Milner-Gulland EJ, Müller B. New horizons for managing the environment: a review of coupled social-ecological systems modelling. Natural Resour Model. 2012;25:219–72.
Schouten MAH, Van der Heide CM, Heijman WJM, Opdam PFM. A resilience-based policy evaluation framework: application to European rural development policies. Ecol Econ. 2012;81:165–75.
Schouten M, Opdam P, Polman N, Westerhof E. Resilience-based governance in rural landscapes: experiments with agri-environmental schemes using a spatially explicit agent-based model. Land Use Policy. 2013;30:934–43.
Schultz L, Folke C, Olsson P. Enhancing ecosystem management through social-ecological inventories: lessons from Kristianstads Vattenrike, Sweden. Environ Conserv. 2007;34:140–52.
Scolozzi R, Morri E, Santolini R. Delphi-based change assessment in ecosystem services values to support strategic spatial planning in Italian landscapes. Ecol Ind. 2012;21:134–44.
Seppelt R, Dormann CF, Eppink FV, Lautenbach S, Schmidt S. A quantitative review of ecosystem services studies: approaches, shortcomings and the road ahead. J Appl Ecol. 2011;48:630–6.
Sheppard SRJ, Meitner M. Using multi-criteria analysis and visualisation for sustainable forest management planning with stakeholder groups. For Ecol Manage. 2005;207:171–87.
Sherrouse BC, Clement JM, Semmens DJ. A GIS application for assessing, mapping and quantifying the social values of ecosystem services. Appl Geogr. 2011;31:748–60.
Singh RK, Murty HR, Gupta SK, Dikshit AK. An overview of sustainability assessment methodologies. Ecol Ind. 2009;9:189–212.
Southern A, Lovett A, O'Riordan T, Watkinson A. Sustainable landscape governance: lessons from a catchment based study in whole landscape design. Landscape Urban Plan. 2011;101:179–89.
Steingröver EG, Geertsema W, Van Wingerden WKRE. Designing agricultural landscapes for natural pest control: a transdisciplinary approach in the HoekscheWaard (The Netherlands). Landscape Ecol. 2010;25:825–38.
Stephenson J. The cultural values model: an integrated approach to values in landscapes. Landscape Urban Plan. 2008;84:127–39.
Syrbe R-U, Walz U. Spatial indicators for the assessment of ecosystem services: providing, benefiting and connecting areas and landscape metrics. Ecol Ind. 2012;21:80–8.
Tallis H, Polasky S. Mapping and Valuing ecosystem services as an approach for conservation and natural-resource management. The year in Ecology and Conservation Biology, Ann NY Acad Sci. 2009;1162:265–283.
Taylor Lovell S, Johnston DM. Designing landscapes for performance based on emerging principles of landscape ecology. Ecol Soc. 2009;14(1):44. http://www.ecologyandsociety.org/vol14/iss1/art44/.
Termorshuizen J, Opdam P. Landscape services as a bridge between landscape ecology and sustainable development. Landscape Ecol. 2009;24:1037–52.
Turner MG. Landscape ecology: the effect of pattern on process. Annu Rev Ecol Syst. 1989;20:171–97.
Tzoulas K, Korpela K, Venn S, Yli-Pelkonen V, Kazmierczak A, Niemela J, James Ph. Promoting ecosystem and human health in urban areas using Green Infrastructure: a literature review. Landscape Urban Plan. 2007;81:167–78.
Van Berkel DB, Carvalho-Ribeiro S, Verburg PH, Lovett A. Identifying assets and constraints for rural development with qualitative scenarios: a case study of Castro Laboreiro, Portugal. Landscape Urban Plan. 2011;102:127–41.
Van Bommel S, Röling NG, Aarts MNC, Turnhout E. Social learning for solving complex problems: a promising solution or wishful thinking? A case study of multi-actor negotiation for the integrated management and sustainable use of the Drentschea area in the Netherlands. Environ Policy Gov. 2009;19:400–12.

Vikervaara P, Kumpula T, Tanskanen A, Burkhard B. Ecosystem services-A tool for sustainable management of human-environment systems. Case study finnish forest Lapland. Ecol Complex. 2010;7:410–20.

Vos CC, Verboom J, Opdam PFM, TerBraak CJF. Towards ecologically scaled landscape indices. Am Nat. 2001;157:24–51.

Wallace KJ. Classification of ecosystem services: problems and solutions. Biol Conserv. 2007;139:235–46.

Walker B, Salt D. Resilience thinking: sustaining ecosystems and people in a changing world. Washington: Island Press; 2006.

Walker B, Holling CS, Carpenter SR, Kinzig A. Resilience, adaptability and transformability in social–ecological systems. Ecol Soc. 2004;9(2):5. http://www.ecologyandsociety.org/vol9/iss2/art5.

Walker BH, Anderies JM, Kinzig AP, Ryan P. Exploring resilience in social-ecological systems through comparative studies and theory development: introduction to the special issue. Ecol Soc. 2006;11(1):12. http://www.ecologyandsociety.org/vol11/iss1/art12/.

Ward Thompson Ch. Landscape and health: the recurring theme. Landscape Urban Plan. 2011;99:187–95.

Wilson MA, Howarth RB. Discourse-based valuation of ecosystem services: establishing fair outcomes through group deliberation. Ecol Econ. 2002;41:431–43.

Wu J, Hobbs R. Landscape ecology: the state of the science. In: Wu J, Hobbs R, editors. Key topics in landscape ecology. Cambridge: Cambridge University Press; 2007.

Chapter 6
Effects of Global Household Proliferation on Ecosystem Services

Jianguo Liu

Abstract Population sizes and growth rates are two major factors used by ecologists in assessing human impacts on ecosystems and landscapes. However, the numbers of households have been increasing much faster than population sizes. As households are basic socioeconomic units (e.g., in consumption of ecosystem services) and key components of coupled human and natural systems, household proliferation has important implications for ecosystem services. On one hand, more households consume more ecosystem services. On the other hand, more households have more impacts on the supply of ecosystem services. So far, most impacts have been negative. As a result, ecosystem services have continued to degrade. It is important to use ecosystem services more efficiently, turn households from consumers to producers of ecosystem services, and incorporate household proliferation into ecosystem service research and management.

Keywords Households · Population · Ecosystem services · Impact · Human activities · Landscape · Coupled human and natural systems · Policy · Management · Housing

6.1 Introduction

Ecosystems and landscapes are coupled human and natural systems (Liu et al. 2007), in which humans interact with natural components. In the past, human population sizes and growth rates were usually used by ecologists in studying relationships between humans and natural systems. However, household numbers

J. Liu (✉)
Center for Systems Integration and Sustainability,
Michigan State University, East Lansing MI 48823, USA
e-mail: liuji@msu.edu

B. Fu and K. B. Jones (eds.), *Landscape Ecology for Sustainable Environment and Culture*, DOI: 10.1007/978-94-007-6530-6_6,

and growth rates were largely overlooked even though households are basic socioeconomic units and are key components of coupled human and natural systems.

Households are major consumers of ecosystem services and play important roles in ecological change. For example, a basic need for each household is a housing unit (e.g., house, apartment), which drives land-use and land-cover changes and subsequently changes in ecosystem services. The household sector is the major consumer of energy in China (Lu et al. 2007). Direct and indirect energy consumption by U.S. households makes up 85 % of national energy use (Bin and Dowlatabadi 2005) and U.S. households emit about 38 % of national carbon emissions through their direct actions (Dietz et al. 2009). On the other hand, households in many areas are vulnerable to threats induced by land change and other types of environmental change (McGranahan et al. 2007). To restore and protect ecosystem services, many countries have implemented payments for ecosystem services (Daily and Matson 2008; Liu et al. 2008). Many of these programs, such as the Grain-to-Green Program of China (Liu et al. 2008) and the Silvopastoral Project in Colombia, Costa Rica, and Nicaragua (Pagiola et al. 2007), occur at the household level.

Given the importance of households, in this chapter we first illustrate global household proliferation (growth in household numbers). Then, we discuss effects of household proliferation on ecosystem services. And finally, we provide suggestions for ecosystem service research and management in the context of household proliferation.

6.2 Global Household Proliferation

Among the 172 countries with relevant data (United Nations Centre for Human Settlements (Habitat) 2001; United Nations Human Settlements Programme 2007), 136 countries (79 %) had faster increases in household numbers than population sizes during 1985–2000 (Fig. 6.1). Over the period of 2000–2030, an even higher percentage of countries (91 %) are projected to have faster growth in household numbers than population sizes (Fig. 6.1).

At the global level, household intensity (number of households per 100 persons) increased 12.6 % from 1985 to 2000. At the country level in 1985, the average household intensity was 22.9 households per 100 persons, and Jordan had the lowest intensity (7.9) while Sweden had the highest density (43.9, Fig. 6.2a). By 2000, the average household intensity increased to 25.8 households per 100 persons. The lowest and highest intensities also increased. Sweden still held the highest spot (48.1), but the country with the lowest intensity had switched to Liberia (9.7) (Fig. 6.2b). The trends of increases in household intensity are projected to continue into the future (Fig. 6.2c).

Over time, a country can have fewer people but more households. During 1985–2000, population declined in 12 countries, but their household numbers increased

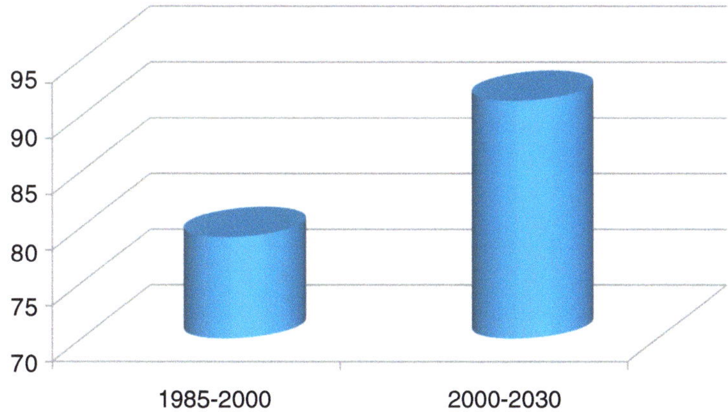

Fig. 6.1 Percentages of countries with faster growth in household numbers than population sizes (actual: 1985–2000, and projected: 2000–2030)

(Fig. 6.3a). For example, Ukraine had a reduction of 1.8 million people but an increase of 1.3 million households. Over the period of 2000–2030, it is projected that 20 countries will experience lower population sizes but higher household numbers (Fig. 6.3b). Russia is projected to have the largest population decline (approximately 21.2 million) but an increase of more than 10.3 million households.

The differences in rates of growth in household numbers and population sizes were due to reductions in household sizes (number of people per household), as a result of such factors as increased nu divorces and declined multigeneration families (Liu et al. 2003). If the average household size in 2000 (3.9 people per household) had remained at the 1985 level (4.4 people per household), there would have been 172 million fewer households in all countries combined by 2000. In other words, there were 172 million "extra" households due to the decline in the average household size alone. It is projected that household sizes will continue to reduce during the period of 2000–2030 and there will be 756 million additional households by 2030 due to reduction in household size alone (with an average household size of 3.1 people per household).

While the discussion above focused on household proliferation at the global and country levels, household proliferation is also common at the regional and local levels. For example, in Wolong Nature Reserve of southwestern China for the conservation of giant pandas, human population size rose from 2,560 in 1975 to 4,550 in 2005, while the number of households jumped from 421 to 1,156 during the same period. In other words, the increase in the number of households was more than twice (174.6 % increase) the increase in the number of people (77.7 % increase). In many regions such as New Zealand (Liu et al. 2003), the numbers of people declined, but the numbers of households continued to increase because household sizes decreased (Liu et al. 2003).

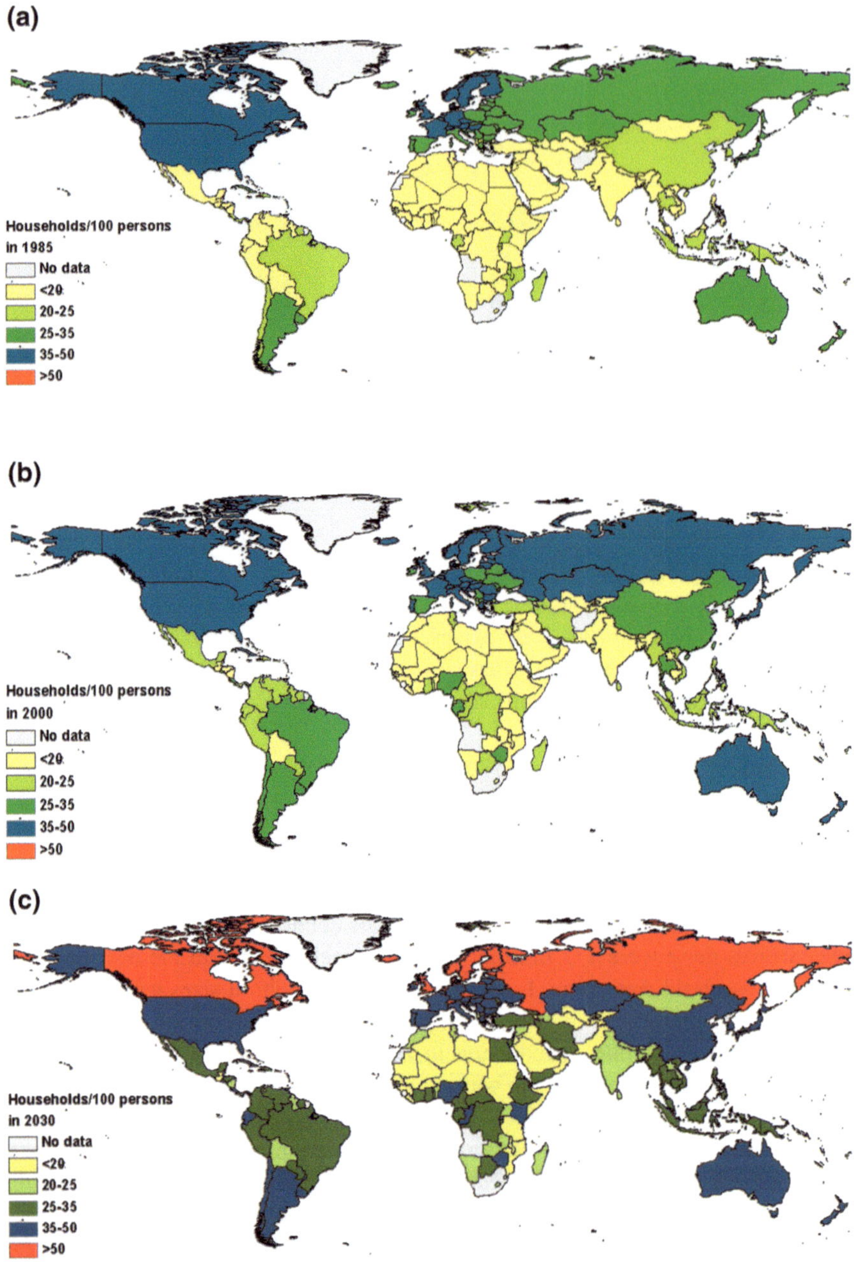

Fig. 6.2 Household intensity in **a** 1985, **b** 2000, and **c** 2030 (projection)

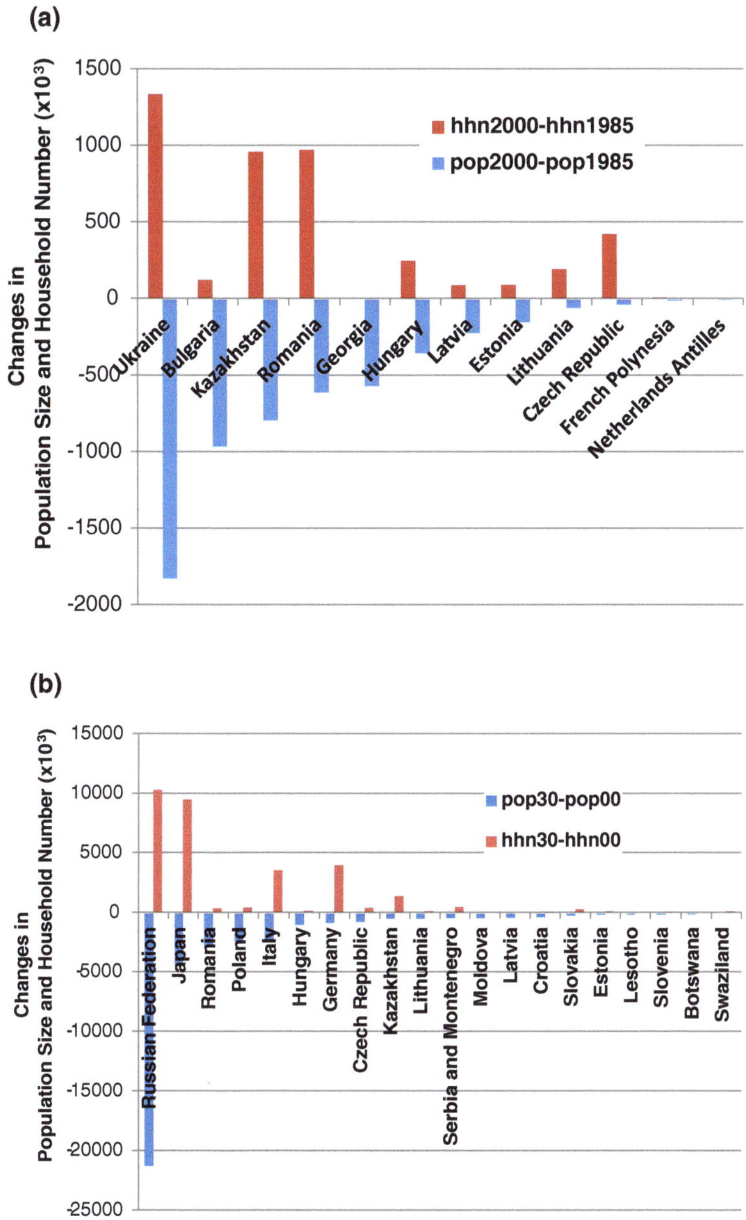

Fig. 6.3 Countries with declined population sizes but increased household numbers during **a** 1985–2000, and **b** 2000–2030 (projection). "pop00–pop85" (and "hhn00–hhn85") indicate the differences between population sizes (numbers of households) in 2000 and 1985. Similarly, "pop30–pop00" (and "hhn30–hhn00") are the projected differences between population sizes (and numbers of households) in 2030 and 2000

6.3 Effects of Household Proliferation on Ecosystem Services

The effects of household proliferation on ecosystem services (provisioning, supporting, regulating, and cultural services) may differ from those of population growth because patterns of household proliferation and population growth vary (Table 6.1).

6.3.1 Demand and Supply of Ecosystem Services

As the number of households increases, so does demand for ecosystem services (or consumption of ecosystem services). This is because more households need more ecosystem services and the efficiency of using ecosystem services is lower per capita in smaller households. For example, more households demand more timber

Table 6.1 Actual and hypothetical impacts of household proliferation on ecosystem services

Type of ecosystem service	Impact of household proliferation (examples)
Provisioning services	
Food (e.g., grains, seafood, spices)	Reduces area for food production (e.g., cropland and other areas suitable for wild foods and spices) through land conversion to residential area (Fazal 2000; Matuschke 2009)
Fresh water	Pollutes water through release of household waste and changes hydrological cycles through land-use change (e.g., application of chemical fertilizers and pesticides for lawn maintenance) (Office for Official Publications of the European Communities 2001; Robbins et al. 2001; Adedeji and Ako 2009; Natural Way 2011)
Fuel, wood, and fiber	Reduces area for production of fuel, wood, and fiber (e.g., fuelwood) through land conversion to residential area (FAO 2002; Carrero and Fearnside 2011)
Pharmaceuticals (e.g., herbal plants, and wildlife)	Destroys plants directly and indirectly (through changes to habitat) (An et al. 2006)
Regulating services	
Carbon sequestration, and climate regulation	Emits CO_2 (Dietz et al. 2009); plants trees and protects forests that sequester CO_2 (Liu et al. 2007)
Flood regulation	Reduces areas (e.g., wetlands) for flood regulation because of land conversion (Schuyt 2005)
Waste decomposition and detoxification	Destroys organisms and habitat of organisms that can decompose waste and toxins (Alavanja 2009)
Purification of water and air	Harms organisms that can purify water and air (Sládeček 1983); creates habitat for biodiversity (e.g., plants, wildlife) that can purify water and air (Liu et al. 2007)

(continued)

Table 6.1 (continued)

Type of ecosystem service	Impact of household proliferation (examples)
Crop pollination	Reduces habitat for pollinators (Hansen et al. 2005); raises honey bees that can enhance pollination (Ogaba 2002)
Pest and disease control	Reduces habitat for natural enemies, spreads pests and diseases (e.g., by introducing garden plants) (Schöller et al. 1997), and creates habitat for pests and diseases; creates habitat for natural enemies and destroys habitats for pests and diseases (Altieri 1993)
Supporting services	
Nutrient cycling	Disrupts nutrient cycling through land conversion (to houses and infrastructure such as roads and other buildings) and creation of barriers (Kaye et al. 2006)
Soil	Uses soil as household construction material, and affects chemical and physical properties of soils through construction of associated infrastructure (e.g., roads, buildings) (Graf 1975)
Seed dispersal	Prevents seed dispersal by forming impermeable surfaces (e.g., houses, roads) (Coffin 2007); facilitates seed dispersal through travel and shipping (Lodge et al. 2006)
Primary production	Damages and occupies areas for primary production through land conversion (Liu et al. 2001)
Cultural services	
Cultural, aesthetic, intellectual and spiritual inspiration	Destroys areas and remnants of cultural and spiritual significance through construction of housing and associated infrastructure (Marsh 1992)
Recreational experiences (including ecotourism)	Destroys through construction and occupies areas suitable for ecotourism (Anderson and Potts 1987)

The impacts of household proliferation are different from those of population growth because patterns of household proliferation and population growth are not the same. For the sake of simplicity, the impacts are phrased in a linear manner, but the actual relationships are much more complex and are often nonlinear with thresholds

for house construction and furniture (Liu et al. 2005), and more fuelwood for heating and cooking. As to fuelwood consumption, a decrease in household size increases fuelwood consumption per capita (An et al. 2001) (Fig. 6.4). This is because houses with different numbers of people used similar amounts of fuelwood for heating. In terms of cooking, more fuelwood is consumed in a large household because more food needs to be cooked for more people, but the efficiency per capita is still higher in a larger household if other conditions are similar (Liu et al. 2005).

On the other hand, households can be ecosystem service producers. For example, some households raise honey bees that are major pollinators (Ogaba 2002), while some other households cultivate plants and flowers in their yards to feed pollinators that can help enhance food production. Some households create habitat for wildlife species and enhance biodiversity, which can generate a variety

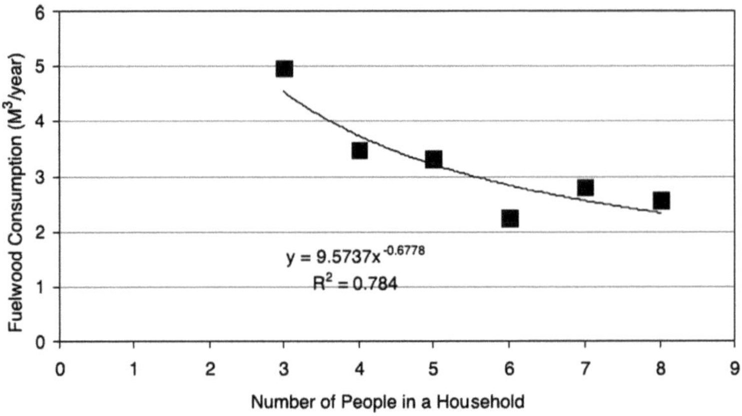

Fig. 6.4 Fuelwood consumption per capita under different household sizes (Liu et al. 2005)

of ecosystem services. Some households also plant trees and protect forests that sequester CO_2, such as those who reduce greenhouse emissions from deforestation and forest degradation (REDD) (The United Nations Collaborative Programme on Reducing Emissions from Deforestation and Forest Degradation in Developing Countries 2012), and who monitor forests from illegal harvesting, such as in China's National Forest Conservation Program (Liu et al. 2008). However, the supply of ecosystem services from households is much less than the demand for ecosystem services. As a result, ecosystem services continue to degrade rapidly (MA 2005).

6.3.2 Impacts on Ecosystem Services

Household impacts on ecosystem services are enormous (Table 6.1). In this section, two examples are given to illustrate the impacts.

6.3.2.1 Impacts of Household Proliferation on Forests and Panda Habitat

Household proliferation is an important contributor to the significant changes in forests and panda habitat in Wolong Nature Reserve. From 1965 to 1997, forest cover and suitable panda habitat in Wolong was substantially reduced (Liu et al. 2001) (Fig. 6.5) because people had used ecosystem services (e.g., fuelwood and timber) in areas that pandas use. Both people and pandas prefer areas that are not too steep. The suitable panda habitat has been much fragmented by human activities (e.g., fuelwood collection, timber harvesting, road construction, and home building). With increases in the total amount of fuelwood consumption and

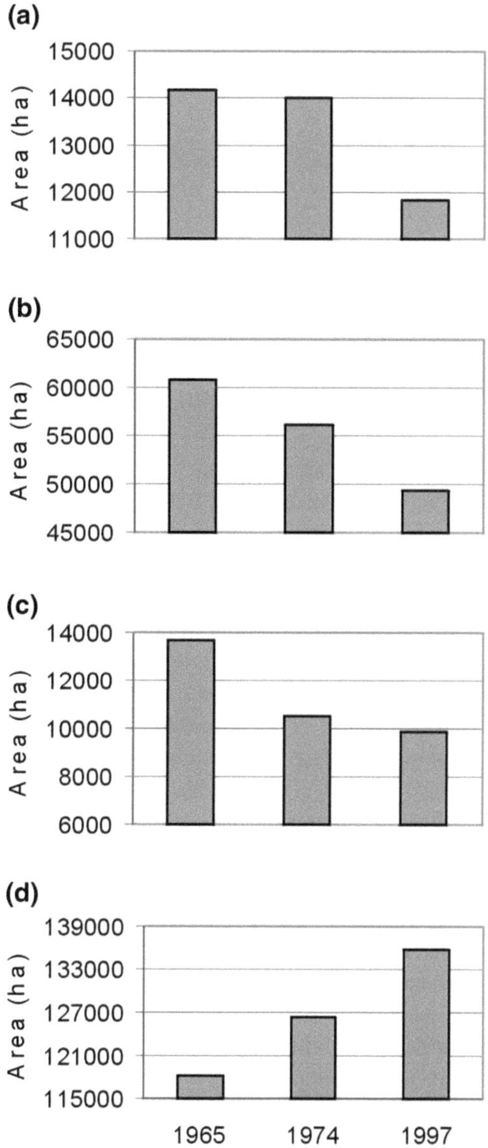

Fig. 6.5 Change in the amount of panda habitat in Wolong Nature Reserve before and after the reserve was established in March 1975. **a** Highly suitable habitat, **b** suitable habitat, **c** marginally suitable habitat, and **d** unsuitable habitat (Liu et al. 2001)

exhaustion of forests near households, local residents went to areas far away from their homes to collect more fuelwood. Consequently, the average distance between homes and locations of fuelwood collection increased over time (He et al. 2009). Fuelwood collection in those remote areas is more damaging to pandas because they are the most suitable panda habitat (Fig. 6.6).

The quantity of panda habitat is more sensitive to factors related to household numbers than to population sizes (An and Liu 2010). Simulations using an agent-

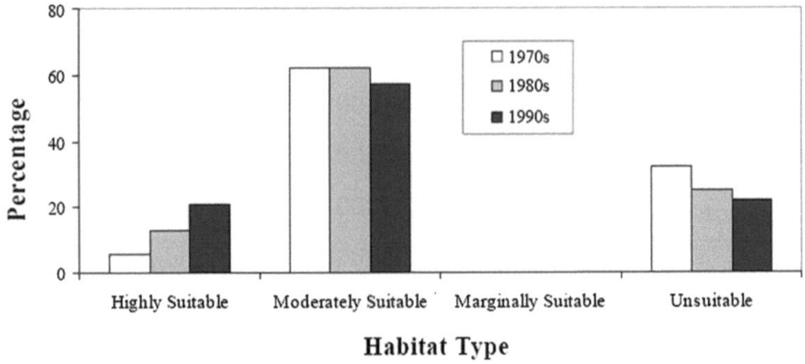

Fig. 6.6 Percentages of fuelwood collection sites in three decades (1970s, 1980s, and 1990s) falling in four types of habitat (He et al. 2009)

based model indicated that household numbers varied very differently than population sizes (An and Liu 2010). Fertility-related factors (e.g., fertility rate, spacing between births, and upper child-bearing age) caused almost instant changes in population size. All the factors except age at the first marriage had time lags of approximately 20 years before they affected household numbers. Age at the first marriage changed household numbers most quickly. A reduction of age at the first marriage from 38 to 18 could lead to a difference of 90 households at year 5, 150 households at year 10, and approximately 220 households at year 20. This is largely because of the household lifecycle: delayed marriage usually postpones the formation of new households and births of babies. It takes more time for other factors to take effect in changing household numbers. For example, increasing fertility rate increases the number of children, but the children still stay with their parents until they establish their own households. This is why there is a time lag of approximately 20 years.

Changing household numbers through age at the first marriage is the most effective and fastest way to lower panda habitat loss. Panda habitat is more influenced by household numbers than population size. This is partly due to how fuelwood is consumed. A major proportion of fuelwood is used for heating, which changes little when an existing household has one more or one fewer person. In terms of cooking, adding or removing one person does not change fuelwood use much (An et al. 2001).

6.3.2.2 Impacts of Household Proliferation on Food Production

Household proliferation also has substantial impacts on other ecosystem services, such as food production (Table 6.1). Because household proliferation requires more areas for housing and associated infrastructures (e.g., roads and sewer services), much agricultural land has been converted into residential areas around the world. Although there are no accurate statistics at the global level, there are numerous reports at the local level. Here are three examples from Africa, Asia, and South America. In Accra (Ghana, Africa), 2,600 ha of agricultural land per year were converted into residential areas (Maxwell 2000). From 1995 to 2005, Ho Chi Minh City of Vietnam lost more than 10,000-ha agricultural land to housing, roads, and other built-up areas (Van 2008). Similar patterns are common in China and Indonesia (Verburg et al. 1999; Weng 2002). In the Pampas ecoregion of Argentina, 39,187 ha of farm land have been converted to exurban use (Matteucci and Morello 2009). An immediate impact of housing expansion is the loss of peri-urban agriculture, which is usually significant in providing perishable food to the urban areas (Matuschke 2009). As a result, agricultural production may be forced to shift to less productive areas and result in yield losses and increased cost of transport.

Food production is further compromised by the use of water by more households because more households require more water for daily consumption and reduce the retention of water because of the impervious surfaces. After the surface water and groundwater in the residential areas cannot meet household demand for water, households have drawn water from far places. This creates cascading effects on distant ecosystem services (Liu et al. in press) and reduces the capacity of food production in distant places (by lowering water table and increasing dry zones in soils), in addition to the agricultural areas that have been converted for residential use. All these affect food security and water security, and ultimately security of all ecosystem services.

Historical trends in household size suggest that there will be many more households even if human population declines. If average household size worldwide were the same as that of the United States (2.5 people per household) in 2010, then the world would have over 40 % more households, or 800 million additional households in the 172 countries with available data (2.7 billion households rather than 1.9 billion households). If each household occupied a 210 m^2 house (the average U.S. house size in 2002), then 168,000 km^2 extra housing area would be required. Even assuming each house has two-stories, then housing needs 89,000 km^2 of additional land area. That would be twice the size of California. Even if the average house globally is half of an average U.S. house, 44,500 km^2 would be needed to accommodate additional households. These estimates have not taken land area for other purposes associated with housing (e.g., infrastructure such as roads, services, yards) into account. Including land for associated functions would require 2–4 times as much land for each home. So the total area for housing would take up nearly half the size of the continental United States (Peterson et al. in press) and severely limit food production.

6.3.3 Payments for Ecosystem Services

Household proliferation has important effects on payments for ecosystem services because many payments for ecosystem services programs are implemented at the household level. For example, China's grassland ecocompensation program distributes 500 yuan (US \$1 = 6.35 yuan as of July 2012) to each household regardless of household size (General Office of the State Council of the People's Republic of China 2010). As there are 2 million households, the total amount of funding needed for all the households is one billion yuan. Thus, the more households, the higher the total amount of payment is needed when the amount of payment for each household is fixed. On the other hand, if the total amount of funding in the program is fixed, each household would receive a smaller amount when there are more households.

More households also can generate a higher amount of funds if they are willing to pay for ecosystem services. For example, Loomis et al. (2000) found that a sampled household would be willing to pay an average of \$252 annually (through a higher water bill) to restore five ecosystem services (dilution of wastewater, natural purification of water, erosion control, recreation, and habitat for fish and wildlife) along a 72 km section of the Platte River in the State of Colorado, USA. Extrapolating the result of the sampled 96 households to all households (281,531) living along the river may reach \$71 million. However, if a quarter of the households are willing to pay, only \$18 million can be collected.

6.4 Research Directions and Management of Ecosystem Services

Household proliferation has been rarely considered in ecosystem services research and management (e.g., valuation) although it may play key roles in ecosystem services and sustainability (Table 6.1). As illustrated above, household dynamics are different from population dynamics because household numbers can increase even though population sizes are stable or even decline. Many questions must be addressed, for example, How do we meet household demands for ecosystem services? How do we reduce household impacts on ecosystem services? How do we determine the most appropriate amounts of payments for ecosystem services in the context of household proliferation? How do households enhance ecosystem services and improve efficiency in using ecosystem services?

New research directions are needed to address questions such as those raised above and test hypotheses such as those listed in Table 6.1. The solutions may include (1) changes in the conceptual frameworks of valuing ecosystem services from static to dynamic processes by incorporating household demand and impacts, (2) changes in research approaches from population-focused to households-focused, and (3) changes from discourse within the ecological and economic communities in valuing ecosystem services to collaborating with researchers in other disciplines (e.g., demography). By collaborating with action-oriented stakeholders and households, the ecological community will be in a stronger position to turn discoveries into actionable knowledge for sustainability of ecosystem services.

The effects of household proliferation may be complex (e.g., with nonlinear relationships and thresholds). Addressing these complexities requires new data and novel tools. Data on household proliferation are not as readily available as population sizes because population sizes are more frequently sampled and widely reported. Obtaining relevant household data is more time-consuming, more complicated, and more costly than research using data on population dynamics.

It is encouraging, however, that new opportunities to address household proliferation are also emerging. More advanced tools for collecting, analyzing, and visualizing data are becoming available. For example, high-resolution remotely sensed data such as QuickBird and IKONOS can help identify locations of housing units (An et al. 2005). A combination of on-the-ground interviews, documents from relevant institutions such as government agencies, and remote sensing data will be helpful in understanding impacts of household proliferation on ecosystem services. Dynamic and interactive web sites (e.g., blogs, social media) and citizen science may provide new tools to understand household demand for ecosystem services.

Current monitoring programs on ecosystem services include indicators of ecosystem services themselves (Table 6.1). To more accurately predict changes in ecosystem services and take proactive adaptive management measures, it is crucial to monitor indicators that affect changes in ecosystem services directly and indirectly, including factors that shape household dynamics. Thus, monitoring efforts should be expanded, especially in areas with severe degradation of ecosystem services, to indicators in human dimensions (e.g., values and attitudes toward household formation and ecosystem services).

Household proliferation generates more complications for ecosystem service management and policy than population growth. In fact, some payments for ecosystem services programs stimulate the formation of new households because the payments are implemented at the household level and dividing a household into two can double the payments (Liu et al. 2007). To achieve sustainability of ecosystem services, current management and stewardship approaches need to adopt a new structure to fully integrate household proliferation and strive to enhance positive and reduce negative effects of household proliferation.

6.5 Conclusions

Global household proliferation provides both challenges and opportunities for research and governance of ecosystem services in coupled human and natural systems across local to global levels. It is projected that household proliferation will intensify even faster than population growth globally in the future. As household proliferation has important implications for demand and impacts on ecosystem services, it should be incorporated into ecosystem services research, monitoring, and scenario analysis. Incorporating household dynamics into research across landscapes around the world would lead to unique new insights. Such research also would generate useful information for managing and governing ecosystem services at a time when the ecological community is faced with unprecedented obligations to address societal needs such as achieving ecological sustainability while improving human well-being worldwide.

Acknowledgments We would like to thank Joanna Broderick and Shuxin Li for assistance, the National Science Foundation and the CAS/SAFEA International Partnership Program for Creative Research Teams of "Ecosystem Processes and Services" for support.

References

Adedeji AA, Ako RT. Towards achieving the United Nations' millennium development goals: the imperative of reforming water pollution control and waste management laws in Nigeria. Desalination. 2009;248(1–3):642–9.
Alavanja MCR. Pesticides use and exposure extensive worldwide. Rev Environ Health. 2009;24(3):303–9.
Altieri MA. Ethnoscience and biodiversity: key elements in the design of sustainable pest management systems for small farmers in developing countries. Agric Ecosyst Environ. 1993;46(1–4):257–72.
An L, Liu J, Ouyang Z, Linderman M, Zhou S, Zhang H. Simulating demographic and socioeconomic processes on household level and implications for giant panda habitats. Ecol Model. 2001;140(1–2):31–50.
An L, Linderman M, Qi J, Shortridge A, Liu J. Exploring complexity in a human-environment system: an agent-based spatial model for multidisciplinary and multiscale integration. Ann Assoc Am Geogr. 2005;95(1):54–79.
An L, He G, Liang Z, Liu J. Impacts of demographic and socioeconomic factors on spatio-temporal dynamics of panda habitat. Biodivers Conserv. 2006;15(8):2343–63.
An L, Liu J. Long-term effects of family planning and other determinants of fertility on population and environment: agent-based modeling evidence from Wolong Nature Reserve, China. Popul Environ. 2010;31:427–59.
Anderson B, Potts DF. Suspended sediment and turbidity following road construction and logging in western montana. JAWRA J Am Water Resour Assoc. 1987;23(4):681–90.
Bin S, Dowlatabadi H. Consumer lifestyle approach to US energy use and the related CO_2 emissions. Energ Policy. 2005;33:197–208.
Carrero GC, Fearnside PM. Forest clearing dynamics and the expansion of landholdings in Apuí, a deforestation hotspot on Brazil's Transamazon Highway. Ecol Soc. 2011;16(2):26.
Coffin AW. From roadkill to road ecology: a review of the ecological effects of roads. J Transp Geogr. 2007;15(5):396–406.
Daily GC, Matson PA. Ecosystem services: from theory to implementation. PNAS. 2008;105(28):9455–6.
Dietz T, Gardner GT, Gilligan J, Stern PC, Vandenbergh MP. Household actions can provide a behavioral wedge to rapidly reduce US carbon emissions. PNAS. 2009;106(44):18452–6.
Fazal S. Urban expansion and loss of agricultural land-a GIS based study of Saharanpur City, India. Environ Urbanization. 2000;12(2):133–49.
FAO. World agriculture: towards 2015/2030. Summary report. rome, food agriculture organization of the United Nations. 2002.
General Office of the State Council of the People's Republic of China. Standing Committee of the State Council decided to establish the conservation of grasslands subsidy incentives. 2010. From http://www.gov.cn/ldhd/2010-10/12/content_1720555.htm.

Graf WL. The impact of suburbanization on fluvial geomorphology. Water Resour Res. 1975;11(5):690.
Hansen AJ, Knight RL, Marzluff JM, Powell S, Brown K, Gude PH, Jones K. Effects of exurban development on biodiversity: patterns, mechanisms, and research needs. Ecol Appl. 2005;15(6):1893–905.
He G, Chen X, Beaer S, Colunga M, Mertig A, An L, Zhou S, Linderman M, Ouyang Z, Gage S, Li S, Liu J. Spatial and temporal patterns of fuelwood collection in Wolong Nature Reserve: implications for panda conservation. Landscape Urban Plan. 2009;92(1):1–9.
Kaye JP, Groffman PM, Grimm NB, Baker LA, Pouyat RV. A distinct urban biogeochemistry? Trends Ecol Evol. 2006;21(4):192–9.
Liu J, An L, Batie SS, Bearer S, Chen X, Groop RE, He G, Liang Z, Linderman MA, Mertig AG, Ouyang Z, Qi J, Zhang H, Zhou S. Beyond Population Size: Examining intricate Interactions among population structure, land Use, and environment in Wolong Nature Reserve (China). In: Entwisle B, Stern P, editors. New research on population and environment. Washington: National Academy of Sciences Press; 2005. p. 217–37.
Liu J, Linderman M, Ouyang ZY, An L, Yang J, Zhang HM. Ecological degradation in protected areas: the case of Wolong Nature Reserve for giant pandas. Science. 2001;292(5514):98–101.
Liu JG, Daily GC, Ehrlich PR, Luck GW. Effects of household dynamics on resource consumption and biodiversity. Nature. 2003;421(6922):530–3.
Liu J, Dietz T, Carpenter SR, Alberti M, Folke C, Moran E, Pell AN, Deadman P, Kratz T, Lubchenco J, Ostrom E, Ouyang Z, Provencher W, Redman CL, Schneider SH, Taylor WW. Complexity of coupled human and natural systems. Science. 2007;317(5844):1513–6.
Liu J, Li S, Ouyang Z, Tam C, Chen X. Ecological and socioeconomic effects of China's policies for ecosystem services. PNAS. 2008;105(28):9477–82.
Liu J, Hull V, Moran E, Nagendra H, Swaffield S, Turner II BL (in press). Telecouplings and global land use. In: Reenberg A, Seto K, editors. Rethinking global land use in an urban era. MIT Press.
Lodge DM, Williams S, MacIsaac HJ, Hayes KR, Leung B, Reichard S, Mack RN, Moyle PB, Smith M, Andow DA, Carlton JT, McMichael A. Biological invasions: recommendations for U.S. policy and management. Ecol Appl. 2006;16(6):2035–54.
Loomis J, Kent P, Strange L, Fausch K, Covich A. Measuring the total economic value of restoring ecosystem services in an impaired river basin: results from a contingent valuation survey. Ecol Econ. 2000;33(1):103–17.
Lu Z, Matsuhashi R, Yoshida Y. Direct and indirect impacts of households by region of China on CO_2 emissions. Environ Inform Arch. 2007;214–23.
MA. Ecosystems and human well-being: Synthesis. Washington (DC): Island Press; 2005. p. 137.
Marsh GA. Walking the spirit trail: repatriation and protection of native American remains and sacred cultural items. Ariz St Law J. 1992;24:79.
Matteucci SD, Silvia D, Morello Jorge. Environmental consequences of exurban expansion in an agricultural area: the case of the Argentinian Pampas ecoregion. Urban Ecosyst. 2009;12: 287–310.
Matuschke I. Rapid urbanization and food security: using food density maps to identify future food security hotspots. 2009. From http://www.fao.org/fileadmin/user_upload/esag/docs/RapidUrbanizationFoodSecurity.pdf.
Maxwell D, Levin C, Armar-Klemesu M, Ruel M, Morris S, Ahiadeke C. Urban livelihoods, and food and nutrition security in Greater Accra, Ghana. IFPRI Research Report 112. Washington DC: International Food Policy Research Institute; 2000.
McGranahan G, Balk D, Anderson B. The rising tide: assessing the risks of climate change and human settlements in low elevation coastal zones. Environ Urbanization. 2007;22(1):17–37.
Natural Way. The Cause and Effects of Water Pollution by Waste Settlement. 2011. from http://miofi.blogspot.com/2011/05/cause-and-effects-of-water-pollution-by.html.
Office for Official Publications of the European Communities. Pollutants in urban waste water and sewage sludge. 2001. From http://ec.europa.eu/environment/waste/sludge/sludge_pollutants.htm .

Ogaba MR. Household poverty reduction through beekeeping amongst Uganda rural women. 2002. From http://www.apimondia.com/apiacta/slovenia/en/ogaba.pdf.
Pagiola S, Ramírez E, Gobbi J, Haan Cd, Ibrahim M, Murgueitio E, Ruíz JP. Paying for the environmental services of silvopastoral practices in Nicaragua. Ecol Econ. 2007;64:374–385.
Peterson N, Peterson T, and Liu J. The Housing Bomb. Johns Hopkins University Press (in press).
Robbins P, Polderman A, Birkenholtz T. Lawns and toxins: an ecology of the city. Cities. 2001;18(6):369–80.
Schöller M, Prozell S, Al-Kirshi AG, Reichmuth C. Towards biological control as a major component of integrated pest management in stored product protection. J Stored Prod Res. 1997;33(1):81–97.
Schuyt KD. Economic consequences of wetland degradation for local populations in Africa. Ecol Econ. 2005;53(2):177–190.
Sládeček V. Rotifers as indicators of water quality. Hydrobiologia. 1983;100(1):169–201.
The United Nations Collaborative Programme on Reducing Emissions from Deforestation and Forest Degradation in Developing Countries. 2012. From http://www.un-redd.org.
Tran TV. Research on the effect of urban expansion on agricultural land in Ho Chi Minh City using remote sensing method. VNU J Sci (Earth Sciences). 2008;24:104–11.
United Nations Centre for Human Settlements (Habitat). Cities in a globalizing world: global report on human settlements 2001. London: Earthscan Publications Ltd; 2001.
United Nations Human Settlements Programme (2007). Global Report on Human Settlements 2007: Enhancing Urban Safety and Security.
Verburg PH, Veldkamp TA, Bouma J. Land use change under conditions of high population pressure: the case of Java. Glob Environ Change. 1999;9:302–12.
Weng W. Land use change analysis in the Zhujiang delta of China using satellite remotesensing, GIS, and stochastic modeling. J Environ Manage. 2002;64:273–84.

Part II
Landscape Modeling

Chapter 7
Bringing Climate Change Science to the Landscape Level: Canadian Experience in Using Landscape Visualisation Within Participatory Processes for Community Planning

Stephen R. J. Sheppard, Alison Shaw, David Flanders, Sarah Burch and Olaf Schroth

Abstract This chapter addresses the role of visualisation tools within participatory processes in bringing climate change science to the local level, in order to increase people's awareness of climate change and contribute to decision-making and policy change. The urgent need to mitigate and adapt to climate change is becoming more widely understood in scientific and some policy circles, but public awareness and policy change are lagging well behind. Emerging visualisation theory suggests that landscape visualisations showing local landscapes in fairly realistic perspective views may offer special advantages in bringing the projected

---

Adapted and updated from a paper (Sheppard et al. 2008) published in Proceedings of "Digital Design in Landscape Architecture 2008", 9th International Conference on IT in Landscape Architecture, May 29–31 2008, Anhalt University of Applied Sciences, Dessau/Bernburg, Germany.

S. R. J. Sheppard (✉) · D. Flanders
Collaborative for Advanced Landscape Planning (CALP), University of British Columbia, 2424 Main Mall, Vancouver, BC V6T1Z4, Canada
e-mail: stephen.sheppard@ubc.ca

D. Flanders
e-mail: david.flanders@ubc.ca

O. Schroth
Department of Landscape, Sheffield University, Arts Tower, Western Bank, Sheffield S10 2TN, UK
e-mail: o.schroth@sheffield.ac.uk

S. Burch
Dept. of Geography and Environmental Management, University of Waterloo, 200 University Ave W. Waterloo, Ontario N2L 361, Canada

A. Shaw
Flipside Sustainability, 3280, W.Broadway, Vancouver, BC, Canada
e-mail: shwali@gmail.com

B. Fu and K. B. Jones (eds.), *Landscape Ecology for Sustainable Environment and Culture*, DOI: 10.1007/978-94-007-6530-6_7,

consequences of climate change home to people in a compelling manner. This chapter draws on and summarizes a unique body of research in Canada, applying and evaluating a local climate change visioning approach in five diverse case study communities across the country. This new participatory process was developed to localize, spatialize, and visualize climate change implications, using landscape visualisation in combination with geospatial and other types of information. The visioning process was successful in raising community awareness, increasing people's sense of urgency, and articulating for the first time holistic community options in mitigating and adapting to climate change at the local level. In some cases the process led to new local policy outcomes and actions. Such methods, if widely implemented in enhanced planning processes, could facilitate uptake of climate change science and potentially accelerate policy change and action on climate change. However, moving from more traditional types of science information and planning to an approach which can engage emotions with visual imagery, will require guidelines and training to address ethical and professional dilemmas in community engagement and planning at the landscape level.

Keywords Climate change visualization · Visioning processes · Landscape visualization · Visual imagery · Landscape planning · Community engagement · Community planning · Decision-support tools · Policy change · Public awareness · Climate change scenarios

7.1 Introduction

Global warming is fundamentally changing the context within which landscapes and communities have traditionally been planned and managed. There is an urgent need to mitigate and adapt to climate change (IPCC 2007), requiring communities to do their part in reducing greenhouse gases (GHGs) and plan for a range of possible energy futures, impacts on ecosystems, risks to infrastructure, and generally unfamiliar circumstances. However, action, policy change and the necessary public support are moving slowly in many parts of the world. Concerns have been raised by climate change scientists over their difficulties in achieving uptake of global modelling by practitioners and decision-makers at the local level (Kriegler et al. 2012). There is therefore a huge need for better communication and decision-support tools to enhance the sense of urgency and help accelerate informed, integrated, and effective responses to climate change.

In this context, realistic landscape visualisations may offer special advantages in bringing the potential consequences of climate change home to local citizens and decision-makers, in a compelling and useful manner. This chapter considers the role of science-based visualisation tools and processes in improving community planning and engagement on climate change; and particularly in making climate change science meaningful at the local level, increasing peoples' awareness of climate change, and possibly affecting policy and collective behaviour change. This chapter reviews first the theoretical basis for such effects, then describes a major research programme conducted in Canada to test the effectiveness of a visualisation-based participatory process in achieving some of these goals. It concludes with implications and recommendations for using such tools and processes in community engagement and planning for climate change internationally.

7.2 Theoretical Background on the Influence of Landscape Visualisation on People in Relation to Climate Change

Landscape visualisation exhibits several characteristics which could be powerful in bringing consequences of climate change home to people. Landscape visualisation attempts to represent actual places and on-the-ground conditions in three-dimensional (3D) perspective views with varying degrees of realism (Sheppard and Salter 2004). This amounts to a unique form of visual communication, conveying information in the dominant form to which the human species is genetically adapted (e.g. visual landscapes), but capable of showing future scenarios and conditions which people may not be able to imagine on their own.

Early evidence from research and practice, and emerging theory on 3D visualisations, provides some reasons for optimism. Human responses to environmental or visual stimuli such as landscape visualisations can be broadly categorized as follows: engagement (level of interest and attention); cognition (related to knowledge, awareness and understanding); affect (related to feelings, perceptions, and emotions); and behaviour (related to changes in behaviour of the viewer) (Appleyard 1977; Zube et al. 1982). In the collective sphere of community planning, related consequences such as capacity-building (to deal with climate change), policy change, and decision-making are also important (Sheppard 2008). There is considerable evidence of the effectiveness of visualisation as a planning tool (e.g. Tress and Tress 2003; Sheppard and Meitner 2005; Salter et al. 2009), and the advantages of interactive systems in particular for engagement, cognition, and awareness building (e.g. Winn 1997; Furness III et al. 1998; Salter 2005; Schroth 2007; Mulder et al. 2007). In this context, potential benefits of landscape visualisation include:

- Attractiveness to lay-people, due to the novelty of the medium, its dynamism and interactivity;
- Combining the predictive capabilities of modelling/GIS with the intuitive and experientially rich media of photo-realistic representation, providing ‘windows into the future’ with changing landscape patterns and meaningful socio-cultural associations;
- The ability to present alternative futures side-by-side and pose ‘what-if’ questions (Ervin 1998); and

- Transparency and flexibility: digital visualisation techniques can be augmented or modified to highlight or simplify almost any aspect of the 3D/4D modelling being conducted, such as underlying meta-data or different levels of realism selected by the user (Bishop and Lange 2005).

Less research has taken place on affective responses (e.g. Bishop and Rohrmann 2003), though there is evidence that visualisations can stimulate positive or negative emotional reactions in observers (e.g. Daniel and Meitner 2001). Nicholson-Cole (2005) documented the influence of popular visual media on people's mental imagery of climate change, and found that respondents were most emotionally affected by national, local, and personal imagery rather than international imagery, in part because it was easier to relate to and more salient (see also Shackley and Deanwood 2002). The ability of visualisations to localize information through detailed depiction of recognizable and familiar sites, as they would be seen by local residents or users (in contrast to a detached plan, aerial view, or abstract diagrams), would seem to tap into people's emotional attachment to place. Nicholson-Cole (2005) describes advantages of visualisation in conveying strong messages quickly and memorably, condensing complex information, and potentially arousing emotional feelings, which may motivate personal action on climate change. The perception literature however, warns that messaging that is too heavy on "doom and gloom" can be counter-productive (Moser and Dilling 2007; Nicholson-Cole 2005).

Very little hard evidence exists on behavioural impacts of landscape visualisation, either during exposure to the visualisation material or afterwards (Sheppard 2005a). Lowe et al. (2006) have evaluated behaviour of people who watched the film "The Day after Tomorrow" which contained extensive visualisations of supposed climate change effects, and found both attitude change and some limited changes in behavioural intent, especially immediately after the viewing. There is evidence from visualisation practice that use of computer visualisations has also led to significant action by decision-makers on policy changes to planning strategies and approvals (Sheppard 2005a; Sheppard and Cizek 2009).

It therefore seems possible that landscape visualisations, if applied to what is arguably the single greatest environmental issue of all (climate change), may be able to capture public interest, influence attitudes and support for climate action, or help trigger policy change, by "making climate change personal" in people's back yards. The actual effectiveness of visualisation in stimulating these responses may depend on many factors including: the delivery mechanism or process for presenting visual imagery to the public or decision-makers, including the role of other forms of available information; the type of audience; the socio-cultural and environmental context; the media employed; and the nature of the climate-change-related subject matter. It seems likely that a combination of techniques and influences would be required if action and policy on climate change is to be implemented at the local community level.

7.3 Research Results from the Local Climate Change Visioning Process

The issues described above have been explored in a unique research programme conducted over several years by researchers and partners that were coordinated by the Collaborative for Advanced Landscape Planning (CALP) at the University of British Columbia. Researchers worked as a trans-disciplinary team involving climate scientists, social scientists, planners, landscape architects, engineers, agency staff and stakeholder representatives in British Columbia and other locations in Canada. The goal was to develop a new approach that bridges the gap between global climate science and the local level, using realistic landscape imagery of alternative climate futures at the neighbourhood scale.

This body of Canadian research appears to be unique in combining the following attributes:

- Applying landscape visualization systematically to future climate change scenarios in real, specific locations.
- Drawing on hybrid modelling, climate change projections, spatial analysis, other locally available data, and local stakeholder opinion to develop scenarios, mapping, and 3D visualisations.
- Developing holistic alternative scenarios which addressed both adaptation, mitigation, and current land use trends.
- Embedding visualisation within a structured participatory process involving multiple stakeholders.
- Evaluation of the process, products, and their impacts on users such as community members, practitioners, and decision-makers.
- A sustained, coherent body of work conducted over the last decade in geographically diverse locations.

In reviewing a range of other scenario 'visioning' approaches, Sheppard (2012) describes various precedents that incorporate one or sometimes more of these attributes, but none that combine them all. The five case study processes described below all involved government partners from the Federal to the local levels, developed multi-stakeholder working groups, and were conducted by researchers from five universities across the continent. The range of environments and types of communities studied (including mild temperate coastal cities, a dry interior rural community, a major metropolitan centre, and a remote arctic hamlet), suggest that the findings apply to local level planning and community engagement in many regions and settlement types in North America and potentially beyond.

This section summarizes the approach and key findings of the Local Climate Change Visioning (LCCV) process, focusing particularly on the results of effectiveness evaluations on the early case studies carried out in the Metro Vancouver region, and contextualized through a brief review of more recent outcomes of related studies across Canada.

7.3.1 Methods of the Local Climate Change Visioning Process in Metro Vancouver[1]

This research aimed to develop and test a process for local visioning of climate change impacts and responses, using an integrated geomatics/visualization system as a prototype for improved community planning and engagement on climate change. More specific objectives, responding to practical and psychological needs of communities and citizens (e.g. Kriegler et al. 2012; Moser and Dilling 2007) included:

- Making climate change choices more explicit in order to build awareness and capacity for behaviour change, policy development, and decision-making: bringing climate change implications home to people and their local governments
- Addressing questions such as "what would your local landscape look like if everyone met specific carbon-reduction targets?" (e.g. BC's GHG reduction targets).
- Enabling the integration of diverse streams of information from the multiple sources and disciplines needed to address climate change somewhat holistically.
- Illustrating various adaptation and mitigation strategies that can be assessed against carbon reduction targets and other key sustainability/feasibility criteria.

The approach was to bring climate change science down to the local level through spatializing, localizing, and visualizing information on climate change, within an enhanced participatory process. The visioning approach harnesses the power of 3D landscape visualization of climate change, supported by Geographic Information Systems (GIS) data, downscaled climate scenarios, and environmental/land-use modelling. The process builds upon early precedents addressing more limited aspects of climate change (e.g. Cohen 1997; Dockerty et al. 2005; Snover et al. 2007), and other scenario-based, modelling-assisted planning processes using visualization (see Sheppard 2012, Chap. 13 for a review). It draws on the best available data, science, and best practices, as well as local knowledge and multidisciplinary expertise, through workshops with scientists, practitioners and community stakeholders.

Products included computer visualizations produced at a scale that matters to decision makers and the community: their neighbourhoods and backyards. These pictures of alternative climate scenarios over time show different levels of climate change causes, impacts, adaptive responses, and mitigation measures in combination. Through the images, people can see, for example, the effects on their community of unmitigated climate change (e.g. sea-level rise, drought, increased fire risk) in their lifetime, or of "complete" resilient low-carbon communities with renewable energy, walkable and more self-contained neighbourhoods, local food supply, and adaptations to more intense rainstorms (Fig. 7.1).

[1] Funded primarily by the GEOIDE National Centres of Excellence research network.

Fig. 7.1 **a** Existing conditions in a high-carbon urban landscape. **b** Conceptual visualisation of a low-carbon future with intensive mitigation (e.g. energy-generating buildings, increased transit, walkable neighbourhoods, live-work buildings, local shops, multi-family housing, etc.) and increased resilience (increased stormwater retention, local food production, local employment, etc.). *Credit* (a) *Photo* S. Sheppard. (b) Visualization: J. Laurenz, CALP, UBC. Reproduced from Sheppard (2012) "Visualizing Climate Change", Earthscan/Routledge

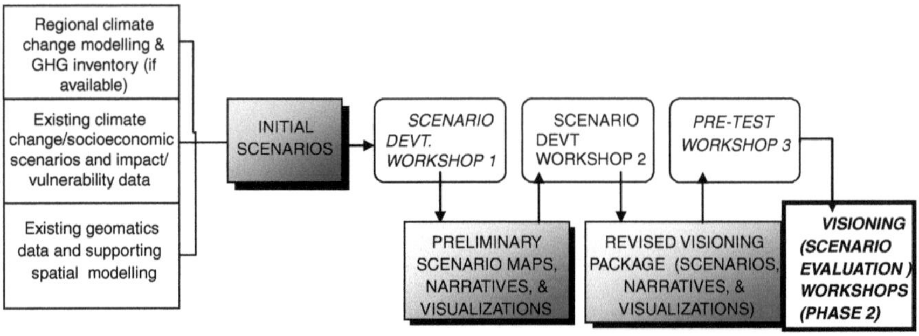

Fig. 7.2 Flowchart for the local climate change visioning process (Scenario Development). Reproduced from Sheppard (2008), "Local Climate Change Visioning", Plan Canada, with permission of Canadian Institute of Planners; and from Sheppard et al. (2008), "Can Visualization Save the World?" Digital design in landscape architecture 2008, 9th international conference Anhalt

The research team initially worked with two communities in south-west BC that represent different climate change challenges: the low-lying coastal community of Delta facing sea-level rise, and the urban fringe on the Northshore mountains, affected by reduced snowpack and increasing natural hazards. In Phase 1, we prepared visioning packages which illustrate four alternative scenarios or "worlds" out to 2100 in each case study area; these were based on assumed local conditions against the backdrop of global climate scenarios and regional modelling of integrated socio-economic and land-use factors in the Georgia Basin QUEST model (Shaw et al. 2009). Steps in this LCCV process (Fig. 7.2) included:

- Downscaling of global climate projections with regional climate change data from Environment Canada.
- Data collection on key land-use/environmental issues and natural hazards at the local level.
- Developing an initial set of plausible alternative climate change scenarios and storylines in the community, through research and workshops with a local working group to define and prioritize GHG sources, potential climate change impacts and vulnerabilities (e.g. snowpack reductions, increased fire and slope stability hazards), adaptation measures, and mitigation measures (e.g. biomass production, neighbourhood retrofitting).
- Mapping impacts and appropriate locations for mitigation and adaptation measures, using spatial analysis with GIS and remote sensing data, interpretation of available urban planning or resource management models, and hybrid modelling to link together various models addressing, for example, climate impacts, land uses, sea level, and energy use.
- Developing 3D models and visualization imagery/animations (in ArcScene, Visual Nature Studio, and Photoshop) for selected neighbourhoods (Fig. 7.3),

Fig. 7.3 3D visualisations of alternative conditions in a BC coastal community at risk of flooding: **a** Existing conditions in 2000. **b** Projection of same neighbourhood in 2100 with a +4 °C global warming scenario, storm surge, and no effective adaptation measures. **c** Projection of same area in 2100 with a low-carbon resilient scenario, storm surge, raised sea-wall, flood-proofed buildings, and on-site energy generation (intensive mitigation). *Credit* David Flanders, CALP, UBC; sea level data provided by Natural Resources Canada. Reproduced from Sheppard (2008), "Local Climate Change Visioning", Plan Canada, with permission of Canadian Institute of Planners; and from Sheppard et al. (2008), "Can Visualization Save the World?" Digital design in landscape architecture 2008, 9th international conference Anhalt

using available 3D datasets such as LiDAR data,[2] and following procedures to ensure accuracy to the data, representative view locations, and themes to be visualized (based on input from the local working group and experts).
- Generating visioning packages for working group review, illustrating possible future neighbourhood conditions through visualizations, profiles using key indicators, GIS mapping, and photographs of best practice precedents.

In Phase 2, the products of this process were tested with local policy-makers and representatives of the public. The evaluation was conducted with approximately 120 community members[3] in Delta and North Vancouver, plus a sample of Lower Mainland planners and engineers, and some local council members. Participants were recruited mainly through posters, web-postings, and key informants using the snowball contact technique. Visioning (scenario evaluation) workshops of 2–3 hours were held in each case study community, using a multi-media PowerPoint presentation with visualisations on two large screens, side-by side. Evaluation methods used standard social science assessment techniques (e.g. pre-post survey written questionnaire, written qualitative comments, participant observations, and some follow-up interviews) to determine changes in participants' attitudes and knowledge due to the presentations, and users' opinions on the process.

7.3.2 Results on Effectiveness of Visioning Tools and Process in Metro Vancouver

The project demonstrated that an integrated, visualization-based process is workable and effective in two very different BC communities. Compelling 3D visualizations of local climate change scenarios can be developed defensibly, despite the multi-disciplinary data/modelling needs, complexity and uncertainty involved. The results of the study suggest that participatory planning processes supported by geo-visualization and visual imagery can have a significant effect on both awareness and affective response. Key findings drawn from initial data analysis are described next (for more details, see Tatebe et al. 2010; Cohen et al. 2011; Sheppard et al. 2011).

---

[2] LIDAR: Light Detection and Ranging techniques using laser-scanning of landscape surfaces to create detailed 3D models.

[3] Approximately half of the Delta public sample were first shown a version of the visioning packages without visualisations, in order to distinguish the results of the overall visioning process from the specific impacts of the visualisations. The results of this comparison are reported elsewhere; in this chapter, the results described apply primarily to the visioning process including visualisations.

7.3.2.1 Participant Engagement

Based on observational data of audience response recorded during the evaluation workshops, it is clear that the extensive use of realistic visualisations maintained a high level of engagement among the public participants over a long and intense visioning session. Some of the most effective images were an animation of rising lea-levels and visualisation sequences with animated slide transitions (wipes) showing time-lapse effects over the 21st Century. Notable examples of the latter included flooding of an island depicted in an aerial view of a LiDAR/terrain model draped with an aerial photograph, and the transition to a higher sea wall blocking views from a back-garden in the community. Participants were also interested in certain other graphics, e.g. illustrative charts and pictographs showing the key indicators differentiating the four future worlds. Interest tended to flag somewhat with longer temporal visualisation sequences.

7.3.2.2 Credibility and Effectiveness

Credibility of the visualisation tools and effectiveness of the visioning process were rated generally as high, though some recommendations for enhanced or additional products were received. Planning and engineering professionals had generally similar responses to those of the public in finding the visualisations credible. Even where the visualisations showed peak events such as storm surge flooding of entire communities, it appears that we did not approach the limits of permissible drama with the participants. Some people commented that the visualisations were too benign, relative to actual storm/flooding conditions which they had experienced.

7.3.2.3 Cognition and Awareness

Professionals registered a substantial increase in the urgency of responding to climate change, after seeing the visioning packages. Using visualizations of alternative climate futures in local and familiar places substantially increased the public's awareness of local climate change impacts and of the response options available to communities. A number of participants remarked on the way the imagery and content of the Local Climate Change Visioning presentation demonstrated the local impacts, making "global warming more immediate, more real". Another participant made the impact of the visuals clear, "I learned how climate change could affect my community in a very graphic way. Numbers may not stay with me but visuals will", The use of photographs from precedents for adaptation or mitigation solutions implemented elsewhere seemed to work to suggest feasibility of future conditions; this and the range of response options visualized seemed to leave people with a sense of the constructive actions that can be taken.

7.3.2.4 Emotions

The analysis of the pre and post questionnaires suggest that, despite a fairly high prior knowledge of global climate change, many respondents' concern about the effects of climate change significantly increased. Many respondents noted that having information locally contextualized and visualized in alternative futures made the climate change information "hit home".

7.3.2.5 Motivation and Behaviour

Results show a significant increase among respondents in the belief that action taken can significantly reduce the impacts of climate change in the future. The visioning material increased stated motivations for behaviour change and altered community participants' attitudes. There was a significant increase in the number of respondents who personally plan to do something about climate change. Analysis of comments revealed that the majority focused on changes to personal auto use (e.g. use car less, walk/bike, use public transport, carpool, buy a hybrid) and the household (e.g. changing light bulbs, using less energy, upgrading appliances), rather than collective responses such as voting with an emphasis on a climate change platform or joining a community group. Willingness to support climate change policies (both mitigation and adaptation) at the local scale increased substantially. One participant from Delta's Environment Committee called the sessions "empowering".

This exploratory research thus offers compelling evidence to support the use of alternative climate change scenarios, downscaled climate information, and geomatics-based visualization techniques to generate significant cognitive and affective responses in community participants, and increase policy support on climate change action. It is difficult to disentangle the effects on participants of the visualisations from those effects arising from the overall participatory process. In a control group of Delta residents who were exposed to the visioning presentation without landscape visualisation but with otherwise similar content, many responses trended in the same direction as those of participants seeing the full presentation, but support for mitigation and adaptation policies was stronger with those seeing the visualisations (Sheppard 2012); engagement and interest levels appeared higher with visualisations also. In subsequent interviews with practitioners involved in the process, the majority of images most vividly recalled were landscape visualisations (Burch et al. 2010).

Broadly similar results from the full visioning process were obtained with both members of the public and practitioners (Tatebe et al. 2010). However, the self-selected nature of the citizen participation may mean that these participants represented an "interest" sample, and the incidence of recent climate change-related events (e.g. flooding in Delta in 2006) may help explain the fairly high levels of awareness on climate change. Accordingly, it cannot be assumed that other types of community would react similarly.

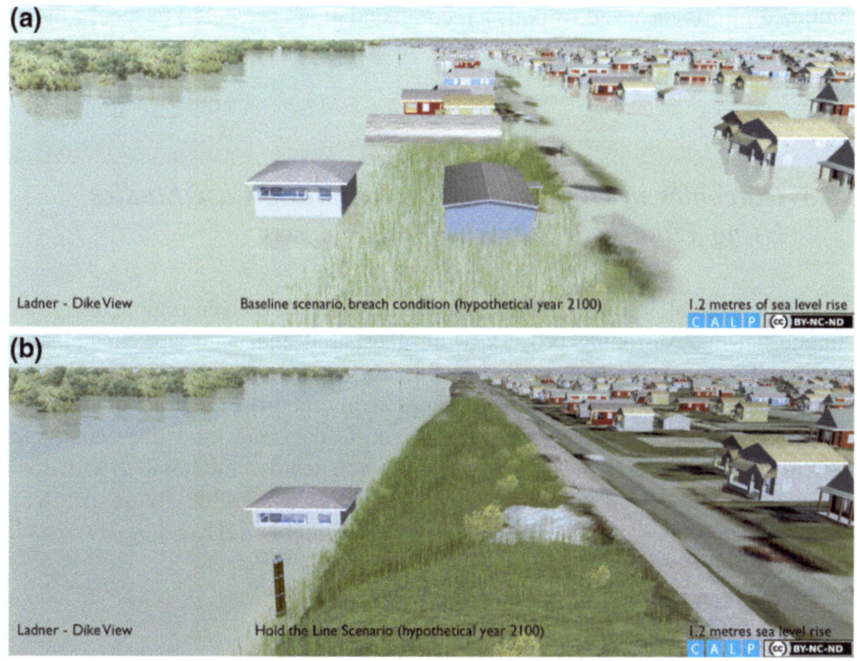

Fig. 7.4 Landscape visualisations showing projected sea level rise impacts and hypothetical adaptations in the year 2100 with 1.2 m sea level rise in South Delta, BC. **a** View in 2100 with current dike conditions and projected flooding from a nearby dike breach. *Visualisation Credit* D. Flanders, CALP, UBC. **b** View with raised dike in the "Hold the Line" scenario, *Visualisation Credit* D. Flanders, CALP, UBC

Beyond the formal evaluation, the visioning project has been well received by the public, politicians, planners, engineers, and international scientists. The resulting visual products have been sought after by local to national media, providing an expanded opportunity for public education and awareness building. The steadily growing number of invited presentations on the visioning methods and results, coming from local, provincial, national and international audiences, suggests a hunger for techniques and information of this kind. Longitudinal studies on the long-lasting impact of the visualisation-aided process are now underway.

Collaboration with the municipality of Delta in particular have continued to this day, drawing on the data and trust relationships previously established. Recent work adapting the LCCV process has focused on developing a range of adaptation scenarios responding to sea level rise though 2100, tied both to spatial analysis of land values and other outcomes, and to detailed landscape visualisations explaining specific adaptation measures and their implications (Fig. 7.4).[4] These

[4] Funded primarily by Natural Resources Canada and BC Ministry of Environment.

combined efforts have led to policy recommendations presented to local officials (Barron et al. 2012), and ongoing efforts to engage a wider community on these critical issues.

7.3.3 Outcomes of Other Canadian Visioning Case Studies Using Landscape Visualisations

Since the original visioning studies in the Metro Vancouver region, the Local Climate Change Visioning process using geospatial and landscape visualization tools has been adapted and applied in three other contexts across Canada.[5] The projects[6] all addressed climate-related issues at community to regional scales; used spatially-based approaches to integrate scientific data and modeling; conducted participatory processes where academic research teams collaborated with local stakeholders and inter-disciplinary experts; explored possible future pathways using scenarios or landscape design options; and applied 2D and/or 3D visualization tools (Pond et al. 2012).

7.3.3.1 A Case Study at Landscape Scale

In Kimberley in the BC Kootenays, the CALP research team applied the visioning process within a small, rural, less well-resourced community watershed (Schroth et al. 2009). This study was embedded in a local process within the Kimberley Climate Adaptation Project (KCAP), a community-driven project to identify local climate change impacts and vulnerabilities, and develop adaptation planning recommendations. The scenario method was simplified to two stakeholder-driven qualitative scenarios (integrated mitigation and adaptation versus adaptation only), supported by quantitative modeling, spatial analysis and 3D visualization of forest fire risks and mountain pine beetle susceptibility under climate change (Pond et al. 2009; Schroth et al. 2009).[7] Google Earth (Fig. 7.5) was chosen as one of the main presentation media because it allowed user interaction and is widely accessible to other smaller communities. Stakeholder interviews and feedback from process participants confirmed that the

[5] Funded primarily by the GEOIDE National Centres of Excellence research network.

[6] A fourth visioning study was conducted for the Elbow River drainage in Alberta, by a University of Calgary research team and partners, using an integrated set of geospatial modeling tools (see Pond et al. 2012); however, this project did not employ landscape visualisation and so is not discussed further here.

[7] Funded by BC Real Estate Foundation and BC Ministry of Community development, with support from the Columbia River Trust and City of Kimberley.

Fig. 7.5 Semi-realistic landscape visualization of the community of Kimberley showing high (*orange*) and moderate (*cream*) susceptibility to Pine Beetle infestation in the surrounding forested watershed due to warmer winters, with potential increased flood hazard (*yellow circles*) in town. *Credit* O. Schroth, CALP, UBC. Mountain Pine Beetle Data source: ILMB, BC Government. Background Image: © 2009 Google Earth. Image © 2009 Province of British Columbia; Image © 2009 Digital Globe; Image © 2009 Terra Metrics. Reproduced from Sheppard (2012) "Visualizing Climate Change", Earthscan/Routledge

visualizations raised awareness and facilitated understanding of climate change impacts and mitigation/adaptation options. Rankings of the various kinds of visualizations in order of importance to users (if involved in commenting on a planning proposal) revealed a bi-modal distribution for Google Earth, where the virtual globe was ranked first 16 times and ranked last 11 times, suggesting that some people really like the interactive tools whereas others preferred the more static media. Overall, the stakeholder workshops produced more than 70 recommendations for climate change mitigation and adaptation, some of which have since been implemented in policy (Schroth et al. 2011).

Based on the Kimberley project and the earlier Metro Vancouver projects, CALP produced a Guidance Manual on the LCCV process and tools (Pond et al. 2010) for interested practitioners. This was used during the next series of projects in a comparative national study (Pond et al. 2012). Over four years, researchers at the Universities of British Columbia, Toronto, Waterloo, and Calgary have collaborated with local partners to adapt LCCV processes to other contexts—from downtown Toronto, to a regional watershed in Alberta, to a Hamlet in Nunavut.

7.3.3.2 A Case Study at Community Scale

The Clyde River project in Nunavut used spatial planning, scenarios, mapping and SketchUp 3D visualizations in a participatory process with translators to bring together local and scientific knowledge, build social learning around planning issues, and visualize potential future resilient pathways for the community. Scenario development was based on four dominant concerns: landscape hazards; housing shortages (as well as how to plan for future population growth); walkability within the community and quality of life issues; and energy resilience (Fig. 7.6). Following review by Hamlet staff and members of the community, two spatially divergent scenarios were developed to explore spatially distinct development alternatives while incorporating more resilient energy production and

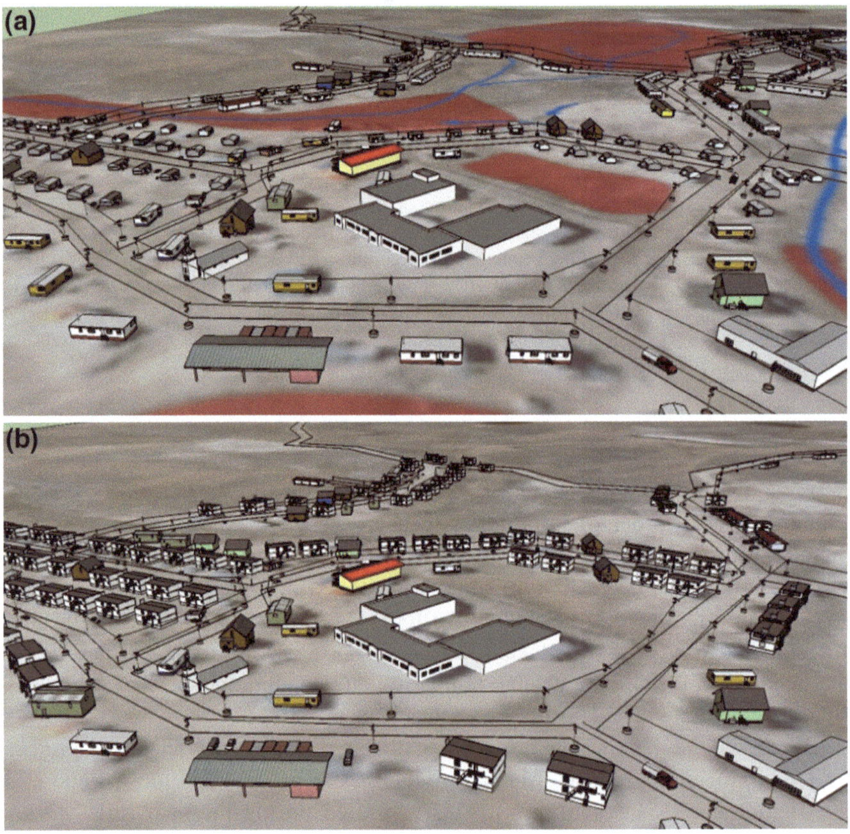

Fig. 7.6 Alternative scenarios and supporting analysis for development of the Arctic community of Clyde River, Nunavut, showing: **a** official plan overlaid with hazard areas in *red*, **b** potentially resilient redevelopment to increase housing, avoid hazard areas, improve walkability, and reduce dependence on imported diesel through renewable energy. *Credit* N. Sinkewicz, D. Flanders, K. Tatebe, and E. Pond, CALP, UBC. Reproduced from Sheppard (2012) "Visualizing Climate Change", Earthscan/Routledge

quality of life concerns in building design and arrangement. "Feedback from community partners suggests that the mapping exercises and 3D visualizations have fostered new conversations and understanding around the community's future and growth options" (Pond et al. 2012), though significant challenges remain in terms of separation between official decision-makers and local community concerns, due to distance and limitations of conventional planning methods still in place.

7.3.3.3 A Case Study at Regional Scale

In Toronto, the case study applied geovisualization methods and tools to help policy-makers and, ultimately, the public, explore where planning policy and mitigation efforts can best be targeted. Two main research foci were identified: reducing heat island effects and increasing green energy production through rooftop photovoltaics. The visualization approaches were driven by a need to help decision makers (e.g. City staff, individual homeowners, etc.) to interactively explore spatial variability in heat and rooftop PV suitability, and to identify tangible linkages between policies and action strategies across multiple scales (Pond et al. 2012). This involved: mapping variations in surface heat using surface temperature variations represented as topographic surfaces on which orthophotos were draped to highlight correspondence between land use and heat effects (Fig. 7.7); development of a web-GIS application and solar modelling to explore

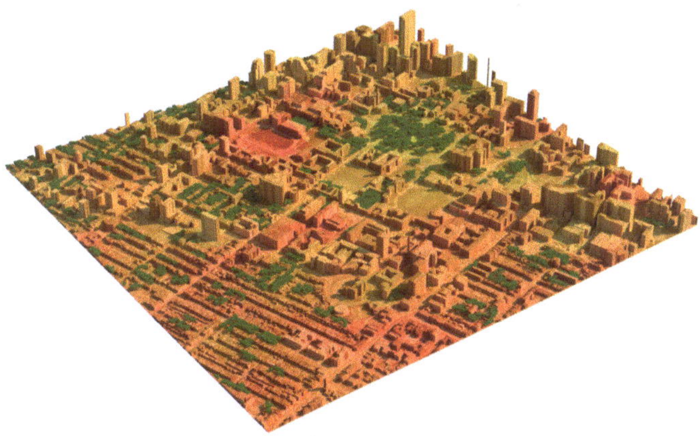

Fig. 7.7 3D model of urban heat island in Toronto: *colours* represent surface temperatures, showing the cooling effect of vegetation (tree canopy shown in *green*). *Credit* J. Danahy. *Data source:* Maloley, MJ. 2010. Thermal remote sensing of Urban Heat Island Effects: Greater Toronto Area, Geological Survey of Canada, open file 6283, 40 pages. doi:10.4095/26339; Behan, KJ, Mate, D, Maloley, MJ, Penney J 2011. Using strategic partnerships to advance urban heat island adaptation in the greater Toronto Area. Geological Survey of Canada, open file 6865. 1 CD-ROM. doi:10.4095/288755. Reproduced from Sheppard (2012) "Visualizing Climate Change", Earthscan/Routledge

solar panel feasibility on individual buildings. This collaboration has led to on-the-ground decision-making in the design and construction of cooler, green parking lots as a heat island mitigation measure, centered on Pearson International Airport; and partnering with the Toronto and Region Conservation Authority (TRCA) to leverage the solar and heat mapping work for their Sustainable Neighbourhood retrofit Action Plan (SNAP) sites.

Local climate change visioning as an applied research program has thus contributed to longer-term outcomes on climate change awareness, policy, and action in several Canadian communities. Early evidence suggests that it may have contributed to a culture of change in thinking about and planning for climate change among the practitioners and policy-makers involved. The next section discusses the implications for practice and further research, within and beyond Canada.

7.4 Implications for the Use of Visualization in Responding to Climate Change

Participatory, iterative processes, involving stakeholders throughout, have led to credible outputs (Moser 2009), based both on underlying science, local knowledge (Rantanen and Kahila 2009), and trust relationships with the research team. Participatory processes have also provided local capacity on the local impacts and response options related to climate change (Shaw et al. 2009). However, evaluation has shown that participatory processes which systematically incorporate visualisations can effectively convey salient information, and help to encourage discussion and informed consideration of local climate change issues, risks, trends, and response/policy options (Schroth 2007; Sheppard et al. 2008; Tatebe et al. 2010; Burch et al. 2010). The process and the tools taken together have had measurable impacts on participants, including increased awareness, understanding, and motivation to support policy change (Sheppard et al. 2008; Schroth et al. 2009; Cohen et al. 2011).

Experience to date with Local Climate Change Visioning processes in Canada suggests a hunger among communities for more detailed and salient information on climate change impacts and response options, a void not being filled at present by any one discipline or conventional planning processes. There are few actual climate change scientists available to communities, and very few if any local governments are using visual learning tools to accelerate awareness and build capacity on climate change among practitioners, policy-makers, or the public. Local planners and engineers are increasingly tasked with forging a response on climate change, but in North America at least there are as yet very few roadmaps and no proven planning processes in place enabling communities to address, holistically and practically, multiple climate change issues at the local level. Adaptation and mitigation responses are often dealt with separately in silos within responsible agencies. How then will these challenges and associated climate change targets be met?

Clearly, there is major potential for the use of visualizations with participatory processes in filling this void. The Local Climate Change Visioning processes described above move beyond current practice. They represent a prototype for improved planning and engagement processes, which could be used to operationalize climate change science into conventional decision-making and design procedures. They show that it is possible to carry out an inclusive, multi-criteria process based on visualizing holistic future conditions, backed by scientific spatial modelling, and supporting structured decision-making. The initial process that was successful in Metro Vancouver has indeed been adapted to fit diverse conditions, data constraints and team capabilities (see Pond et al. 2012) in three other locations. What began as an experimental stand alone research technique has since been embedded in municipal climate change planning processes (Kimberley), applied to several planning and design interventions across a major city (Toronto), and used to formulate alternative northern development strategies that are more sustainable and culturally responsive than official community plans (Clyde River).

Such tools and processes could be central to resolving difficult dilemmas in contested landscapes, e.g. the impacts of wind power on areas of sensitive landscape character, or the opposition to increased urban density in existing neighbourhoods. We will need good communications, fair processes, and excellent design if we are to resolve problems and preserve quality of life during the transition away from high carbon energy sources and as climate uncertainties increase.

However, using powerful techniques like landscape visualisation as both an objective decision-making tool and a way to motivate changing attitudes and policy, requires a credible and ethical stance with sound methodologies for preparing valid visualisations of climate change. Sheppard (2001, 2005b) has proposed a Code of Ethics for use of landscape visualisation, which identifies the following principles or criteria that may be relevant to climate change applications:

- accuracy of visualisation relative to expected conditions;
- representativeness of views in space and time, relative to the context;
- visual clarity of presentations;
- interest and engagement of the audience/users;
- legitimacy or accountability of the visualisation, including transparency of data and the production process; and
- accessibility of the visualisations to the public and potential users.

Mulder et al. (2007) expanded these to include a range of quality criteria for visualisation, and Sheppard (2012) has outlined principles for defensible use of various visual media, including landscape visualisation, in engaging people on climate change specifically. Development and presentation of visualisations by a trusted source would appear to be an important aspect of defensibility (Nicholson-Cole 2005; Sheppard 2012). A strategy that has worked in the LCCV studies was to secure effective stakeholder participation in the development of socioeconomic scenarios and the application of decision-rules for visualising the scenarios.

Another possible strategy would be to allow the doubtful or skeptical user to freely navigate, interact with, or interrogate the visualisation imagery and underlying databases (Furness III et al. 1998; Sheppard and Salter 2004).

Capacity-building and training of practitioners on structured and ethical preparation and use of visualisations in participatory planning processes for climate change is needed across a range of climate change issues affecting communities, working with other extension agents in the government, regional agencies, industry, and NGOs to transfer knowledge to and from communities. Initial guidance on learning, planning, and implementing these processes already exists, in the form of a Visioning Manual (Pond et al. 2010) and review of visual media techniques, principles, and examples in Sheppard (2012). It is also possible that the audience or receiver of information from visualisation-based processes (e.g. the public or local councils) may need training in order to absorb and correctly interpret the meaning of sometimes novel or unfamiliar visual imagery or tools.

Visual imagery is to some extent a universal language, transcending linguistic and sometimes cultural differences. Landscape visualisations have been shown to be effective in a variety of social groups and community types, including aboriginal communities in BC (Lewis and Sheppard 2006) and Nunavut (Pond et al. 2012). This suggests that the visioning processes tested in diverse Canadian communities may also be applicable in other countries and community types.

Further research is therefore needed to test the effectiveness of visioning processes in various environments and community types internationally. It is also important to conduct more evaluations of the impact of visualisations relative to the larger participatory process. Lastly, it would be instructive to evaluate and compare visioning processes which are researcher-driven versus those conducted by practitioner and embedded in official planning processes.

7.5 Conclusions

Communities all over the world face an urgent need to choose between possible climate change strategies with far-reaching consequences, while keeping their public involved and supportive. Experience gained through a decade of Canadian research suggests that landscape ecologists and practitioners can employ powerful science-based visual tools capable of improving understanding, influencing people's perceptions, and helping to motivate action at the landscape or community level. Such approaches could help fill the void in developing practical, holistic, collective solutions to climate change problems, using defensible visual imagery of future low-carbon resilient communities. Participatory visioning processes can dramatically bring the impacts of climate change home to people, making it 'personal' through realistic visualizations of their familiar landscape under future scenarios informed by climate change projections. The Canadian community case studies described in this chapter suggest that this novel approach, combining various scientific, geomatic, communication and psychological techniques, represents a better way to inform community dialogue and decision-making on climate change mitigation and adaptation. Practitioners and professional associations internationally should consider enhancing conventional community planning methods along these lines, in order to test or adapt methods such as those applied in Canada, to address growing issues of climate change at the local level everywhere.

If visioning processes and visualisations are to be used more systematically in planning and engagement on climate change, training and guidance are needed. Because landscape visualisations help to engaging emotional responses, strong ethical procedures will be key. Visualisation tools are too powerful to be ignored, but also to be used without careful consideration of defensibility. Scientists and practitioners should adopt better standards for using visualisation and visioning processes to convey the science, acknowledge the uncertainties, engage stakeholders, and ultimately help local communities to develop their own solutions to climate change.

Acknowledgments The co-authors wish to acknowledge the following people whose work on previous studies and papers on climate change visioning has directly contributed to this chapter: Adelle Airey, Kristi Tatebe, Ellen Pond, Sara Muir-Owen, Sara Barron, Glenis Canete, Jon Laurenz, Stewart Cohen, John Robinson, Jeff Carmichael, Sonia Talwar, Rob Feick, John Danahy, and Rob Harrap. We also wish to acknowledge the vital support from our many partners on these projects, including agency staff and stakeholders from the communities of Delta, BC; District of North Vancouver; Kimberley, BC; Clyde River Hamlet, Nunavut; and Toronto, Ontario.

References

Appleyard D. Understanding professional media: issues, theory, and a research agenda. In: Altman I, Wohlwill JF, editors. Human Behavior and Environment, vol. 1. New York: Plenum Press; 1977. p. 43–88.

Barron S, Canete G, Carmichael J, Flanders D, Pond E, Sheppard SRJ, Tatebe K. A climate change adaptation planning process for low-lying, communities vulnerable to sea level rise. Sustainability. 2012;4:2176–208.

Bishop ID, Lange E, editors. Visualization in landscape and environmental planning. London: Taylor and Francis; 2005.

Bishop ID, Rohrmann B. Subjective responses to simulated and real environments: a comparison. Landscape Urban Plann. 2003;65:261–77.

Burch S, Sheppard SRJ, Shaw A, Flanders D. Planning for climate change in a flood-prone community: municipal barriers to policy action and the use of visualisations as decision-support tools. J Flood Risk Manage. 2010;3(2):126–39.

Cohen SJ. Scientist-stakeholder collaboration in integrated assessment of climate change: lessons from a case study of Northwest Canada. Environ Model Assess. 1997;2:281–93.

Cohen S, Sheppard SRJ, Shaw A, Flanders D, Burch S, Taylor B, Hutchinson D, Cannon A, Hamilton S, Burton B, Carmichael J. Downscaling and visioning of mountain snow packs and other climate change implications in North Vancouver, Br Columbia. J Mitig Adapt Strat Glob Change. 2011;17(1):25–49.

Daniel TC, Meitner MJ. Representational validity of landscape visualizations: the effect of graphical realism on perceived scenic beauty of forest vistas. J Environ Psychol. 2001;21(1):61–72.

Dockerty T, Lovett A, Sunnenberg G, Appleton K, Parry M. Visualising the potential impacts of climate change on rural landscapes. Comput Environ Urban Syst. 2005;29(3):297–320.
Ervin SM. Answering the 'what ifs'. Landscape Archit Mag. 1998;10:64–73.
Furness III TA, Winn W, Yu R. The impact of three dimensional immersive virtual environments on modern pedagogy: global change, VR and learning. In: Proceedings of workshops in Seattle, Washington, and Loughborough, England in May and June 1997. Seattle, WA: Human Interface Technology Lab, University of Washington. 1998. Available online at http://www.hitl.washington.edu/publications/r-97-32/. Accessed 1 June 2004.
IPCC. Climate change 2007: synthesis report. In: Core Writing Team, Pachauri RK, Reisinger A, editors. Contribution of working groups I, II and III to the fourth assessment report of the intergovernmental panel on climate change. Geneva: IPCC; 2007. p. 104.
Kriegler E, O'Neill B, Hallegatte S, Kram T, Lempert R, Moss R, Wilbanks T. The need for and use of socio-economic scenarios for climate change analysis: a new approach based on shared socio-economic pathways. Global Environmental Change. 2012;22:807–22.
Lewis JL, Sheppard SRJ. Culture and communication: can landscape visualisation improve forest management consultation with indigenous communities? Landscape Urban Plan. 2006;77: 291–313.
Lowe T, Brown K, Dessai S, de França Doria M, Haynes K, Vincent K. Does tomorrow ever come? Disaster narrative and public perceptions of climate change. Public Underst Sci. 2006; 15:435–57.
Moser S. Making a difference on the ground: the challenge of demonstrating the effectiveness of decision support. Climatic Change. 2009;95(1–2):11–21.
Moser S, Dilling L, editors. Creating a climate for change: communicating climate change and facilitating social change. Cambridge: Cambridge University Press; 2007.
Mulder J, Sack-da-Silva S, Bruns D. Understanding the role of 3D visualisations: the example of Calden Airport expansion, Kassel, Germany. In: van den Brink A, van Lammeren R, van de Velde R, Dane S, editors. Geo-visualisation for participatory spatial planning in Europe: imaging the future. vol. 3. Mansholt Publication Series. Wageningen: Wageningen Academic Publishers; 2007. p. 75–88.
Nicholson-Cole SA. Representing climate change futures: a critique on the use of images for visual communication. Comput Environ Urban Syst. 2005;29(3):255–73.
Pond E, Schroth O, Sheppard SRJ, Feick R, Marceau D, Danahy J, Burch S, Cornish L, Cohen S, Pooyandeh M, Wijesekara N, Flanders D, Tatebe K, Barron S. Collaborative processes and geo-spatial tools in support of local climate change visioning and planning. In: Chrisman and Wachowicz, editors. The added value of scientific networking: perspectives from the GEOIDE network members 1998–2012. GEOIDE Network, Québec: Cité Universitaire; 2012.
Pond E, Schroth O, Sheppard SRJ, Muir-Owen S, Liepa I, Campbell C, Salter J, Tatebe K, Flanders D. Local climate change and landscape visualizations guidance manual. Collaborative for Advanced Landscape Planning, UBC. 2010.
Pond E, Schroth O, Sheppard SRJ, CALP. Visioning and visualizations, Kimberley Climate Adaptation Project. Vancouver: Real Estate Foundation; 2009. p. 28.
Rantanen H, Kahila M. The SoftGIS approach to local knowledge. J Environ Manage. 2009;90(6):1981–90.
Salter JD. Designing and testing a prototypical landscape information interface for lay-people. Unpublished Masters thesis, Faculty of Forestry. Vancouver: University of British Columbia; 2005.
Salter JD, Campbell C, Journeay M, Sheppard SRJ. The digital workshop: exploring the use of interactive and immersive tools in participatory planning. J Environ Manage. 2009;90: 2090–101.
Schroth O. From information to participation—interactive landscape visualization as a tool for collaborative planning. Institute for Spatial and Landscape Planning. Dissertation. Zurich: ETH Zürich; 2007.

Schroth O, Pond E, Campbell C, Cizek P, Bohus S, Sheppard SRJ. Tool or Toy? virtual globes in landscape planning. Future Internet. 2011;3:204–27.
Schroth O, Pond E, Muir-Owen S, Campbell C, Sheppard SRJ. Tools for the understanding of spatio-temporal climate scenarios in local planning: Kimberley BC case study. Zurich: Swiss National Sciences Foundation; 2009.
Shackley S, Deanwood R. Stakeholder perceptions of climate change impacts at the regional scale: implications for effectiveness of regional and local responses. J Environ Plan Manage. 2002;45(3):381–402.
Shaw A, Sheppard SRJ, Burch S, Flanders D, Wiek A, Carmichael J, Robinson J, Cohen S. Making local futures tangible—synthesizing, downscaling, and visualizing climate change scenarios for participatory capacity building. Global Environ Change. 2009;19(4):447–63.
Sheppard SRJ. Guidance for crystal ball gazers: developing a code of ethics for landscape visualization. Landscape Urban Plan. 2001;54(1–4):183–99.
Sheppard SRJ. Landscape visualisation and climate change: the potential for influencing perceptions and behaviour. Environmental Science and Policy. vol. 9. New York: Elsevier; 2005a. p. 637–54.
Sheppard SRJ. Validity, reliability, and ethics in visualization. In: Bishop ID, Lange E, editors. Visualization in landscape and environmental planning. Chapter 5. London: Taylor and Francis; 2005b. p. 79–97.
Sheppard SRJ. Local climate change visioning: a new process for community planning and outreach using visualization tools. Plan Canada. 2008;48(1):36–40.
Sheppard SRJ. Visualizing climate change: a guide to visual communication of climate change and developing local solutions. Abingdon: Earthscan/Routledge; 2012.
Sheppard SRJ, Cizek P. The ethics of Google-Earth: crossing thresholds from spatial data to landscape visualization. J Environ Manage. 2009;90:2102–17.
Sheppard SRJ, Meitner MJ. Using multi-criteria analysis and visualisation for sustainable forest management planning with stakeholder groups. For Ecol Manage. 2005;207(1–2):171–87.
Sheppard SRJ, Salter J. The role of visualization in forest planning. In: Encyclopedia of Forest Sciences. Oxford: Academic Press/Elsevier; 2004. p. 486–98.
Sheppard SRJ, Shaw A, Flanders D, Burch S. Can visualization save the world? Lessons for landscape architects from visualizing local climate change. Conference Proceedings, digital design in landscape architecture 2008, 9th international conference, 29–31 May 2008, Germany: Anhalt University of Applied Sciences Dessau/Bernburg; 2008.
Sheppard SRJ, Shaw A, Burch S, Flanders D, Wiek A, Carmichael J, Robinson J, Cohen S. Future visioning of local climate change: a framework for community engagement and planning with scenarios and visualization. Futures. 2011;43(4):400–12.
Snover AK, Whiteley Binder L, Lopez J, Willmott E, Kay J, Howell D et al. Preparing for climate change: a guidebook for local, regional, and State Governments. Prepared by climate impacts group (Centre for Science in the Earth System), University of Washington, and King County, Washington, in association with and published by ICLEI—Local Governments for Sustainability, Oakland. 2007.
Tatebe K, Shaw A, Sheppard SRJ. Technical report on local climate change visioning for Delta: findings and recommendations. The Collaborative for Advanced Landscape Planning. Vancouver: University of British Columbia; 2010.
Tress B, Tress G. Scenario visualisation for participatory landscape planning: a study from Denmark. Landscape Urban Plan. 2003;64:161–78.
Winn W. The impact of three-dimensional immersive virtual environments on modern pedagogy. HITL Technical Report R-97-15. Human Interface Technology Laboratory, Seattle: University of Washington; 1997.
Zube EH, Sell JL, Taylor JG. Landscape perception: research, application, and theory. Landscape Plan. 1982;9:1–33.

Chapter 8
Watershed Scale Physically Based Water Flow, Sediment and Nutrient Dynamic Modeling System

Billy E. Johnson, Zhonglong Zhang and Charles W. Downer

Abstract Non-point source (NPS) runoff of pollutants is viewed as one of the most important factors causing impaired water quality in freshwater and estuarine ecosystems and has been addressed as a national priority since the passage of the Clean Water Act. To control NPS pollution, state and federal agencies developed a variety of programs that rely heavily on the use of watershed management in minimizing riverine and receiving water pollution. Watershed models have become critical tools in support of watershed management. Lumped, empirical models such as HSPF do not account for spatial heterogeneity within subwatersheds and the simulations of the actual processes are greatly simplified. This chapter describes a distributed water flow, sediment and nutrient dynamic modeling system developed at U.S. Army Engineer Research and Development Center. The model simulates detailed water flow, soil erosion, nitrogen (N) and phosphorus (P) cycling at the watershed scale and computes sediment transport across the landscape, nutrient kinetic fluxes for N and P species. The model consists of three distinct parts: (1) watershed hydrology, (2) soil erosion and sediment transport, and (3) nitrogen and phosphorus transport and cycling. The integrated watershed model was tested and validated on two watersheds in Wisconsin (French Run and Upper Eau Galle Watersheds). The model performed well in predicting runoff, sediment, nitrogen and phosphorus. This chapter presents the model development and validation studies currently underway in Wisconsin.

B. E. Johnson (✉)
Environmental Laboratory, U.S. Army Engineer Research and Development Center, 3909 Halls Ferry Rd, Vicksburg, MS 39180, USA
e-mail: Billy.E.Johnson@usace.army.mil

Z. Zhang
Badger Technical Services, U.S. Army Engineer Research and Development Center, 3909 Halls Ferry Rd, Vicksburg, MS 39180, USA

C. W. Downer
Coastal and Hydraulics Laboratory, U.S. Army Engineer Research and Development Center, 3909 Halls Ferry Rd, Vicksburg, MS 39180, USA

B. Fu and K. B. Jones (eds.), *Landscape Ecology for Sustainable Environment and Culture*, DOI: 10.1007/978-94-007-6530-6_8,

Keywords Watershed model • Hydrology • Sediment transport • Nonpoint source • Nutrient transformations

8.1 Introduction

Nutrient pollution is a leading cause of water quality impairment in lakes and estuaries and is also a significant issue in rivers (USEPA 2007). Non-point source (NPS) pollution, especially from nitrogen and phosphorus, has consistently ranked as one of the top causes of degradation in some U.S. waters. Nutrient problems can exhibit themselves locally or much further downstream leading to degraded estuaries, lakes and reservoirs, and to hypoxic zones where fish and aquatic life can no longer survive. The growing concern about the environmental impact of NPSs has enhanced a "watershed approach" to reduce NPS pollution and coordinate the management of water resources. The concept of this watershed approach is based on multi-purpose, multi-objective management, and in examining all water needs in the watershed and receiving water bodies. A watershed scale flow, sediment and water quality modeling system has been developed at U.S. Army Engineer Research and Development Center (ERDC) in support of the U.S. Army Corps of Engineers (USACE)' watershed approach.

This chapter describes the on-going watershed water quality modeling development and integration with the Gridded Surface Subsurface Hydrologic Analysis (GSSHA) model. The major chemical and physical processes influencing sediments and nutrients in the soil, overland flow and stream have been accounted for in the GSSHA. The hydrological variables required to drive the sediment and nutrient simulation were provided using the existing GSSHA model. Integrated physically based hydrologic models with sediment and nutrient transport across the landscape give more realistic descriptions of the sediment and nutrient dynamics in watersheds. This is especially important for agricultural watersheds where the sediment and nutrient play important roles and their occurrence are highly variable both in time and space. Hence, hot-spots with high contaminant loading sources can be more accurately identified and watershed management practices to reduce sediment and nutrient transport can be made more confidently.

8.2 Watershed Scale Water Flow and Sediment Model

The U.S. Army Corps of Engineer's Gridded Surface Subsurface Hydrologic Analysis (GSSHA) is a physically-based, distributed-parameter, structured grid, hydrologic model that simulates the hydrologic response and sediment transport of a watershed subject to given hydrometeorological inputs. The watershed is divided into grid cells that comprise a uniform finite difference grid. GSSHA is a

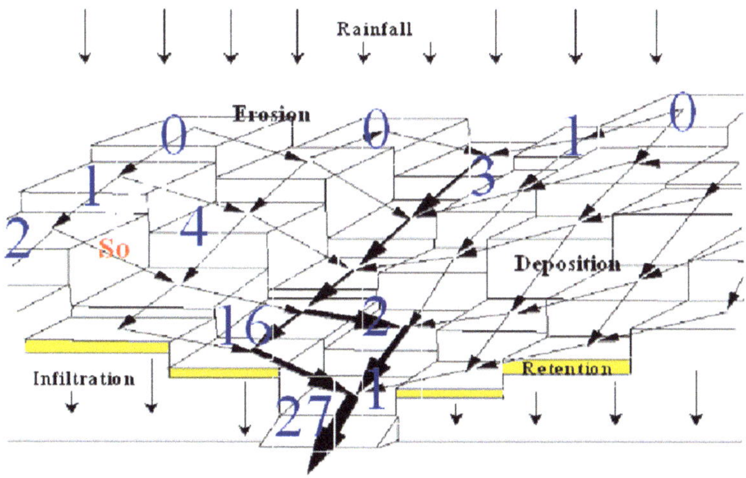

Fig. 8.1 Topographical representation of overland flow and channel routing schemes within a watershed

reformulation and enhancement of the CASC2D (Fig. 8.1). The model incorporates 2D overland flow, 1D stream flow, 1D unsaturated flow and 2D groundwater flow components. Within GSSHA, sediment erosion and transport processes take place both on the land and within the channel. The GSSHA model employs mass conservation solutions of partial differential equations and closely links the hydrologic components to assure an overall mass balance. GSSHA had already been tested and applied for hydrologic response and sediment transport in several watersheds across the United States and achieved satisfactory results (CHL 2012). A brief introduction is given as follows however details of the GSSHA model can be found in Downer and Ogden (2004). A review of hydrologic and sediment erosion and transport process descriptions is informative to illustrate the physics behind individual process representations and specific to those needed to drive full nitrogen and phosphorus cycling at the watershed scale.

8.2.1 Hydrologic Simulation

The modeling of hydrologic processes begins with rainfall being added to the watershed, some of which is intercepted by the canopy cover, evapotranspirated or infiltrated. Hydrologic processes that can be simulated and the methods used to approximate the processes with the GSSHA model are listed in Table 8.1.

GSSHA uses two-step, finite-volume schemes to route water for both 2D overland flow and 1D channel flow where flows are computed based on heads and volumes are updated based on the computed flows. Several modifications were made to both the GSSHA channel routing and the overland flow routing schemes

Table 8.1 Processes and approximation techniques in the GSSHA model

Process	Approximation
Precipitation distribution	Thiessen polygons (nearest neighbor) Inverse distance-squared weighting
Snowfall accumulation and melting	Energy balance
Precipitation interception	Empirical two parameter
Overland water retention	Specified depth
Infiltration	Green and Ampt (GA) Multi-layered GA Green and Ampt with Redistribution (GAR) Richard's equation (RE)
Overland flow routing	2-D diffusive wave
Channel routing	1-D diffusive wave, 1-D dynamic wave
Evapo-transpiration	Deardorff Penman-Monteith with seasonal canopy resistance
Soil moisture in the vadose zone	Bucket model RE
Lateral groundwater flow	2-D vertically averaged
Stream/groundwater interaction	Darcy's law
Exfiltration	Darcy's law

to improve stability, and allow interaction between the surface and subsurface components of the model. The combination of improvements in the stability of the overland and channel routing schemes has allowed significant increases in model computational time steps over previous versions.

8.2.1.1 Overland Flow Routing

Water flow across the land surface is shallow, unsteady, and non-uniform. This flow regime can be described by the Saint-Venant equations which are derived from physical laws regarding the conservation of mass and momentum. Overland flow routing in GSSHA employs the 2D diffusive wave equation, which allows for backwater and reverse flow conditions. The 2D (vertically integrated) continuity equation for gradually-varied flow over a plane in rectangular (x, y) coordinates is (Julien et al. 1995):

$$\frac{\partial h}{\partial t} + \frac{\partial q_x}{\partial x} + \frac{\partial q_y}{\partial y} = i_e \tag{8.1}$$

where h = surface water depth [L], q_x, q_y = unit discharge in the x- or y-direction = Q_x/B_x, Q_y/B_y [L^2/T], Q_x, Q_y = flow in the x- or y-direction [L^3/T], B_x, B_y = flow width in the x- or y-direction [L], i_e = excess net precipitation rate [L/T].

The diffusive wave momentum equations for the x- and y-directions are written as:

$$S_{fx} = S_{0x} - \frac{\partial h}{\partial x} \tag{8.2a}$$

$$S_{fy} = S_{0y} - \frac{\partial h}{\partial y} \tag{8.2b}$$

where S_{fx}, S_{fy} = friction slope (energy grade line) in the x- or y-direction, S_{0x}, S_{0y} = ground surface slope in the x- or y-direction.

8.2.1.2 Channel Flow Routing

Channel flow routing in GSSHA employs the 1D diffusive wave equation. The 1D (laterally and vertically integrated) continuity equation for gradually-varied flow along a channel is (Julien et al. 1995):

$$\frac{\partial A}{\partial t} + \frac{\partial Q}{\partial x} = q_l \tag{8.3}$$

where A = cross sectional area of channel flow [L^2], Q = total discharge [L^3/T], and q_l = lateral flow into or out of the channel [L^2/T].

8.2.2 Sediment Simulation

Sediment erosion and transport are potentially very important processes in water quality modeling. Excess sediment affects water quality directly by itself. Sediment transport also influences chemical transport and fate. Suspended sediments act as a carrier of chemicals in the watershed flow. Many chemicals sorb strongly to sediment and thus undergo settling, scour, and sedimentation. Sorption also affects a chemical's transfer and transformation rates. The amount of chemicals transported by the sediments depends on the suspended sediment concentration and the sorption coefficient. Both sediment transport rates and concentrations must be estimated in most toxic modeling studies. The sediment algorithm is included as a sub-model in GSSHA and invoked only when sediment simulation is required. The sediment sub-model is designed for estimating sediment delivery and channel transport in watersheds. It consists of four primary components: (1) sediment transport; (2) erosion; (3) deposition; and (4) bed processes (bed elevation dynamics).

8.2.2.1 Sediment Transport

The sediment transport model is based on the suspended sediment mass conservation equation (advection-diffusion equation with the sink-source term describing sedimentation resuspension rate) and the equation of bottom deformation. For the overland plane in 2D, the concentration of particles in a flow is governed by conservation of mass (sediment continuity) (Julien 1998):

$$\frac{\partial C_{ss}}{dt} + \frac{\partial \hat{q}_{tx}}{\partial x} + \frac{\partial \hat{q}_{ty}}{\partial y} = \hat{J}_e - \hat{J}_d + \hat{W}_s = \hat{J}_n \tag{8.4}$$

where C_{ss} = concentration of sediment particles in the flow [M/L^3], $\hat{q}_{tx}$, $\hat{q}_{ty}$ = total sediment transport areal flux in the x- or y-direction [M/L^{2}T], $\hat{J}_e$ = sediment erosion volumetric flux [M/L^{3}T], $\hat{J}_d$ = sediment deposition volumetric flux [M/L^{3}T], $\hat{W}_s$ = sediment point source/sink volumetric flux [M/L^{3}T], $\hat{J}_n$ = net sediment transport volumetric flux [M/L^{3}T].

The total sediment transport flux in any direction has three components, advective, dispersive (mixing), and diffusive, and may be expressed as (Julien 1998):

$$\hat{q}_{tx} = u_x C_{ss} - (R_x + D)\frac{\partial C_{ss}}{\partial x} \tag{8.4a}$$

$$\hat{q}_{ty} = u_y C_{ss} - (R_y + D)\frac{\partial C_{ss}}{\partial y} \tag{8.4b}$$

where u_x, u_y = flow (advective) velocity in the x- or y-direction [L/T], R_x, R_y = dispersion (mixing) coefficient the x- or y-direction [L^2/T], D = diffusion coefficient [L^2/T].

Note that both dispersion and diffusion are represented in forms that follow Fick's Law. However, dispersion represents a relatively rapid turbulent mixing process while diffusion represents a relatively slow a Brownian motion, random walk process (Holley 1969). In turbulent flow, dispersive fluxes are typically several orders of magnitude larger than diffusive fluxes. Further, flow conditions for intense precipitation events are usually advectively dominated as dispersive fluxes are typically one to two orders smaller than advective fluxes. As a result, both the dispersive and diffusive terms may be neglected.

Similarly, the suspended sediment transport in channels is described by the 1-D advection-diffusion equation that includes a source-sink term describing sedimentation and resuspension rates and laterally distributed inflow of sediments. The concentration of particles in flow is governed by the conservation of mass (Julien 1998):

$$\frac{\partial C_{ss}}{\partial t} + \frac{\partial \hat{q}_{tx}}{\partial x} = \hat{J}_e - \hat{J}_d + \hat{W}_s = \hat{J}_n \tag{8.5}$$

Individual terms for the channel advection-diffusion equation are identical to those defined for the overland plane. Similarly, the diffusive flux term can be neglected. The dispersive flux is expected to be larger than the corresponding term for overland flow. However, the channel dispersive flux still may be neglected relative to the channel advective flux during intense runoff events. The distributed runoff inflow to the channel and the suspended sediment concentration in the runoff are simulated by the overland component.

8.2.2.2 Sediment Erosion and Deposition

In the overland plane, sediment particles can be detached from the bulk soil matrix by raindrop impact and entrained into the flow by hydraulic action when the exerted shear stress exceeds the stress required to initiate particle motion (Julien and Simons 1985). The overland erosion process is influenced by many factors including precipitation intensity and duration, runoff length, surface slope, soil characteristics, vegetative cover, exerted shear stress, and sediment particle size. In channels, sediment particles can be entrained into the flow when the exerted shear stress exceeds the stress required to initiate particle motion. For non-cohesive particles, the channel erosion process is influenced by factors such as particle size, particle density and bed forms. For cohesive particles, the erosion process is significantly influenced by inter-particle forces (such as surface charges that hold grains together and form cohesive bonds) and consolidation. The surface erosion algorithm represents the mechanisms by which sediment is eroded from hillslopes and transported to the stream or channel network. Entrainment rates may be estimated from site-specific erosion rate studies or, in general, from the difference between sediment transport capacity and advective fluxes:

$$
\begin{aligned}
v_r &= \frac{J_c - v_a C_{ss}}{\rho_b} \; J_c > v_a C_{ss} \\
v_r &= 0 \; J_c \le v_a C_{ss}
\end{aligned}
\tag{8.6}
$$

where v_r = resuspension (erosion) velocity [L/T], J_c = sediment transport capacity areal flux [M/L^2/T], v_a = advective (flow) velocity (in the x- or y-direction) [L/T].

The rate of sediment deposition is proportional to the sediment concentration and settling velocity. If the sediment transport capacity is lower than the sediment load, then sediment deposition occurs. The process of sediment deposition is highly selective, the settling velocity of an aggregate or particle being a function of its size, shape, and density. Coarse particles (>62 μm) are typically non-cohesive and generally have large settling velocities under quiescent conditions. Numerous empirical relationships to describe the non-cohesive particle settling velocities are available. For non-cohesive (fine sand) particles with diameters from 62 to 500 μm, the settling velocity can be computed as (Cheng 1997):

$$
v_{sq} = \frac{\nu}{d_p}\left[\left(25 + 1.2 d_*^2\right)^{0.5} - 5\right]^{-1.5} \tag{8.7a}
$$

$$
d_* = d_p \left[\frac{(G-1)g}{\nu^2}\right]^{1/3} \tag{8.7b}
$$

where v_{sq} = quiescent settling velocity [L/T], ν = kinematic viscosity of water [L^2/T], and d_* = dimensionless particle diameter.

Fine particles often behave in a cohesive manner. If the behavior is cohesive, flocculation may occur. Floc size and settling velocity depend on the conditions under which the floc was formed (Krishnappan 2000; Haralampides et al. 2003). As a result of turbulence and other factors, not all sediment particles settling through a column of flowing water will necessarily reach the sediment-water interface or be incorporated into the sediment bed. Beuselinck et al. (1999) suggested this process also occurs for the overland plane. When flocculation occurs, settling velocities of cohesive particles can be approximated by relationship of the form (Burban et al. 1990):

$$v_s = a \cdot d_f^m \tag{8.8a}$$

$$v_{se} = P_{dep} v_s \tag{8.8b}$$

where $v_s =$ floc settling velocity [L/T], $a =$ experimentally determined constant, $d_f =$ median floc diameter [L], $m =$ experimentally determined constant, $v_{se} =$ effective settling (deposition) velocity [L/T], and $P_{dep} =$ probability of deposition.

8.2.2.3 Upper Sedimentation Processes

The upper soil and sediment bed play important roles in the transport of contaminants. Once a particle erodes, it becomes part of the flow and is transported downstream within the watershed. The fluxes of the channel erosion and sedimentation control the dynamics of the upper most contaminated layer. Particles and associated contaminants in the surficial sediments may enter deeper sediment layers by burial or be returned to the water column by scour. Whenever burial/scour occurs, particles and associated contaminants are transported through each subsurface sediment segment within a vertical stack. In response to the difference between bed form transport, erosion, and deposition fluxes, the net addition (burial) or net loss (scour) of particles from the bed causes the bed surface elevation to increase or decrease. The rise or fall of the bed surface is governed by the sediment continuity (conservation of mass) equation, various forms of which are attributed to Exner equation (Simons and Sentürk 1992). Julien (1998) presents a derivation of the bed elevation continuity equation for an elemental control volume that includes vertical and lateral (x- and y-direction) transport terms. Neglecting bed consolidation and compaction processes, and assuming that only vertical mass transport processes (erosion and deposition) occur, the sediment continuity equation for the change in elevation of the soil or sediment bed surface may be expressed as:

$$\rho_b \frac{\partial z}{dt} + v_{se} C_{ss} - v_r C_{sb} = 0 \tag{8.9}$$

where z = elevation of the soil surface [L], ρ_b = bulk density of soil or bed sediments [M/L^3], C_{sb} = concentration of sediment at the bottom boundary [M/L^3].

8.3 Nutrient Cycling Simulation

There are two components to simulate water quality. The first component is for transport of reactive or nonreactive materials throughout the watershed, both insoluble and dissolved. The second component is a flexible biogeochemistry that addresses the water quality state variables and transformation processes. Water quality state variables included in GSSHA can either be transported by advection-dispersion processes or storage routing depending on the water engines. Conceptually, three hydrologic domains and associated nutrient pathways in the watershed were modeled: (1) subsurface soils, (2) overland flow, and (3) channel flow. Currently GSSHA includes: (1) subsurface soil nitrogen module, (2) subsurface soil phosphorus module, (3) soil plant dynamic module, (4) overland flow nitrogen module, (5) overland flow phosphorus module, and (6) in-stream water quality module.

8.3.1 Nitrogen Cycle

The nitrogen cycle represents one of the most important nutrient cycles found in terrestrial ecosystems which includes stores of nitrogen found in the atmosphere, where it exists as a gas (mainly N_2) and other major stores of nitrogen including organic matter in soil and the oceans. Nitrogen in soil and water can be present in organic or inorganic forms and in either dissolved or particulate forms. The inorganic forms of nitrogen include nitrate (NO_3^-), nitrite (NO_2^-), exchangeable ammonium ($_4^+$), and fixed ammonium. The activities of humans have severely altered the nitrogen cycle. Some of the major processes involved in this alteration include: the application of nitrogen fertilizers to crops and increased deposition of nitrogen from atmospheric sources. A schematic representation of the watershed nitrogen transport and transformation processes involved in the nitrogen cycle is given in Fig. 8.2a.

8.3.1.1 Nitrogen Transformations in Soils

Once in the soil, the nitrogen will transform through the processes of mineralization, immobilization, volatilization, nitrification, denitrification, plant uptake, nitrogen fixation, and sediment sorption. Most plants can only take up nitrogen in two solid forms: ammonium ion (NH_4^+) and the nitrate ion (NO_3^-). Ammonium is

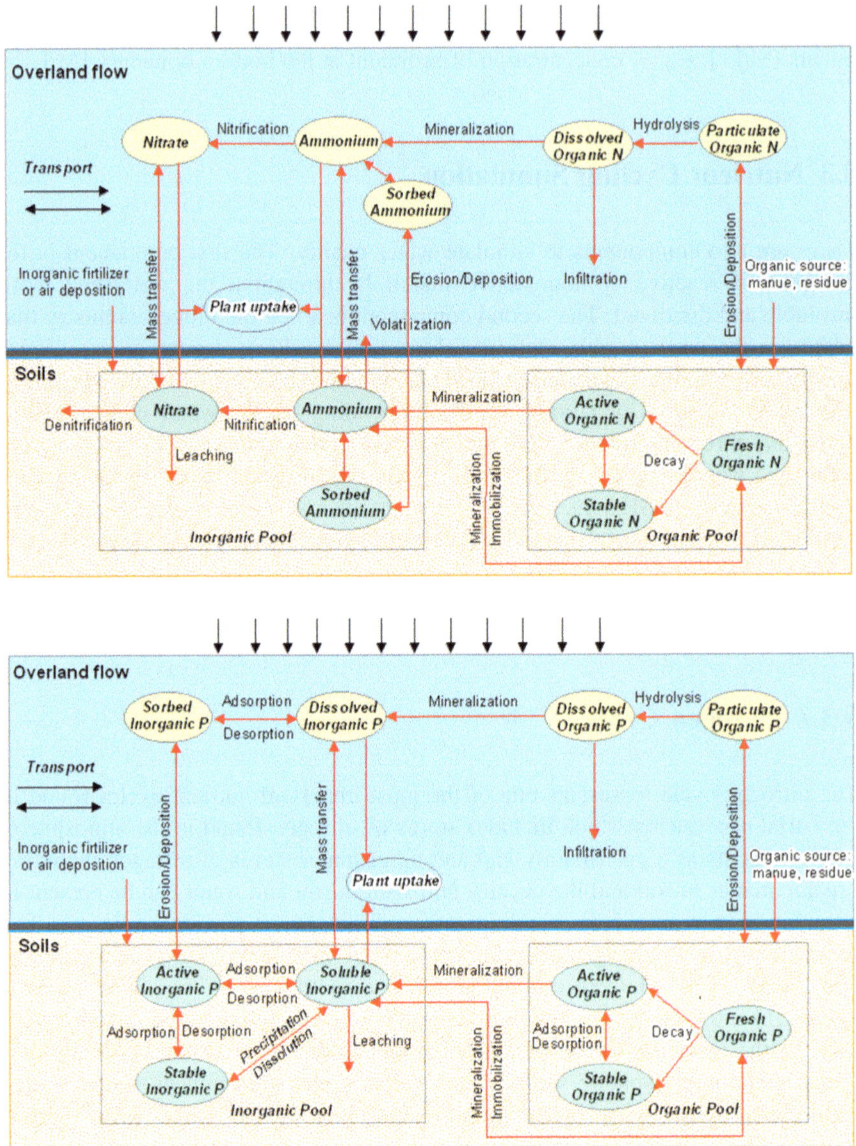

Fig. 8.2 Simplified **a** nitrogen, and **b** phosphorus cycle in soil and water

used less by plants for uptake because in large concentrations it is extremely toxic. Mathematical models of the soil nitrogen are generally in the form of storage (pool) accounting procedures. Soil nitrogen cycling is simulated in NSM for the five pools for each of the soil layers. The mass balance equations used to describe the nitrogen cycle in soils are summarized in Table 8.2.

Table 8.2 Mathematical expressions for soil nitrogen transformations

N pool	N transformation equation
Fresh organic N ($orgN_{frs}$)	$\frac{d(\Delta z \cdot orgN_{frs})}{dt} = -ON_{min\|imb} - ON_{dec} - ON_{frs,e} + ON_{frs,s}$
Active organic N ($orgN_{act}$)	$\frac{d(\Delta z \cdot orgN_{act})}{dt} = ON_{dec} - ON_{trn} - ON_{min} - ON_{act,e} + ON_{act,s}$
Stable organic N ($orgN_{sta}$)	$\frac{d(\Delta z \cdot orgN_{sta})}{dt} = ON_{dec} + ON_{trn} - ON_{sta,e} + ON_{sta,s}$
Ammonium N (NH_4^+)	$\frac{d(\Delta z \cdot NH_4^+)}{dt} = NH_{\min} - NH_{nit\|vol} - NH_{up} - R_{NH4,e} + NH_s$
Nitrate N (NO_3^-)	$\frac{d(\Delta z \cdot NO_3^-)}{dt} = NH_{nit} - NO_{dnit} - NO_{up} - R_{NO3,f} - R_{NO3,e} + NO_s$

Table 8.3 Mathematical expressions for overland flow nitrogen transformations

N species	N transformation equation
Particulate organic N (*PON*)	$\frac{\partial PON_{ov}}{\partial t} = L(PON_{ov}) - k_{hn}PON_{ov}$
Dissolved organic N (*DON*)	$\frac{\partial DON_{ov}}{\partial t} = L(DON_{ov}) + k_{hn}PON_{ov} - \frac{r}{h}DON_{ov} - k_{mn}DON_{ov}$
Ammonium N (NH_4^+)	$\frac{\partial NH_{4\,ov}^+}{\partial t} = L(NH_{4\,ov}^+) + k_{mn}DON_{ov} - \frac{r}{h}NH_{4\,ov}^+ - k_{nit}NH_{4\,ov}^+ - R_{NH4,up}$
Nitrate N (NO_3^-)	$\frac{\partial NO_{3\,ov}^-}{\partial t} = L(NO_{3\,ov}^-) + k_{nit}NH_{4\,ov}^+ - \frac{r}{h}NO_{3\,ov}^- - R_{NO3,up}$

8.3.1.2 Nitrogen Transformations in Surface Runoff

The dominant nitrogen species in waters are dissolved inorganic nitrogen $\left(NH_4^+, NO_2^-, NO_3^-\right)$, dissolved organic nitrogen (DON), particulate organic nitrogen (PON) and particulate inorganic nitrogen (PIN) (Burt and Haycock 1993). Models may consider particulate nitrogen as a single variable, or, alternately, represent from one to many particle types or fractions. In NSM, dominant nitrogen transformation processes in surface runoff are simulated for *PON*, *DON*, NH_4^+, and NO_3^-. Transformation processes include mineralization of DON to NH_4^+, nitrification of NH_4^+, to, NO_3^-, plant uptake of NH_4^+, and, NO_3^-, soil mass transfer of NH_4^+, NO_3^-, and DON, sediment sorption of NH_4^+, and hydrolysis of PON to DON. The mass balance equations used to simulate the nitrogen cycle in surface runoff are summarized in Table 8.3.

8.3.2 Phosphorus Cycle

The phosphorus cycle differs from the other major biogeochemical cycles in that it does not include a gas phase. The largest reservoir of phosphorus is in sedimentary rock. When it rains, phosphates are removed from the rocks via weathering and are distributed throughout both soils and water. Plants take up the phosphate ions from the soil. Phosphorus is not highly soluble, binding tightly to molecules in soil. Therefore it mainly reaches waters by traveling with runoff soil particles.

Table 8.4 Mathematical expressions for soil phosphorus transformations

P pool	P transformation equation
Fresh organic P ($orgP_{frs}$)	$\frac{d(\Delta z \cdot orgP_{frs})}{dt} = -OP_{min\|imb} - OP_{dec} - OP_{frs,e} + OP_{frs,s}$
Active organic P ($orgP_{act}$)	$\frac{d(\Delta z \cdot orgP_{act})}{dt} = OP_{dec} - OP_{min} - OP_{trn} - OP_{act,e} + OP_{act,s}$
Stable organic P ($orgP_{sta}$)	$\frac{d(\Delta z \cdot orgP_{sta})}{dt} = OP_{dec} + OP_{trn} - OP_{sta,e} + OP_{sta,s}$
Labile (soluble) P (P_{sol})	$\frac{d(\Delta z \cdot P_{sol})}{dt} = IP_{\min} - IP_{sol\|act} - IP_{up} - R_{DIP,e} + IP_s$
Active inorganic P ($minP_{act}$)	$\frac{d(\Delta z \cdot minP_{act})}{dt} = IP_{sol\|act} - IP_{act\|sta} - IP_{act,e} + IP_{act,s}$
Stable inorganic P ($minP_{sta}$)	$\frac{d(\Delta z \cdot minP_{sta})}{dt} = IP_{act\|sta} - IP_{sta,e} + IP_{sta,s}$

A schematic representation of the watershed phosphorus transport and transformation processes involved in the phosphorus cycle is given in Fig. 8.2b.

8.3.2.1 Phosphorus Transformations in Soils

Phosphorus can exist in the soil as phosphate $\left(HPO_4^{-2}, \text{ or, } H_2PO_4^{-}\right)$, particulate phosphorus, organic phosphorus, or in phosphorus minerals. Many reactions and mechanisms regulate and control the composition and forms of phosphorus present in the soil. Phosphorus is generally much less mobile than nitrogen, being strongly adsorbed to soil particles as well as organic matter. Phosphorus transformations in the soil include decomposition and mineralization of organic phosphorus, immobilization of labile phosphorus, and sorption of labile phosphorus to/from soil particles, and plant uptake. Soil phosphorus cycling is simulated by NSM for the six pool state variables for each of the soil layers. The mass balance equations used to describe the phosphorus cycle in soils are summarized in Table 8.4.

8.3.2.2 Phosphorus Transformations in Surface Runoff

The same process occurs within the aquatic ecosystem as for that in soils. Phosphorus is not highly soluble, binding tightly to soil particles. Therefore it mostly reaches waters by traveling with runoff soil particles. Phosphorus enters surface water primarily as particulate matter and secondarily as dissolved inorganic phosphorus (phosphate and its conjugate base forms). In NSM, dominant transformation processes are simulated for Particulate Organic Phosphorus (*POP*), Dissolved Organic Phosphorus (*DOP*), Particulate Inorganic Phosphorus (*PIP*), and Dissolved Inorganic Phosphorus (*DIP*). Transformation processes in surface runoff include mineralization of DOP to DIP, plant uptake of DIP, soil mass transfer of DIP and DOP, adsorption/desorption of DIP onto suspended sediments, and hydrolysis of POP to DOP. The mass balance equations used to simulate the phosphorus cycle in surface runoff are summarized in Table 8.5.

Table 8.5 Mathematical expressions for overland flow phosphorus transformations

P species	P transformation equation
Particulate organic P (*POP*)	$\frac{\partial POP_{ov}}{\partial t} = L(POP_{ov}) - k_{hp}POP_{ov}$
Dissolved organic P (*DOP*)	$\frac{\partial DOP_{ov}}{\partial t} = L(DOP_{ov}) + k_{hp}POP_{ov} - \frac{r}{h}DOP_{ov} - k_{mp}DOP_{ov}$
Dissolved inorganic P (*DIP*)	$\frac{\partial DIP_{ov}}{\partial t} = L(DIP_{ov}) + k_{mp}DOP_{ov} - \frac{r}{h}DIP_{ov} - R_{DIP,up}$

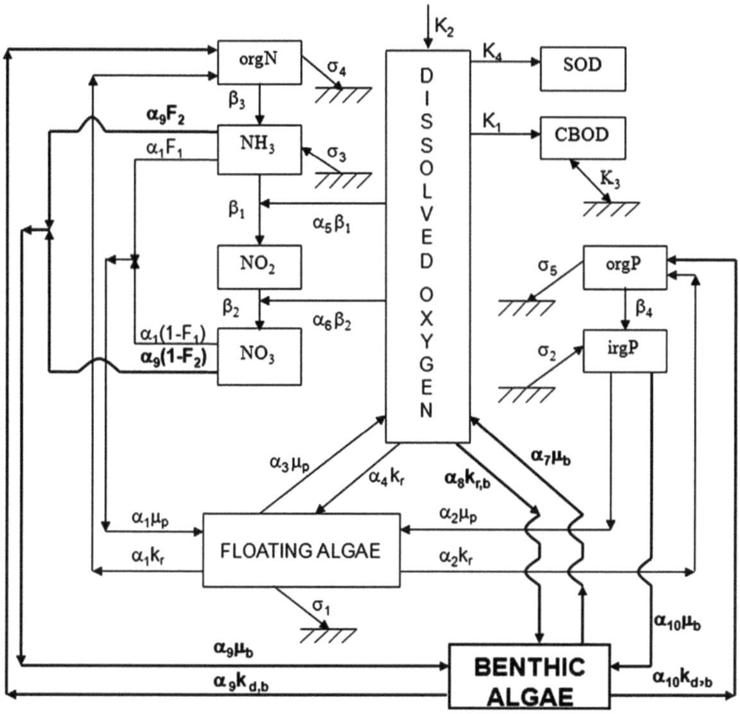

Fig. 8.3 Schematic representation of GSSHA in-stream water quality modeling

8.3.3 In-Stream Water Quality

For in-stream water quality modeling it is assumed that longitudinal and temporal changes (1D transport) are applicable. Water quality is affected in streams due to physical transport and exchange processes and biological, chemical, and biochemical kinetic processes along with changes due to benthic sediments. Currently, the in-stream water quality module includes a set of nutrient simulation kinetics. In-stream water quality kinetics computes algal biomass, organic and inorganic nitrogen and phosphorus species, CBOD and DO. The schematic representation of GSSHA in-stream water quality processes is shown Fig. 8.3.

The nutrient transport and transformation equations in the water column are summarized in Table 8.6.

Table 8.6 Mathematical expressions for in-stream water quality processes

Water quality variables	Transformation equation
Particulate organic N	$\frac{dON}{dt} = (\alpha_1 \cdot k_r \cdot A_p - \beta_3 \cdot ON - \sigma_4 \cdot ON) + \alpha_9 \cdot k_{db} \frac{A_b}{h} F_b$
Ammonium N	$\frac{dNH4}{dt} = \left(\beta_3 \cdot ON - \beta_1 \cdot NH4 + \frac{\sigma_3}{h} - F_1 \cdot \alpha_1 \cdot \mu_p \cdot A_p\right) - F_2 \cdot \alpha_9 \cdot \mu_b \frac{A_b}{h} F_b$
Nitrate N	$\frac{dNO3}{dt} = (\beta_2 \cdot NO2 - (1 - F_1)\alpha_1 \cdot \mu_p \cdot A_p) - (1 - F_2)\alpha_9 \cdot \mu_b \frac{A_b}{h} F_b$
Particulate organic P	$\frac{dOP}{dt} = (\alpha_2 \cdot k_r \cdot A_p - \beta_4 \cdot OP - \sigma_5 \cdot OP) + \alpha_{10} \cdot k_{db} \frac{A_b}{h} F_b$
Dissolved inorganic P	$\frac{dDIP}{dt} = \left(\beta_4 \cdot OP + \frac{\sigma_2}{h} - \alpha_2 \cdot \mu_p \cdot A_p\right) - \alpha_{10} \cdot \mu_b \frac{A_b}{h} F_b$
Dissolved oxygen	$\frac{dDO}{dt} = (k_2(DO_s - DO) + (\alpha_3 \cdot \mu_p - \alpha_4 \cdot k_r)A_p - k_1 \cdot CBOD - \frac{k_4}{h}$ $-\alpha_5 \cdot \beta_1 \cdot NH4 - \alpha_6 \cdot \beta_2 \cdot NO3) + (\alpha_7 \cdot \mu_b - \alpha_8 \cdot k_{rb}) \frac{A_b}{h} F_b$
Algae (phytoplankton)	$\frac{dA_p}{dt} = \left(\mu_p - k_r - \frac{\sigma_1}{h}\right) A_p$
Algae (bottom algae)	$\frac{dA_b}{dt} = (\mu_b - k_{rb} - k_{db}) A_b$

8.3.4 Nutrient and Interaction with Flow and Sediment Transport

Surface runoff can remove large quantities of nutrients from the soil in both dissolved and particulate forms. The loss of dissolved nutrients in surface runoff is the result of rainfall mixing with the dissolved nutrients in the upper portion of the soil. Dissolved nutrients interact with surface runoff and once in water, they are transported. Suspended nutrients, which are assumed to be either organic or adsorbed inorganic components, attach to eroded sediment material derived from erosion, and are transported with water. The process of erosion is selective for finer particles. The finer particles, particularly clay, have larger surfaces of adsorption, and the clay fraction contains much of the organic matter and hydrous oxides (iron and aluminum) that can bind nitrogen and phosphorus (Nelson and Logan 1983). Runoff has two roles in the transport of nutrients: particle detachment and transport. Most of the nitrogen leaving watersheds through surface runoff is attached to finer soil particles. Novotny and Chester (1981) reported enrichment ratios for organic nitrogen or nitrogen adsorbed onto organic matter ranging from 2 to 4. Surface runoff also contains dissolved forms of nitrogen including NH_4^+ and NO_3^-. Phosphorus is most commonly assumed to be transported predominantly in particulate forms through soil erosion by surface runoff. Particulate phosphorus is attached to mineral and organic sediment as it moves with the runoff. The enrichment ratio for particulate phosphorus varies from 1 to 10 depending on watershed size and soil characteristics. However, where soil erosion is limited, the majority of phosphorus transported by surface runoff may be in dissolved forms.

8.3.4.1 Dissolved Mass Transfer from the Upper Soil

The complicated nature of the flux at the soil or sediment surface is usually characterized through the use of a mass transfer coefficient, an empirical coefficient that relates the concentration gradient to mass transport (Choy and Reible 1999). The transfer rate of dissolved species from the soil to the water column is affected by the concentration gradient across the water-soil interface as well as flow conditions in the water column. This rate is computed with the NSM and then incorporated into the GSSHA-NSM integration as an external source/sink flux. Mass transfer theory states that the mass flux of a given species under a given set of flow conditions can be expressed as:

$$S_d = k_e(C_{d2}/\phi - C_d) \tag{8.10}$$

where S_d is mass transfer flux of a dissolved species [$ML^{-2}T^{-1}$], C_d is dissolved concentration of a species in the water column, C_{d2} is dissolved concentration of a species in the soil layer in terms of mass of the substance per bulk volume of the soil layer [M/L^3], k_e is mass transfer coefficient between water column and soil layer [L/T], and ϕ is porosity of the soil layer.

8.3.4.2 Leaching

Nitrate is highly mobile as discussed previously and subject to leaching losses when both soil NO_3^- content and water movement are high. NO_3^- leaching from soils must be carefully controlled because of the serious impact that it can have on the groundwater. Movement of NO_3^- through soil is governed by bulk flow which results in the movement of nitrate with the flow of water, molecular diffusion which results in the movement of nitrate due to the concentration gradient, and hydrodynamic dispersion in the soil due to the heterogeneity and internal structure of the soil. Leaching of NH_4^+ is usually insignificant.

Phosphorus is mainly bound to the fine soil particles. Only a small fraction of the phosphorus in the soil is present in the dissolved phase. However, some dissolved phosphorus is transported with runoff, and small amounts of phosphorus can reach the ground water through leaching. The amount of percolating phosphorus is controlled by the phosphorus adsorptive capacity of the soils above the aquifer (Nelson and Logan 1983). Transport of dissolved phosphorus involves the same processes as those described for N: convection, diffusion and hydrodynamic dispersion. These flux terms are computed through the GSSHA-NSM integration.

8.3.4.3 Erosion and Sedimentation

The erosion and sedimentation of sediments and associated pollutants are two important processes in water quality modeling. Sediment detachment by surface runoff is usually simulated in terms of a generalized erosion-deposition theory proposed by Smith et al. (1995). This assumes that the transport capacity concentration of the runoff reflects a balance between the two continuous counteracting processes of erosion and deposition. In general, the insoluble forms of nitrogen and phosphorus far exceed their soluble forms, the physical transport rate of both inorganic and organic forms of nitrogen and phosphorus with sediment is computed through NSM integration with GSSHA. The transport of nutrient particulates from the soil surface to the water column via erosion occurs at a rate that is proportional to the rate at which sediment particles are eroded (resuspended).

$$S_r = \sum_{n=1}^{N} f_p^n v_r^n C_{s2}^n \tag{8.11}$$

where S_r is total erosion rate of a nutrient [$ML^{-2}T^{-1}$], C_{s2}^n is particulate concentration associated with particle "n" in the soil layer [M/L^3], f_p^n is fraction of the total chemical in the sorbed phase associated with particle "n", and v_r^n can be defined as a erosion velocity associated with particle "n" [L/T].

The magnitude of the deposition flux of a contaminant is equal to the product of the rate of sediment deposition and the contaminant concentration associated with the settling particles. Settling velocity depends not only on the size, shape, and density of particles, but also on the concentration of the particles. The deposition of particulate nutrients from the water column is computed as:

$$S_s = \sum_{n=1}^{N} f_p^n v_s^n C_s^n \tag{8.12}$$

where S_s is total deposition rate of a nutrient [$ML^{-2}T^{-1}$], C_s^n is particulate concentration associated with particle "n" in water column [M/L^3], and v_s^n is settling velocity associated with particle "n" [L/T].

8.4 Water Flow, Sediment and Nutrient Modeling Validation

Model validation is important in verifying that the proper processes are represented adequately. Currently there are two case studies underway to validate the nutrient cycling processes at Eau Galle Watershed. The Eau Galle Watershed encompasses a 402 km^2 area in northwest Wisconsin, Fig. 8.4. The lower portion of the watershed is relatively data poor. The upper portion of the watershed, that portion above Spring Valley Dam, has been the subject of intensive past studies, and is relatively data rich. In addition to the concern about agricultural effects on water quality in the lake and river, there are concerns about the effects of land use change on hydrologic and water quality conditions in the larger Eau Galle River system. Hydrology, sediment transport, nutrient cycling and export were examined

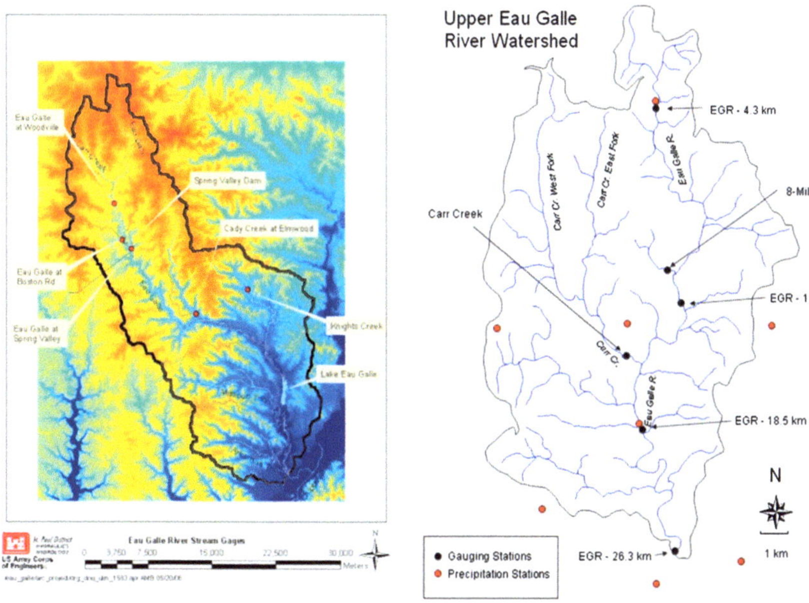

Fig. 8.4 Eau Galle River watershed and sampling locations

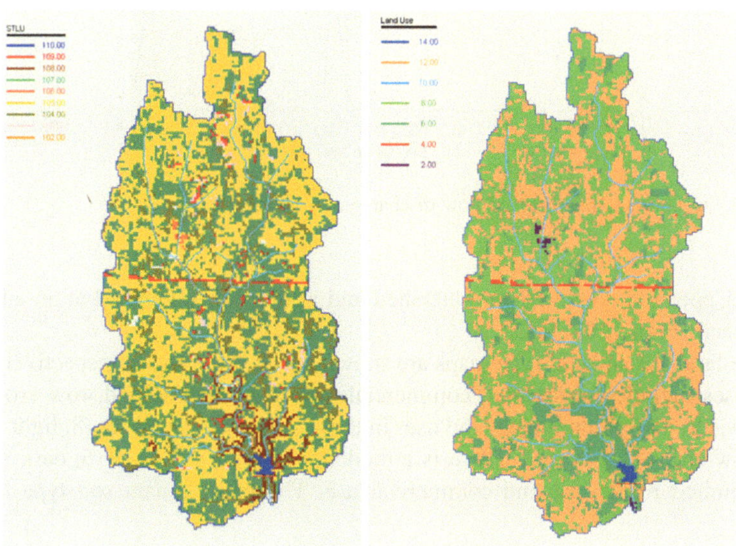

Fig. 8.5 Eau Galle River watershed land use and soil distribution

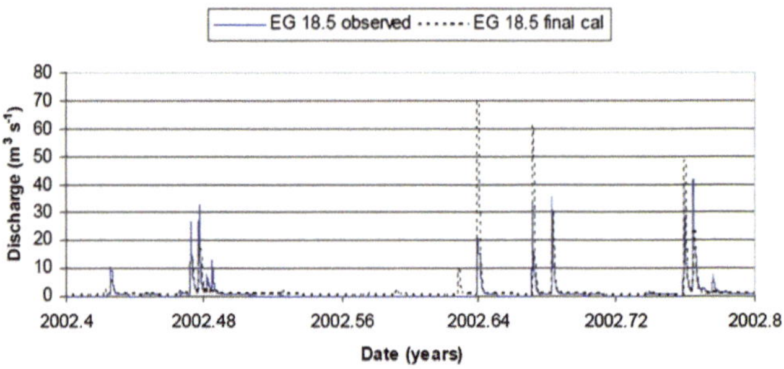

Fig. 8.6 Calibrated and observed flow discharge at EG 8.5

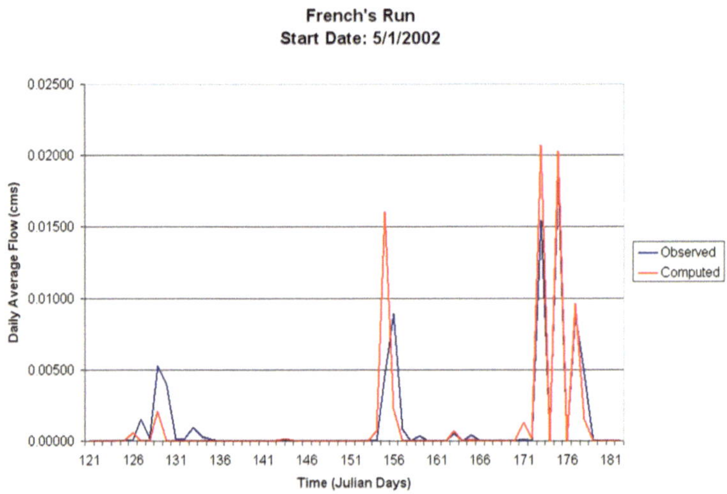

Fig. 8.7 Calibrated and observed flow discharge at Franch's Run

at the Upper Eau Galle River watershed and at one sub-watershed at an adjacent watershed (French Creek).

The land use and soil type maps are shown in Fig. 8.5a and b, respectively. The land uses includes residential, commercial, forest, grass, wetland, row crop, and open water. The predominate land uses in the watershed are pasture (8, light green) and row crops (12, beige). There is a moderate amount of forest (6, dark green), with limited residential and commercial use. The predominant soil type is silty loam.

The hydrologic model of GSSHA was developed and calibrated to observed flows at the USGS gauge (EG 8.5 in Fig. 8.4) because this site was believed to provide the most reliable data for model calibration. Flows from the other sites are considered less reliable. Results of the model calibration during the period June through October 2002 are shown in Fig. 8.6. The mean absolute error (MAE) of

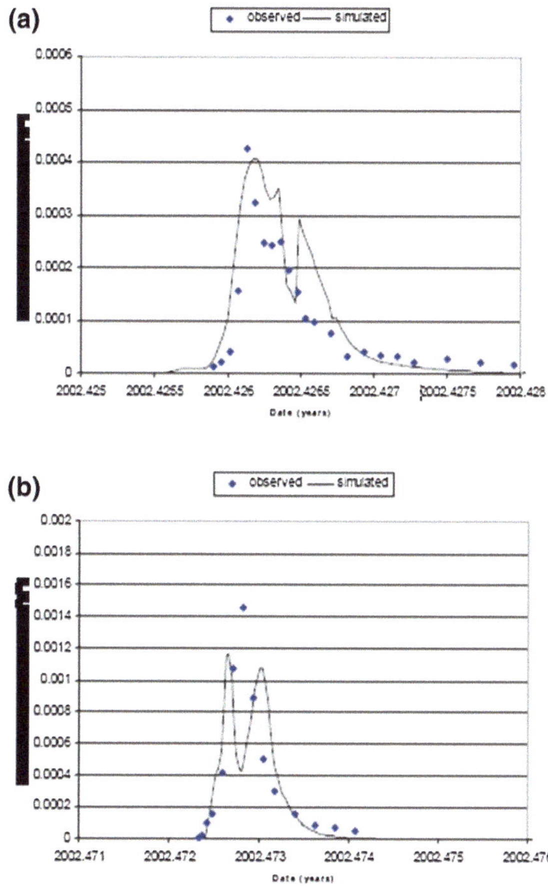

Fig. 8.8 Calibrated and observed total suspended sediment at EG 8.5

the larger two peaks is 3 percent of the observed. The error in total discharge is 1.5 percent of observed. The hydrograph shapes and base flow are accurately reproduced. The GSSHA model was able to adequately simulate hydrology as seen by the above calibration.

The French's Run study site was located in the headwaters of French Creek watershed, which is adjacent to and just south of the Upper Eau Galle River basin. The only defined channels in French's Run were ditches located on either side of two roads that bisected the watershed. Land use in the watershed was dominated by corn production during the study period and flows during storm events occurred as overland runoff from the field that drained directly into the ditches and a culvert that passed under a road. Runoff was exacerbated by contouring crop rows parallel to the slope of the field to promote better field drainage (James et al. 2003). Calibrated and observed flow discharge at French's run gage is given in Fig. 8.7.

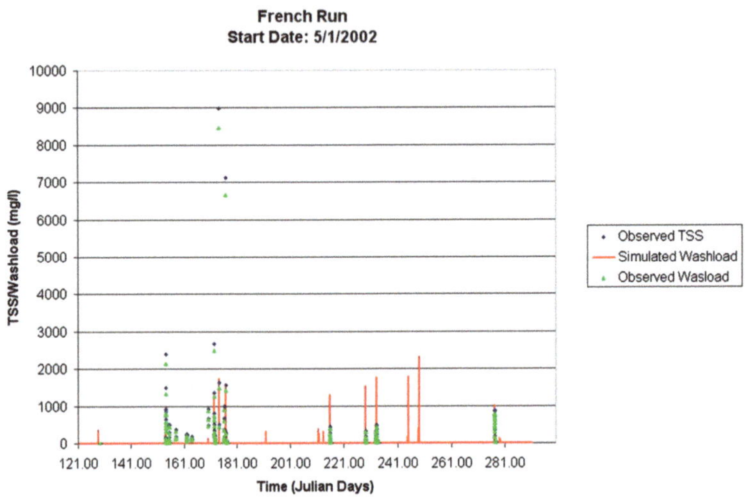

Fig. 8.9 Calibrated and observed total suspended sediment at Franch's Run

Three sediment size fractions were simulated, sand, silt, and clay. The model was calibrated to two observed events that occurred in June 2002 using the hydrologic parameters from calibration 1. Observed values of total suspended solids (TSS) and flow were combined according to USGS standards to produce sediment discharge ($m^3\ s^{-1}$) and compared to the model stream values of wash load, which is composed of clay and silt size fractions. The sand is expected and assumed to move as bed load and not be in TSS measurements. The calibration results are shown in Fig. 8.8. The MAE for the total sediment discharge (m^3) for the two events was 12 and 4 percent of the observed, respectively. In general, the sediment calibration and verification results are good.

For the field scale, calibrated and observed suspended sediment at French's run gage is given in Fig. 8.9.

The nutrient cycling simulation within GSSHA was tested and validated for the same watershed. In preparing the nutrient loadings, total nitrogen and total phosphorous were measured, and the soluble forms were estimated from these totals as inputs. No continuously N and P concentration measurement data were available. The model calibration for water quality was conducted only for dissolved N and P at gages where observed data were available. Based on multiple GSSHA runs, Fig. 8.10 shows the comparison between observed and modeled nitrate N and dissolved inorganic P for the same simulation period with the hydrology. The figure indicates that the trend of modeled nutrient concentration match with the trend of the measured data.

For the field scale, calibrated and observed flow inorganic phosphorus at French's run gage is given in Fig. 8.11.

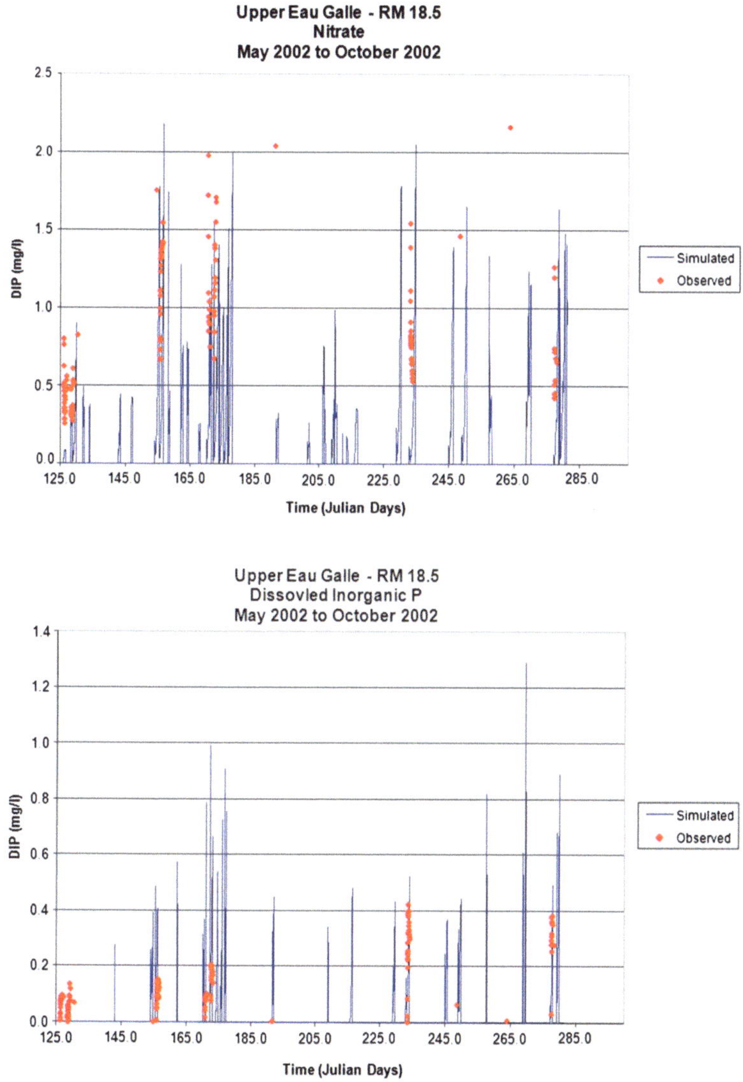

Fig. 8.10 Modeled and observed nitrogen and phosphorus results at EG 8.5

As the results of the demonstrative application show, the model presently developed achieves detailed analysis of the water quality aspects of a watershed including nutrient transport and fate across the landscape. Due to limited data issues it is inferred that current GSSHA modeling system needs further testing and validation at the watershed scale.

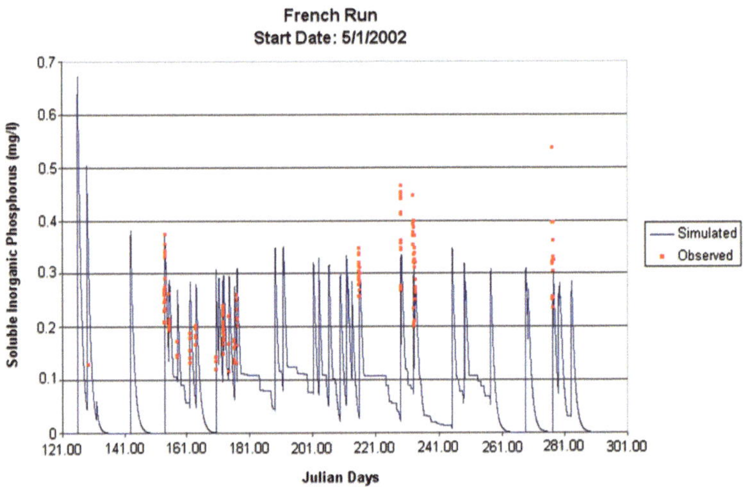

Fig. 8.11 Modeled and observed inorganic phosphorus results at Franch's Run

8.5 Summary

The GSSHA model has the capability to simulate hydrology, hydraulics, and sediment transport in variety of watershed applications. A distributed nutrient transport and transformation modeling has been integrated with GSSHA. The nutrient modeling described in this chapter is able to simulate both soil, surface runoff and channel processes of nitrogen and phosphorus cycling. Now the model is composed of cascade-linked three sub-models. An integrated model serves to evaluate evapotranspiration, infiltration, soil water content, discharge, soil erosion, nutrient transport and fate. The GSSHA was applied to a real watershed—upstream of the Spring Valley Dam located on the Eau Galle River in Wisconsin. The GSSHA model was able to adequately simulate hydrology and sediment transport as seen by the calibration. The model closely reproduces discharge and sediment transport during the calibration. This application also demonstrated its capabilities in simulating the fate and transport of nutrients in watersheds as well. The ability to simulate spatially distributed nutrient concentrations within the watershed has not been evaluated due to a lack of field data at this time.

Use of the structured grid in space is a great convenience in preparation of the input data including maps of soil texture, land use and nutrient loading. The GSSHA model generates time series outputs of model state variables at specified points in space over time. The model also provides the temporal variation and spatial distribution of sediment and nutrient transport. The GSSHA model is capable of predicting runoff depth, soil moistures, discharge, soil erosion, sediment and nutrient (nitrogen and phosphorus) transport and fate. The model is useful in determining the relative contributions of various sources of surface runoff and base flow and corresponding sediment and nutrient sources. Furthermore, the GSSHA model has the advantage of being fully distributed and physically based. All simulation parameters employed have physical or well-established empirical meaning, and all are within the bounds of published or field measured values. It is thus concluded that the model could be a powerful tool to investigate and assess the time-varying hydrology and sediment and nutrient transport and fate of a watershed. Especially, in agricultural sector, it would absolutely be an invaluable tool for quantifying water quality impacts by agricultural farming and practices. However, model physically based formulation will require application on a data set that includes detailed rainfall, soil moisture, and distributed source observations.

Appendix: Water Quality Parameters

$orgN_{frs}$	concentration of soil layer fresh organic N pool [M/L^3]
$ON_{min\|imb}$	net mineralization/immobilization rate of soil layer fresh organic N pool [M/L^2/T]
ON_{dec}	decomposition rate of soil layer fresh organic N pool [M/L^2/T]
$ON_{frs,e}$	net surface erosion/deposition rate of soil layer fresh organic N pool [M/L^2/T]
$ON_{frs,s}$	external sources [M/L^2/T]
$orgN_{act}$	concentration of soil layer active organic N pool [M/L^3]
ON_{trn}	rate transferred between the active and stable organic N pools [M/L^2/T]
ON_{min}	mineralization rate of soil layer active organic N pool [M/L^2/T]
$ON_{act,e}$	net surface erosion/deposition rate of soil layer active organic N pool [M/L^2/T]
$ON_{act,s}$	external sources [M/L^2/T]
$ON_{sta,e}$	net surface erosion/deposition rate of soil layer stable organic N pool [M/L^2/T]
$ON_{sta,s}$	external sources added to the soil layer stable organic N pool [M/L^2/T]
NH_4^+	concentration of soil layer NH_4^+ pool [M/L^3]
NH_{min}	total mineralization processes rate of soil layer organic N pools [M/L^2/T]
$NH_{nit\|vol}$	net nitrification/volatilization processes rate in the soil layer [M/L^2/T]
NH_{up}	plant uptake rate of soil layer NH_4^+ pool [M/L^2/T]
$R_{NH4,e}$	mass transfer rate of NH_4^+ between the upper soil layer and surface runoff [M/L^2/T]
NH_s	external sources [M/L^2/T]
NO_3^-	concentration of soil layer NO_3^- pool [M/L^3]
NO_{dnit}	denitrification processes rate in the soil layer [M/L^2/T]
NO_{up}	plant uptake rate of soil layer NO_3^- pool [M/L^2/T]

$R_{NO3,e}$	mass transfer rate of NO_3^- between the upper soil layer and surface runoff [M/L^2/T]
$R_{NO3,f}$	infiltration rate of soil layer NO_3^- pool [M/L^2/T]
NO_s	external sources [M/L^2/T]
PON_{ov}	concentration of the overland flow PON [M/L^3]
$orgN$	total concentration of organic N in the upper soil layer [M/L^3]
k_{hn}	PON hydrolysis rate constant [1/T]
DON_{ov}	concentration of DON in the overland flow [M/L^3]
k_{mn}	DON mineralization rate constant [1/T]
$NH^+_{4\,ov}$	concentration of NH_4^+ in the overland flow [M/L^3]
k_{en}	effective mass transfer rate constant [L/T]
k_{nit}	nitrification rate constant [1/T]
$R_{NH4,up}$	plant uptake rate of the overland flow NH_4^+ [M/L^3/T]
$NO^-_{3\,ov}$	concentration of NO_3^- in the overland flow [M/L^3]
$R_{NO3,up}$	plant uptake rate of the overland flow NO_3^- [M/L^3/T]
$orgP_{frs}$	concentration of soil layer fresh organic P pool [M/L^3]
OP_{dec}	decomposition rate of soil layer fresh organic P pool [M/L^2/T]
$OP_{min\|imb}$	net mineralization/immobilization rate of soil layer fresh organic P pool [M/L^2/T]
$OP_{frs,e}$	net surface erosion/deposition rate of soil fresh organic P pool [M/L^2/T]
$OP_{frs,s}$	external sources [M/L^2/T]
$orgP_{act}$	concentration of soil layer active organic P pool [M/L^3]
OP_{min}	mineralization rate of soil humic active organic P pool [M/L^2/T]
OP_{trn}	rate transferred between the active and stable organic P pools [M/L^2/T]
$OP_{act,e}$	net surface erosion/deposition rate of soil humic active organic P pool [M/L^2/T]
$OP_{act,s}$	external sources [M/L^2/T]
$orgP_{sta}$	concentration of soil layer stable organic P pool [M/L^3]
$OP_{sta,e}$	net surface erosion/deposition rate of soil humic stable organic P pool [M/L^2/T]
$OP_{sta,s}$	external sources [M/L^2/T]
P_{sol}	concentration of soil layer soluble P pool [M/L^3]
IP_{min}	total mineralization processes rate of soil layer organic P pools [M/L^2/T]
$IP_{sol\|act}$	net sorption rate transferred between the soluble P pool and active inorganic P pool [M/L^2/T]
IP_{up}	plant uptake rate of soil layer soluble P pool [M/L^2/T]
$R_{DIP,e}$	mass transfer rate of soluble P between the upper soil layer and surface runoff [M/L^2/T]
IP_s	external sources [M/L^2/T]
$minP_{act}$	concentration of soil layer active inorganic P pool [M/L^3]

$IP_{act\|sta}$	net slow sorption transfer rate between the active inorganic P pool and the stable inorganic P pool [M/L^2/T]
$IP_{act,e}$	surface erosion/deposition rate of soil active inorganic P detachment [M/L^2/T]
$IP_{act,s}$	external sources [M/L^2/T]
$minP_{sta}$	concentration of soil layer stable inorganic P [M/L^3]
$IP_{sta,e}$	surface erosion/deposition rate of soil stable inorganic P detachment [M/L^2/T]
$IP_{sta,s}$	external sources [M/L^2/T]
POP_{ov}	concentration of POP in the overland flow [M/L^3]
$orgP$	total concentration of organic P in the upper soil layer [M/L^3]
k_{hp}	POP hydrolysis rate constant [1/T]
DOP_{ov}	concentration of DOP in the overland flow [M/L^3]
k_{mp}	DOP mineralization rate constant [1/T]
DIP_{ov}	concentration of DIP in the overland flow [M/L^3]
k_{ep}	DIP mass transfer rate between the upper soil layer and overland flow [L/T]
$R_{DIP,up}$	plant uptake rate of the overland flow DIP [M/L^3/T]
PON_{ch}	concentration of in-stream PON [M/L^3]
k_{dp}	temperature-dependent phytoplankton death rate [T^{-1}]
k_{db}	temperature-dependent bottom algae death rate [T^{-1}]
A_p	stream phytoplankton concentration [M/L^3]
A_b	stream bottom algae concentration [M/L^2]
k_{hn}	temperature-dependent PON hydrolysis rate coefficient [T^{-1}]
DON_{ch}	concentration of in-stream DON [M/L^3]
k_{mn}	temperature-dependent DON mineralization rate coefficient [T^{-1}]
F_{oxmn}	DON mineralization attenuation due to low oxygen
$TNH^+_{4\,ch}$	Total concentration of in-stream NH_4^+ [M/L^3]
k_{rp}	temperature-dependent phytoplankton respiration rate [T^{-1}]
F_{oxna}	nitrification attenuation due to low oxygen
k_{nit}	temperature-dependent NH_4^+ nitrification rate coefficient [T^{-1}]
P_{ap}	preference coefficient of phytoplankton for NH_4^+
P_{ab}	preference coefficient of bottom algae for NH_4^+
$NO^-_{3\,ch}$	concentration of in-stream NO_3^- [M/L^3]
K_{scdn}	DOC half-saturation constant for denitrification [gC/m^3] [M/L^3]
k_{dnit}	temperature-dependent NO_3^- denitrification rate coefficient [T^{-1}]
F_{oxdn}	effect of low oxygen on denitrification
POP_{ch}	concentration of in-stream POP [M/L^3]
DOP_{ch}	concentration of in-stream DOP [M/L^3]
TIP_{ch}	total concentration of in-stream inorganic P [M/L^3]
k_{hp}	temperature-dependent POP hydrolysis rate coefficient [T^{-1}]
k_{mp}	temperature-dependent DOP mineralization rate coefficient [T^{-1}]
F_{oxmp}	DOP mineralization attenuation due to low oxygen
POC_{ch}	concentration of in-stream POC [M/L^3]

DOC_{ch}	concentration of in-stream DOC [M/L^3]
DIC_{ch}	concentration of in-stream DIC (mole/L) [M/L^3]
k_{hc}	temperature-dependent POC hydrolysis rate coefficient [T^{-1}]
F_{oxmc}	DOC mineralization attenuation due to low oxygen
k_{mc}	temperature-dependent DOC mineralization rate [T^{-1}]
k_{ac}	0.923 k_a = temperature-dependent CO_2 deaeration coefficient [T^{-1}]
k_H	Henry's constant [mole/(L atm)]
p_{CO2}	partial pressure of carbon dioxide in the atmosphere [atm]
α_0	fraction of total inorganic carbon in carbon dioxide
DO_{ch}	concentration of in-stream DO [M/L^3]
k_a	temperature-dependent oxygen reaeration coefficient [T^{-1}]
DO_s	saturation concentration of oxygen [mgO_2/L]
S_{SOD}	sediment oxygen demand rate [M/L^3]
A_b	stream bottom algal concentration [M/L^2]
μ_b	benthic algal photosynthesis rate [T^{-1}]
F_{oxb}	attenuation due to low oxygen
k_{rb}	temperature-dependent benthic algal respiration rate [T^{-1}]
k_{db}	temperature-dependent benthic algal death rate [T^{-1}]

References

Beuselinck L, Govers G, Steegen A, Quine TA. Sediment transport by overland flow over an area of net deposition. Hydrol Process. 1999;13(17):2769–82.

Burban PY, Xu Y, McNeil J, Lick W. Settling speeds of flocs in fresh and sea waters. J Geophys Res [Oceans]. 1990;95(C10):18213–20.

Burt TP, Haycock NE. Controlling losses of nitrate by changing land use. In: Burt TP, Heathwaite AL, Trudgill ST, editors. Nitrate: processes, patterns and management. Wiley; 1993. p. 342–67.

Cheng NS. Simplified settling velocity formula for sediment particle. J Hydraul Eng. 1997;123(2):149–152

Choy B, Reible DD. Contaminant transport in soils and sediments: mathematic analysis. Report to the Hazardous Substance Research Center (S&SW), Baton Rouge: Louisiana State University; 1999.

Coastal and Hydraulic Laboratory (CHL). http://chl.erdc.usace.army.mil/gssha 2012.

Downer CW, Ogden FL. GSSHA: a model for simulating diverse streamflow generating processes. J Hydrol Eng. 2004;9(3):161–74.

Haralampides K, McCourquodale JA, Krishnappan BG. Deposition properties of fine sediment. J Hydraul Eng. 2003;129(3):230–4.

Holley ER. Unified view of diffusion and dispersion. J Hydraul Div. 1969;95(2):621–31.

James WF, Eakin HL, Barko JW. Phosphorus forms and export from sub-watersheds in the Upper Eau Galle River basin exhibiting differing land-use practices. Water Quality Technical Notes Collection (ERDC WQTN-PD-15), US. Vicksburg: Army Engineer Research and Development Center; 2003.

Julien PY, Simons DB. Sediment transport capacity of overland flow. Trans ASAE. 1985;28:755–62.

Julien PY, Saghafian B, Ogden FL. Raster-based hydrologic modeling of spatially-varied surface runoff. Water Resour Bull. 1995;31(3):523–36.

Julien PY. Erosion and sedimentation. Cambridge: Cambridge University Press; 1998.
Krishnappan BG. In situ distribution of suspended particles in the Frasier River. J Hydraul Eng. 2000;126(8):561–9.
Nelson DW, Logan TJ. Chemical processes and transport of phosphorus. In: Schaller FW, Bailey GW, editors. Agricultural management and water quality. Iowa State University Press; 1983. p. 65–91.
Novotny V, Chester G. Handbook of nonpoint pollution: sources and management. New York: Van Nostrand Reinhold Co.; 1981.
Simons DB, Sentürk F. Sediment transport technology: water and sediment dynamics (revised edition). Littleton: Water Resources Publications; 1992.
Smith RE, Goodrich D, Quinton JN. Dynamic distributed simulation of watershed erosion: the KINEROS2 and EUROSEM models. J Soil Water Conserv. 1995;50:517–20.
USEPA (US Environmental Protection Agency). Hypoxia in the Northern Gulf of Mexico: an update by the EPA Science Advisory Board, EPA-SAB-08-003; 2007.

Chapter 9
Simulation of River Flow for Downstream Water Allocation in the Heihe River Watershed, Northwest China

Chansheng He, Lanhui Zhang, Li Fu, Yi Luo, Lanhai Li and Carlo DeMarchi

Abstract Multiple demands for water for large agricultural irrigation schemes, increasing industrial development, and rapid urban population growth have depleted downstream flows most of the time in the arid Heihe River Watershed, Northwest China, causing shrinking oases and tensions among different water jurisdictions and ethnic groups. To address this pressing issue, the State Council of the People's Republic of China has issued an executive order to mandate the release of water downstream for ecosystem restoration. This paper describes the adaptation of the Distributed Large Basin Runoff Model to the Heihe Watershed to gain an understanding of the distribution of glacial/snow melt, groundwater, surface runoff, and evapotranspiration in the upper and middle reaches of the watershed. The simulated daily river flows for 1990–2000 show that Qilian

C. He · L. Zhang
Center for Dryland Water Resources Research and Watershed Sciences, Key Laboratory of Western China's Environmental Systems (MOE), Lanzhou University, Lanzhou 730000, China

C. He (✉)
Department of Geography, Western Michigan University, Kalamazoo, MI 49008-5424, USA
e-mail: He@wmich.edu

L. Fu
Institute of Remote Sensing Applications, The Chinese Academy of Sciences, Beijing 100101, China

Y. Luo
Institute of Geographical Science and Natural Resources Research, The Chinese Academy of Sciences, Beijing 100101, China

L. Li
Xinjiang Institute of Ecology and Geography, Chinese Academy of Sciences, Urumqi 830011 Xinjiang, People's Republic of China

C. DeMarchi
Department of Geological Sciences, Case Western Reserve University, Cleveland, OH 44106-7216, USA

B. Fu and K. B. Jones (eds.), *Landscape Ecology for Sustainable Environment and Culture*, DOI: 10.1007/978-94-007-6530-6_9,

Mountain in the upper reach area was the main source of runoff generation in the Heihe Watershed, and annually, the Heihe River discharged about 1×10^9 m^3 of water from the middle reach (at Zhengyixia Gage Station) to the lower reach under the normal climatic conditions (with a likelihood of 50 %). These flows are consistent with the State Council's mandate of delivering 0.95×10^9 m^3 water downstream annually. However, the river flow would be significantly less under the dry climatic conditions, making it much more difficult to deliver the mandated amount of water downstream for ecosystem rehabilitation.

Keywords Multiple water demands · Distributed large basin runoff model (DLBRM) · The Heihe watershed · Northwest China · Water shortage

9.1 Introduction

Proper management of the limited water resources is essential to ensure the welfare of human beings and the sustainability of dry land ecosystems, which support over 38 % of the world population (Reynolds et al. 2007). During the past few decades, however, improper water resource management has resulted in numerous problems worldwide, including poor food security, increased human diseases, conflicts between different users, limitations on economic development and human welfare, desertification, salinization, sand storms, water pollution, and so forth (UN World Water Development Report 2003; Reynolds et al. 2007). In China, the increased withdrawals from the upper and middle reaches of the Yellow River depleted groundwater in much of the basin and contributed to desiccation (i.e. no measurable flow in the river) of the lower reaches in 22 of the years between 1972 and 2000. This desiccation has created serious economic and environmental problems throughout the North-Central China, including water rationing, under-capacity industrial production, reduced crop yields, water pollution, wildlife habitat depletion, coastline recession, and sea water intrusion (He et al. 2005). In arid Northwest China, irrigated farming accounts for more than 80 % of total water usage. Over the past few decades, the increased withdrawals for agricultural irrigation and municipal water supplies in the Hexi Corridor of the Heihe River Watershed (the second largest inland river or terminal lake in the nation, with a drainage area of 128,000 km^2) since the 1970s have depleted much of the river flows to the lower reach, shrinking the East Juyan Lake and drying up the West Juyan Lake, endangering aquatic ecosystems, accelerating desertification, intensifying water conflicts between the middle reach of Gansu Province and lower reach of the Inner Mongolian Autonomous Region (IMAR), and damaging relationships among Han, Mongolian, and Hui ethnic groups (He et al. 2009) (Fig. 9.1). To mitigate the water conflicts and rehabilitate West Juyan Lake, the State Council of the People's Republic of China (the executive branch of the central government) has issued a "Water Allocation Plan for the Heihe Watershed

Mainstream", mandating water allocation to the lower reach each year (Pan and Qian 2001; Feng et al. 2002). While a number of studies have been done in the Heihe Watershed (Cheng et al. 1999; Feng et al. 2002; Pan and Qian 2001; Jia et al. 2005), the magnitude, spatial and temporal distribution, and transfer mechanism of the Heihe hydrological system are still not well understood, especially in the face of climate change and urbanization. This gap, together with the lack of a comprehensive implementation plan has impeded the implementation of the State Council's water allocation plan. This study uses a hydrological modeling approach to address a key research question facing the Heihe River Watershed: how much water flows from the middle reach to the lower reach annually to support competing demands for water for domestic, irrigation, industrial supplies and rehabilitation of the ecosystems under the current climatic conditions?

Specifically, this study simulates the hydrological processes of the upper and middle reaches of the Heihe River Watershed to determine the amount of water flowing downstream annually from the middle reach (at Zhengyixia Gage Station). It describes our collaborative work to adapt the Distributed Large Basin Runoff Model (DLBRM) to the Heihe Watershed for understanding the hydrological processes of the river system and thereby provides partial basis for implementation of the central government's water allocation plan. We first describe the physical features of the Heihe watershed, then briefly introduce the structure, input, and output of the DLBRM, and finally discuss the simulation results of the DLBRM in the Heihe River Watershed.

9.2 Methods

9.2.1 The Study Area

The boundary of the Heihe River Watershed is shown in Fig. 9.1. The Qilian Mountain is situated at the south of the watershed, with a peak elevation of 5,584 m. Ice and snow cover it year round above 4,500 m. Mixed alpine meadow and permafrost dominate between 3,600 to 4,500 m. While the main vegetation is forest and grassland with a mean annual precipitation of 250–500 mm in the 1,900–3,600 m range, the landscape below 1,900 m is dominated by hilly or grassland desert with a mean annual precipitation of 200–250 mm (Pan and Qian 2001; He et al. 2009). Located in the middle reach of the Heihe Watershed, the Hexi Corridor hosts over 90 % of the total agricultural oases in the watershed and supports more than 97 % of the Heihe River Watershed's 1.8 million inhibits in two metropolitan areas: Zhangye (population 1.25 million in 2000) and Jiuquan (population 0.49 million in 2000). Irrigation supply is from both surface water withdrawals and groundwater pumping. North of the Hexi Corridor is the Alashan Highland (the area north of Zhengyixia Gage Station with mean elevation 1,000 m), an extremely dry desert with an annual precipitation below 50 mm.

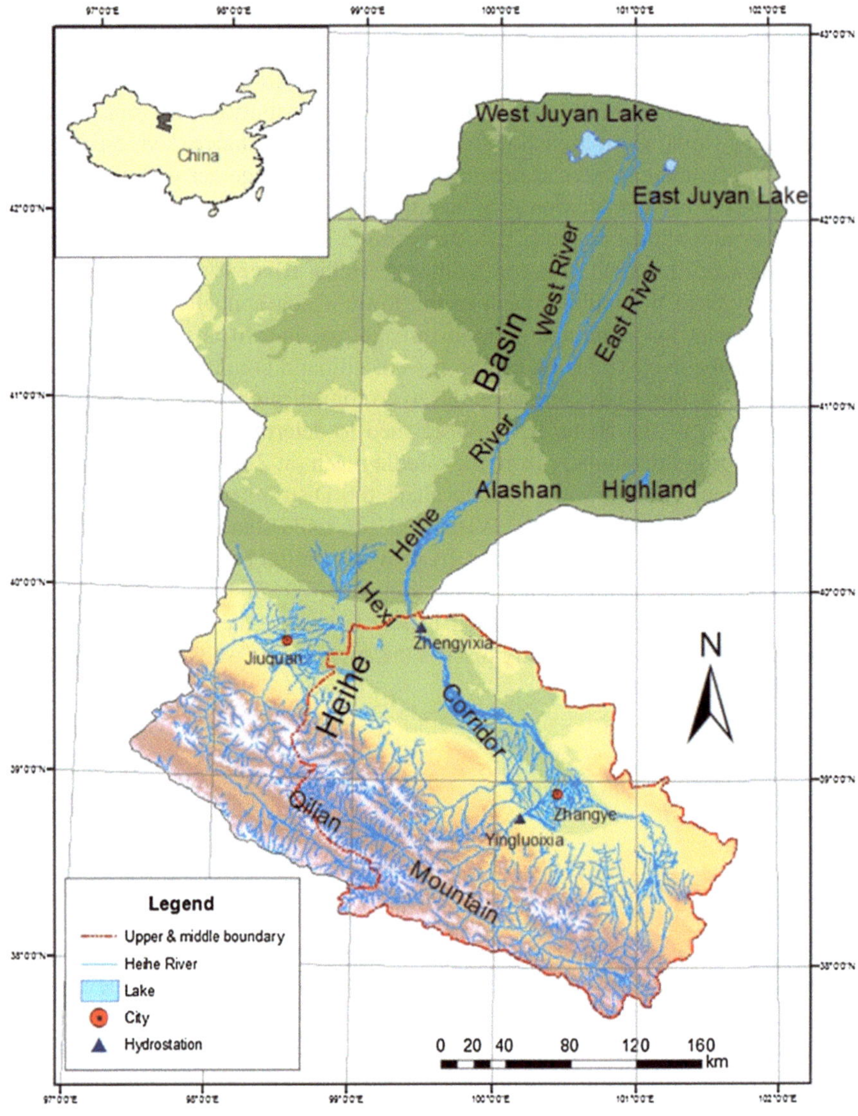

Fig. 9.1 Boundary of the Heihe River watershed

Spotty oases appear intermittently along the streams, lakes, and irrigation ditches. Since it is extremely dry, the Alashan Highland is a large source of frequent sandstorms (Cheng et al. 1999; He et al. 2009).

The total annual water withdrawal in the Heihe in 1995 was about 3.36 billion (10^9) m^3 and 86 % of that was used to irrigate over 288,000 ha of farmland mainly located in the Hexi Corridor (Pan and Qian 2001; He et al. 2009), leaving less than 0.5 billion (10^9) m^3 of annual flow to the lower reach (north of the Zhengyixia Gage

Station). Competitions for the limited water resources have caused intense conflicts between water users in the Hexi Corridor and those in Alashan Highland in the lower reach. To address this pressing issue, the State Council has mandated the allocation of 0.95×10^9 m^3 of water to the lower reach under normal climatic conditions for rehabilitating the downstream ecosystems. Implementation of such a mandate means reducing the current water use by 0.58×10^9 m^3 for agricultural irrigation (potentially taking about 40,000 ha of farmland out of irrigation) and domestic supply in Zhangye City alone in the middle reach (Pan and Qian 2001), a potential loss of about $240 million annually for the city.

9.2.2 Description of the DLBRM

To answer the question, "How much water flows downstream from the upper and middle reaches (at the Zhengyixia Gage Station) in the Heihe Watershed?", this study uses the Distributed Large Basin Runoff Model to simulate the hydrology of the combined upper and middle reaches of Heihe River Watershed at daily intervals over the period of 1978–2000. The DLBRM was developed by the U.S. National Oceanic and Atmospheric Administration (NOAA)'s Great Lakes Environmental Research Laboratory and Western Michigan University. It represents a watershed by using 1 km^2 (or other size) grid cells. Each cell of the watershed is composed of moisture storages of the upper soil zone (USZ), lower soil zone (LSZ), groundwater zone (GZ), and surface, which are arranged as a serial and parallel cascade of "tanks" to coincide with the perceived basin storage structure (Fig. 9.2). Water enters the snow pack, which supplies the basin surface (degree-day snowmelt). Infiltration is proportional to this supply and to saturation of the upper soil zone (partial-area infiltration). Excess supply is surface runoff. Flows from all tanks are proportional to their amounts (linear-reservoir flows). Mass conservation applies for the snow pack and tanks; energy conservation applies to evapotranspiration (ET). The model computes potential ET from a heat balance, indexed by daily air temperature, and calculates actual ET as proportional to both the potential and storage. It allows surface and subsurface flows to interact both with each other and with adjacent-cell surface and subsurface storages. The model has been applied extensively to riverine watersheds draining into the North America's Laurentian Great Lakes for use in both simulation and forecasting (Croley and He 2005, 2006; He and Croley2007a, 2010; DeMachi et al. 2011; Croley et al. 2005). The unique features of the DLBRM include: (1) use of readily available climatological, topographical, hydrological, soil, and land use databases; (2) applicability to large watersheds; and (3) analytical solutions for mass continuity equations, (mathematical equations are not shown here due to space limitations; for details, see Croley and He 2005, 2006; He and Croley 2007a).

The DLBRM requires 16 input variables for each of the grid cells. To facilitate the input and output processing for the DLBRM, an ArcView-DLBRM (AVDLBRM) interface program has been developed to assist with the model

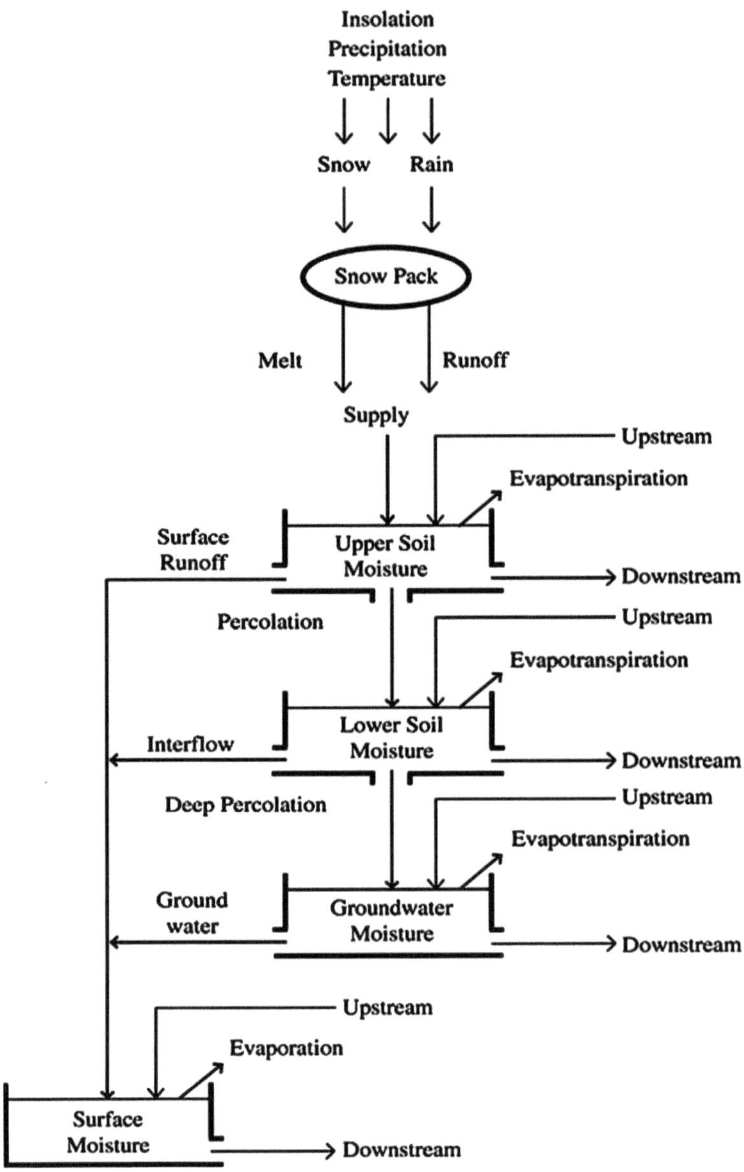

Fig. 9.2 Tank cascade schematic of distributed large basin runoff model

implementation. The interface was written in ArcView Avenue scripts by modifying the ArcView Nonpoint Source Modeling interface by He (2003). It consists of six modules: (1) Soil Processor, (2) DLBRM Utility, (3) Parameter Generator, (4) Output Visualizer, (5) Statistical Analyzer, and (6) Land Use Simulator. Multiple databases of meteorology, soil, DEM, land use/cover, and hydrology and

Table 9.1 Input Variables for the DLBRM, sourced from CAREERI

Variables	Databases
Elevation	Digital elevation model (DEM)
Flow direction	DEM
Slope	DEM
Land use	Land use database
Depth of upper soil zone (USZ)	Compiled soil database
Depth of lower soil zone (LSZ)	Compiled soil database
Available water capacity (%) of USZ	Compiled soil database
Available water capacity of LSZ	Compiled soil database
Permeability of USZ	Compiled soil database
Permeability of LSZ	Compiled soil database
Soil texture	Compiled soil database
Manning's coefficient value	Land use, slope, and soil texture

Table 9.2 Time series meteorological and flow variables for the DLBRM, sourced from CAREERI

Variables	Databases
Daily precipitation	Gansu Bureau of Meteorology
Daily air temperature	Gansu Bureau of Meteorology
Daily solar isolation	Gansu Bureau of Meteorology
Daily flows	Gansu Bureau of Hydrology

hydrography are used by the interface through the draw-down menu to derive input variables for the DLBRM (He et al. 2001; He 2003; He and Croley 2007b). The derived variables for each of the cells include: elevation, flow direction, slope, land use, Manning's coefficient (*n*) values, soil texture, USZ and LSZ depths, available water capacity, and, permeability, as well as daily precipitation, air temperature, and solar isolation. DLBRM outputs include, for every cell, surface runoff, ET, infiltration, percolation, interflow, deep percolation, groundwater flow, USZ, LSZ groundwater, and surface moisture storages, and lateral flows between USZ, LSZ, groundwater, and the surface (Tables 9.1 and 9.2). The outputs can be examined either in tabular or map format using the interface.

9.2.3 DLBRM Input Data

The combined upper and middle reaches of the Heihe River Watershed were discretized into a grid network of 9,790 cells at 4 km^2 resolution. Multiple databases of DEM (at 100 m resolution), land use/cover for the year 2000, meteorological and hydrological databases for 1978–2000 were provided by The Chinese Academy of Sciences (CAS) Cold and Arid Regions Environmental and Engineering Research Institute (CAREERI). These databases were used to derive relevant input variables for the DLBRM using the AVDLBRM interface for each grid cell (Croley and He 2005; He and Croley 2007a). Since the soil database of 1999 (1:250,000) from the Gansu Province only contained soil types, we used SPAW (Soil-Plant-Atmosphere-Water) Field & Pond Hydrology model (developed by the U.S. Department of Agriculture Agricultural Research Service and Natural Resources Conservation Service) to determine relevant soil attributes for each of the soil types based on the limited soil survey data collected by Xiao and his group (2006) Chen and Xiao (2003). Such attributes include soil texture, depth of USZ and LSZ, water holding capacity (%) and permeability (cm/hr). Manning's coefficients were assigned to each cell by the hydrological response units (HRU), which was determined according to the combination of land use, soil texture, and slope (He and Croley 2007b). Average daily river flow rates (in m^3/s) were converted into daily outflow volumes and used to conduct a systematic search of the parameter space to minimize the root mean square errors (RMSE) between actual and simulated daily outflow volumes at the watershed outlet (Croley et al. 2005; Croley and He 2006).

9.2.4 Model Calibration and Verification

The DLBRM was calibrated over the period of 1978–1987 for each of the 9,790 cells (4 km^2) at daily intervals. The calibration shows a 0.69 correlation between simulated and observed watershed outflows and a 0.072 mm/d root mean square error. The ratio of model to actual mean flow was 1.011; and the ratio of model to actual flow standard deviation was 0.68 (Table 9.3). Over a separate verification period (1990–2000), the model demonstrated a 0.71 correlation between simulated and observed watershed outflows and a 0.006 cm/d RMSE; the ratio of model to actual mean flow was 1.409; and the ratio of model to actual flow standard deviation was 0.72, showing the model overestimated the mean daily flows over the period of 1990–2000. The simulated annual water budget (averages of the 1990–2000) shows that annual surface net supply from both rainfall and snow melt

Table 9.3 DLBRM Heihe calibration statistics

					Long-term average ratio to surface net supply				
Calibration period	Correlation	RMSE (cm)	μ_M/μ_A	σ_M^2/σ_A^2	Surface runoff	Interflow	GW	USZ ET	LSZ ET
1978–1987	0.690	0.007	1.014	0.683	0.065	0.061	0.000	0.000	0.884
1978–1987*	0.690	0.007	1.011	0.682	0.065	0.061	0.000	0.000	0.885
1990–2000	0.710	0.006	1.409	0.717	0.070	0.000	0.000	0.000	0.870

*A 10.85 m snow pack was assumed in about 305 km^2 of mountain area (elevation >4500 m) in the simulation

ET represents evaporation

μ_M/μ_A = ration of simulated mean annual outflow to the actual annual outflow was about 8.92 billion (10^9) m^3 (Fig. 9.3), which mainly came from Qilian Mountain in the upper reach area. The USZ stored about 391 billion m^3 of water, the largest storage among all the four storage tanks (USZ, LSZ, groundwater zone, and surface storage). Surface runoff from the USZ averaged about 0.54 billion m^3, while a much larger portion of water (8.37 billion m^3) percolated down to the LSZ. A majority (94 %) of the percolated water evaporated to the atmosphere from the LSZ and the rest flowed to the stream in the form of interflow. There was hardly any deep percolation to the groundwater since the LSZ is up to several hundred meters deep in much of the middle reach area (Pan and Qian 2001). The average annual outflow at the outlet (Zhengyixia) of the middle reach was about 1.05 billion m^3 to the downstream (Fig. 9.3) (He et al. 2009).

Simulation results of daily flows for 1990 are shown in Fig. 9.4. Compared to the observed daily discharge, DLBRM using parameters from both the 1978–1987 and the 1990–2000 calibration periods reasonably depicted the daily variations of the Heihe discharge at the outlet (near Zhengyixia Station), with the former performing better than the latter. However, the simulations underestimated the discharges during the cold season, when there was not much precipitation, and overestimated the discharges during the spring and summer, when there were more storms (except underestimating the discharge from the July 25, 1990 storm of 2.5 cm) (Fig. 9.4) (He et al. 2009).

9.3 Discussion

The Qilian Mountain (with a peak elevation of 5,500 m above sea level) makes up the upper reach of the Heihe River Watershed. Due to the high altitude and steep slope of the mountain area, much of the snow melt and rainfall becomes surface runoff. Once reaching the mountain outlet (Yingluoxia Gage Station), the water quickly percolates to the deep, coarse sandy and loamy soils in the alluvial fan (up to several hundred m deep) which is the main agricultural oasis in the middle reach (between the mountain outlet at Yingluoxia Gage Station and the middle reach outlet at Zhengyixia Gage Station) (Cheng et al. 1999; Pan and Qian 2001; and Wu et al. 2010). As annual precipitation in the oasis is less than 200 mm, the majority of the river flow is used to irrigate crops like spring wheat, corn and rice in the oasis, depleting river flow downstream of Zhangye City (Fig. 9.1). The simulated results show that there was hardly any deep percolation to the groundwater. Instead, a portion of the water in the LSZ was simulated to flow to the river channels though interflow. This is due to the fact that LSZ is several hundred meters deep and could be mixed with groundwater zone. Studies by Cheng et al. (1999), Pan and Qian (2001), and Wu et al. (2010) report similar findings that groundwater recharge is only observable in the middle reach area with groundwater level less than 5 m deep and daily precipitation more than 10 mm, river flow infiltrates to the alluvial fan, and then flows out to the river channel from the aquifer.

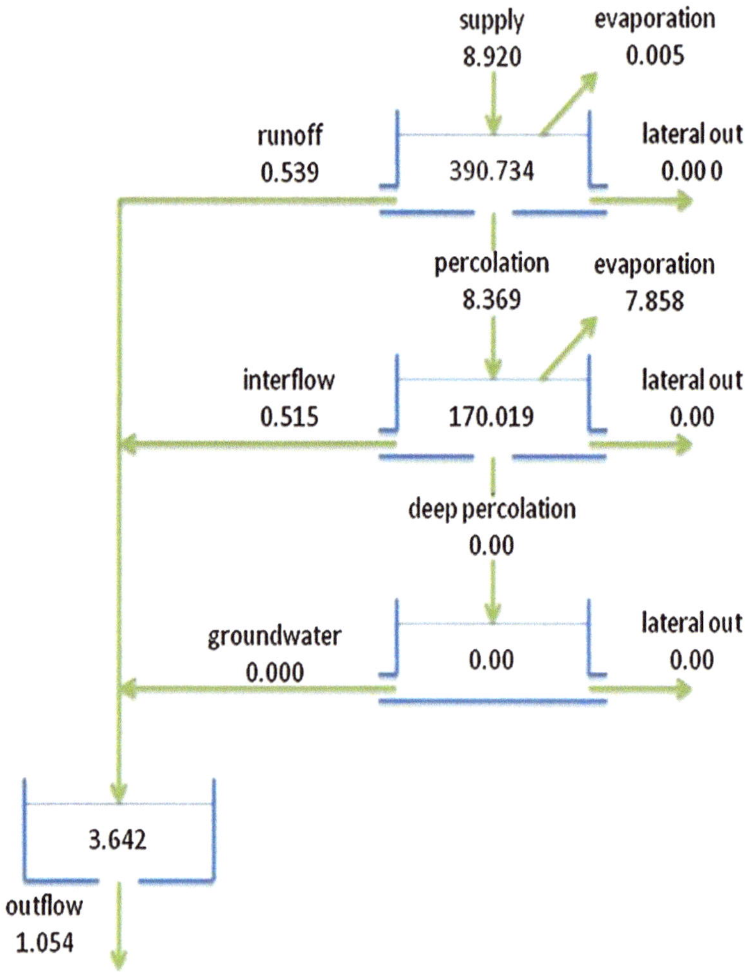

Fig. 9.3 Annual water budget (1990–2000 average in 10^9 m^3) for the upper-middle reaches of the Heihe River watershed

The simulation results show there was little ET from the upper soil zone. Instead, a very high ET rate occurred in the LSZ (Fig. 9.3). This phenomenon is attributable to several factors. First, the USZ is a hypothetical storage layer with a simulated capacity of a few cm to up to 100 cm. Second, between the mountain outlet (Yingluoxia) and middle reach outlet (Zhengyixia), soil is quite coarse and sandy, and thus water from the USZ infiltrates to the LSZ quickly (Cheng et al. 1999; Pan and Qian 2001; and Wu et al. 2010). Third, consumption of groundwater is mainly through ET in the middle reach of the watershed. Since the LSZ is several hundred meters deep and could be mixed with groundwater zone, loss of soil water and groundwater was simulated through the form of ET to the atmosphere in this study (Cheng et al. 1999; Jia et al. 2005; Pan and Qian 2001; and

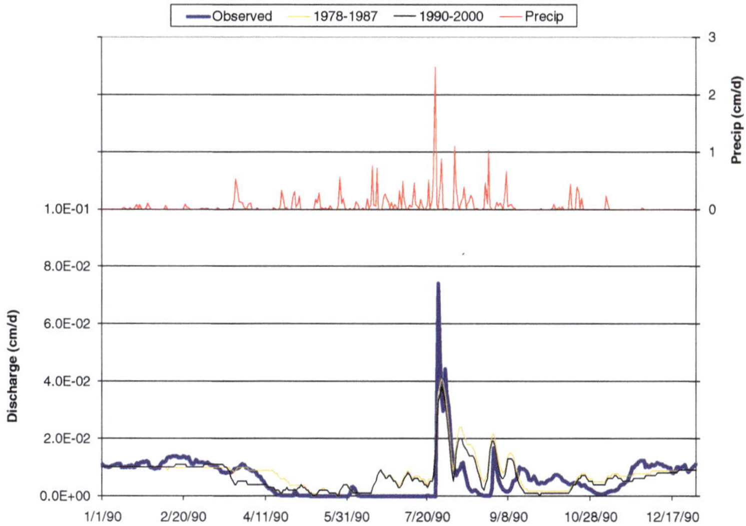

Fig. 9.4 Comparison of the simulated discharges and observed discharge in the Heihe River for 1990

Wu et al. 2010). Fourth, lack of detailed soil attribute and aquifer information led to the failure of the DLBRM to detect the movement of groundwater in the study.

The simulated average annual flow for 1990–2000 was about 1.05×10^9 m^3 at the middle reach outlet (at Zhengyixia Gage Station) under a normal precipitation year (P = 50 %). But the annual river flow was simulated to change from 0.80×10^9 m^3 in 1991 (a dry year, P = 75 %) to 1.27×10^9 m^3 in 1998 (a wet year, P = 20 %). It appears that under normal climatic years, the amount of flow passing the middle reach outlet is slightly over 1×10^9 m^3, barely satisfying the requirement of delivering 0.95×10^9 m^3 downstream annually by the State Council. This amount of flow, however, only has an exceedence probability of 50 %, meaning that the annual river flow is less than 1×10^9 m^3 at Zhengyixia Gage Station 50 % of the time. In addition, the simulations of 1990–2000 overestimated the actual flow by 40 % (Fig. 9.3). The simulated daily flows for 1999 underestimated the discharges during the cold season, and overestimated the discharges during the spring and summer (Fig. 9.4) (He et al. 2009). These discrepancies are related to both the model structure and data availability and accuracy. First, the DLBRM was developed to simulate hydrological processes of North America's Great Lakes watersheds, and does not have explicit modules available to simulate either permafrost processes or river channel infiltrations to aquifer, which leads to its underestimation of discharges during the winter season and overestimation of the discharges in the warm season, respectively. Second, lack of data on detailed soil attributes and aquifer settings made the DLBRM unable to discern the groundwater flows to the river. Third, it seems that large storages of water in the reservoirs for the increased irrigation withdrawals in the middle reach oasis reduced discharges at the Zhengyixia Gage Station. thus contributing to the high ratio of modelled to actual mean annual flow volumes of 1.4 achieved in the model calibration for the 1990–2000 model verification period (Pan and Qian 2001; and Wu et al. 2010).

Implementation of the State Council's water allocation plan requires taking about 0.58×10^9 m^3 water out of irrigation each year in the middle reach in order to deliver 0.95×10^9 m^3 of water at Zhenyixia Gage Station for rehabilitating the downstream ecosystem (Pan and Qian 2001; The City of Zhangye 2004). This goal seems achievable during normal climatic conditions. However, governmental entities (e.g. Cities of Zhangye and Jiuquan) in the middle reach, while coping with an annual economic loss of about $240 million, must take into account the uncertainties associated with the simulated mean annual flows while taking a number of actions such as adjusting crop patterns, water pricing, and market transfer to deliver more water downstream. Under dry years, a significantly lesser amount of flow would be available at the Zhengyixia Gage Station, making it much more difficult to deliver the targeted 0.95×10^9 m^3 of water downstream.

9.4 Conclusions

Simulations of the hydrology of the combined upper and middle reaches of the Heihe River Watershed in Northwest China show that Qilian Mountain in the upper reach area is the main runoff production area for the entire watershed. On average, surface runoff and interflow contributed 51 and 49 % of the river flow respectively for the period of 1990–2000. Annually the river was simulated to discharge slightly more than 1×10^9 m^3 of water from the middle reach (at Zhengyixia Gage Station) downstream under normal climatic conditions. While requiring a significant reduction in water withdrawals by water users in the middle reach, this amount seems to meet the mandate of delivering 0.95×10^9 m^3 of water at Zhengyixia Gage Station for rehabilitation of downstream ecosystems by the State Council. However, the amount of the flow at the middle reach outlet has an exceedence probability of 50 % under normal climatic conditions, and is much less under dry climatic conditions, making it much harder to deliver the required 0.95×10^9 m^3 of water downstream. In addition, climate change and rapid urban expansion will further intensify the water shortage problem in the Heihe River Watershed. Thus, how to develop a comprehensive water management plan to address the competing demands for water among agricultural irrigation, industrial development, urban supplies, and ecosystem protection remains a long term challenge between water users in the upper, middle, and lower reaches of the Heihe River Watershed. As limitations in the current structure of the DLBM and input data led to significant uncertainties in the simulated results, our future work includes further model refinement and field work to support water resource decision making in the study watershed and other similar areas of Northwest China.

Acknowledgments Partial support for this research is provided by the National Natural Science Foundation of China (Grant No: 91125010), the National Basic Research Program of Global Change (Grant No. 2010CB951002), and the CAS/SAFEA International Partnership Program for Creative Research Teams of"Ecosystem Processes and Services. The Chinese Academy of Sciences (CAS) (Grant No: GJHZ0948 and CXTD-Z2005-2), and Scherer Endowment Fund of Western Michigan University Department of Geography. The data set used in the study is provided by "Environmental & Ecological Science Data Center for West China, National Natural Science Foundation of China" (http://westdc.westgis.ac.cn).

References

Chen L, Xiao H. Heixi mountain soils and their applications. Beijing: Ocean Press; 2003.

Cheng GD, Chang BY, Qu YG. Assessment and Rational Use of Water Resources for Coordinated Economic Development and Ecologic and Environmental Protection in the Heihe Watershed. The Chinese Academy of Sciences, Lanzhou: Lanzhou Institute of Glaciology and Geocryology; 1999.

Croley TE II, He C. Distributed-parameter large basin runoff model I: model development. J Hydrologic Eng. 2005;10(3):173–181.

Croley TE II, He C, Lee DH. Distributed-parameter large basin runoff model II: application. J Hydrologic Eng. 2005;10(3):182–191.

Croley TE II, He C. Watershed surface and subsurface spatial intraflows. J Hydrologic Eng. 2006;11(1):12–20.

DeMarchi C, Xing F, Croley TE II, He C, Wang Y. Application of a distributed large basin runoff model to lake erie: model calibration and analysis of parameter spatial variation. J Hydrologic Eng. 2011;16(3):193–202.

Feng Q, Cheng GD, Kunihiko E. Towards sustainable development of the environmentally degraded river Heihe basin. China Hydrological Sci J. 2002;46(5):647–58.

He C, Shi C, Yang C, Agosti BP. A Windows-based GIS-AGNPS interface. J Am Water Resour Assoc. 2001;37(2):395–406.

He C. Integration of GIS and simulation model for watershed management. Environ Model Softw. 2003;18(8–9):809–13.

He C, Cheng S, Luo Y. Desiccation of the yellow river and the South water Northward diversion project. Water Int. 2005;30(2):261–8.

He C, Croley TE II. Application of a distributed large basin runoff model in the great lakes basin. Control Eng Pract. 2007a;15(8):1001–11.

He C, Croley TE II. Integration of GIS and visualization for distributed large basin runoff modeling of the great lakes watersheds. In: Scarpati and Jones, editors. Environmental change and rational water use. Orientación Gráfica Editora S.R.L. Argentina: Buenos Aires; 2007b. p. 247–60.

He C, DeMarchi C, Croley TE II, Hunter T, Feng Q. Modeling the hydrology of the Heihe Watershed in Northwestern China. J Glaciol Geocryol. 2009;31(3):420–1.

He C, Croley TE II. Hydrological resource sheds and the U.S. great lakes applications. J Res Ecol. 2010;1(1):25–30.

Jia Y, Wang H, Nin G, Yang D, Wang J, Qin Q. Principles and applications of distributed watershed hydrological models. Beijing, China: Water Resources Pressing; 2005.

Pan QM, Qian SL. Water resources in the Heihe Watershed. The Yellow River Press: Zhengzhou;2001.

Reynolds JF, Smith DMS, Lambin EF, et al. Global desertification: building a science for dryland development. Science. 2007;316(2007):847–51.

Wu Y, Zhang Y, Wen X, Su J. Hydrologic cycle and water resource modeling for the Heihe River basin in Northwestern China. Science Press:Beijing (in Chinese);2010.
UN World Water Development Report. Water for People. Water for Life: UNESCO Publishing and Berghahn Books; 2003.
Xiao H. Analysis of types and mechanical composition of soil samples (unpublished data). Lanzhou: The Chinese Academy of Sciences Cold and Arid Regions Environmental and Engineering Research Institute; 2006.

Chapter 10
Evaluating the Transport and Fate of Nutrients in Large Scale River Basins Using an Integrated Modeling System

Zhonglong Zhang and May Wu

Abstract Watershed management is essential in minimizing riverine and receiving water pollution. No watershed model is currently available that adequately model the large river basin system over long periods of time in a satisfied level of detail. Watershed models must be linked to riverine or other detailed receiving water models in order to adequately represent the intricacies of the physical system under study. Watershed models are used to provide "boundary conditions," both hydrologic/hydraulic and nonpoint source loading fluxes, to the receiving water models. The objectives of this study were to develop an integrated watershed and riverine modeling system using Soil and Water Assessment Tool (SWAT) model and Hydrologic Engineering Center-River Analysis System (HEC-RAS) model to evaluate the transport and fate of nutrients and water quality impacts in large scale river basins. The SWAT model was constructed for the Upper Mississippi River Basin (UMRB) in an attempt to account for key elements associated with crop production and land use changes. The model was calibrated and validated by using 18 years of observed United States Geological Survey (USGS) streamflow discharge and water quality data. The SWAT model was used to predict flow and nutrient exports from each tributary within the watershed. The results were used as HEC-RAS model's inputs through an interface. The HEC-RAS model was able to simulate the transport and fate of nutrients and dynamic changes in riverine nutrient concentrations. The integrated SWAT and HEC-RAS modeling system provides a systematic approach to modeling nonpoint nutrient sources, transport, and fate in a large scale river basin. The modeling system can be used to predict downstream water quality impacts with land use changes and

---

Z. Zhang (✉)
Badger Technical Services, U.S. Army Engineer Research and Development Center, 3909 Halls Ferry Road, Vicksburg, MS 39180, USA
e-mail: Zhonglong.Zhang@usace.army.mil

M. Wu
Center for Transportation Research, Energy System Division, Argonne National Laboratory, 9700 S. Cass Avenue, Lemont, IL 60439, USA

B. Fu and K. B. Jones (eds.), *Landscape Ecology for Sustainable Environment and Culture*, DOI: 10.1007/978-94-007-6530-6_10,

assess the effectiveness of different watershed management practices whether directed toward nutrient source supply or abatement.

Keywords SWAT · HEC-RAS · Watershed model · Riverine model · Hydrology · Land use change · Sediment · Nutrient transport

10.1 Introduction

Nutrient pollution is a leading cause of water quality impairment in lakes and estuaries and is also a significant issue in rivers (USEPA 2007). The Mississippi River and tributary streams have been greatly impacted by excess nitrogen, phosphorus, and sediment loadings from cropland and other sources. The importance of linkages between watershed processes, and the environmental and ecological responses of aquatic resources are best exemplified by the formation of an oxygen deprived area in Gulf of Mexico. The hypoxic zone in the Gulf varies in size from year to year, in 2007, it was the third-largest hypoxic zone on record (Devine et al. 2008). The hypoxic zone has become a serious threat to commercial fishing, shrimping, and recreation industries. A number of factors contribute to the size of the dead zone, but nutrient pollution substantially drives the problem. The formation of hypoxic zone in the Gulf is a problem caused in large part by excessive inputs of sediments and nutrients arising from industrial and agricultural activities in the Mississippi River Basin (MRB) (Diaz and Rosenberg 2008). Soil erosion that moves sediments and sediment-bound nutrients and pesticides into waterways is another factor influencing water quality in the Gulf. In particular, nitrogen is more abundant in dissolved forms, whereas phosphorus is largely present in particulate forms (either adsorbed or as a constituent of inorganic and organic particles). As a result, there is a strong correlation between suspended sediment and total phosphorus concentrations, and changes to the river system that alter the flow of water or sediment in the system are likely to cause a larger change in the concentration and transport of phosphorus than of nitrogen (Wetzel 2001). The evidence now suggests that both nitrogen and phosphorus affect the size of the dead zone (Devine et al. 2008).

Nutrients enter the Mississippi River and the Gulf from a variety of sources, including fertilizer runoff from farms, golf courses, and lawns; manure disposal; discharge from sewage treatment plants and industrial facilities; nitrogen deposition from the atmosphere; and erosion of nutrient-rich soil. The data provided by USGS shows that areas with significant agricultural uses are the largest contributors of nutrient pollution to the Gulf. The United States Geological Survey (USGS) found that on average, from 2001 to 2005, the Upper Mississippi and Ohio-Tennessee River basins represent about 31 % of the total land area within the MRB, they contribute about 82 % of the nitrate-nitrogen flux, 69 % of the total Kjedahl Nitrogen (TKN), and 58 % of the total phosphorus flux.

The amount of sediments and nutrients transported to the Gulf is not simply a direct function of what is coming off the field but must also include what's lost along the way as water moves through the drainage network (Alexander et al. 2000; Mueller and Spahr 2006). Studies conducted only at fields or small watersheds are not able to capture how sedimentation and nitrogen and phosphorus concentration at multiple scales are influenced by various agricultural practices. Effective management of water quality and ecosystem, whether directed toward nutrient supply or abatement, requires both watershed and riverine models to quantify the transport and fate of nutrients throughout the basin system.

The Soil and Water Assessment Tool (SWAT) model has been widely used to quantify the flow, sediment and nutrient loadings in large river basins (Gassman et al. 2007). The SWAT model can be used to identify the location and magnitude of sediment and nutrient runoff hotspots associated with crop production and land use changes. The flow-routing methods presently used in SWAT are hydrological routing methods that are based on the continuity equation and on empirical relationships to replace the momentum equation. In SWAT, one-dimensional (1D) hydrologic model is used to mathematically represent flow routing along a river reach. In this case, simplified schemes, such as linear reservoirs, Muskingum-Cunge methods may be applied. When dealing with large scale rivers, however, backwater effect and floodplain inundation may become governing factors for flood wave routing, and a 1D-hydraulic model is a more suitable method. The weakness of SWAT models for dynamic flow routing in the large scale watershed areas is well known. It is very important to perform the flow routing process accurately because routed results affect other aspects such as sediment routing and the in-stream nutrient process, both of which are strongly tied to water routing. To address this problem, a 1D unsteady state flow model (UNET) developed for the main stem of the Illinois River was coupled with the HSPF model to perform the flow routing (Lian et al. 2007). SWAT was also coupled with 2D hydrodynamic and water quality model (CE-QUAL-W2) in the Cedar Creek Reservoir study (Debele et al. 2008). The results indicated that the two models are compatible and can be used to assess and manage water resources in complex watersheds comprised of watershed and receiving waterbodies.

In this study, an integrated watershed and riverine modeling system was developed. The water quality impacts of crop production and land use changes in the Upper Mississippi River Basin (UMRB) were evaluated. The UMRB is the main focus for future renewable bioenergy sources and provides a majority of conventional and potential cellulosic feedstock for biofuel production (USDOE 2011). Results of the basin's SWAT model were used as the mainstem Hydrologic Engineering Center-River Analysis System (HEC-RAS) model boundary conditions and inputs for evaluating the transport and fate of sediments and nutrients in the Upper Mississippi River. An integrated modeling system can be used to predict: (1) Spatial and temporal patterns of nutrient export in the watershed; (2) Nutrient cycling and microbial and chemical reactions within the river; (3) Long-term changes in riverine nutrient concentrations and their causes; (4) Effects of watershed nutrient loading on the downstream water quality; and (5) Effectiveness of watershed management scenarios implemented to address problems in the downstream.

10.2 Upper Mississippi River Basin

The Upper Mississippi River Basin (UMRB) is located in the Midwest United States and is one of the seven major basins contributing to the Mississippi River (Fig. 10.1). The Mississippi River is the seventh largest in the world, based on discharge (580 km^3 $year^{-1}$); whereas the MRB is the third largest in the world based on area (3,225,000 km^2). The Mississippi river flows some 3,770 km downstream from Lake Itasca, Minnesota, through ten states and eventually discharges to the Gulf of Mexico. The UMRB includes more than 2,011 km of the Mississippi River extends from the Mississippi River head water at Lake Itasca in Minnesota to the confluence of Mississippi and Ohio Rivers at Cairo in Illinois. The Mississippi River is divided into the upper and lower basin at Cairo, Illinois where the Missouri River enters as the last major tributary. UMRB has the greatest amount of artificially drained soil, the highest percentage of total land in agriculture (corn and soybean) and the highest use of nitrogen fertilizers in the nation. The Upper Mississippi River (UMR) has been defined by the U.S. Congress as active commercial navigation routes within the UMRB. The UMRB has a total drainage area of approximately 490,000 km^2, about 15 % of the entire MRB, including large parts of the states of Illinois, Iowa, Minnesota, Missouri, and Wisconsin. Small portions of Indiana and South Dakota are also within the basin.

Fig. 10.1 Geographic locations of Mississippi River sub-basins

Soil type in the UMRB ranges from heavy, poorly drained clay soil to light, well-drained sands. Mollisols dominate but there are also Alfisols that have high clay subhorizons and lower amounts of organic matter and clay in the surface horizons. These soils are highly productive but have poor internal drainage. Subsurface tile drainage of soils in this region is necessary for crop production. Corn and soybean require a well-drained warm soil for optimum growth. In most parts of the basin, agriculture is the dominant land use (Fig. 10.2). This figure indicates that over 60 % of total land is used for agriculture and pasture. Corn, soybeans, and alfalfa are the major crops in the basin. The UMRB encompasses most of the Corn Belt region of the U.S. Nitrogen and phosphorus are abundant in the UMRB because of the widespread use of commercial and animal-manure fertilizers. In fact, the quantity

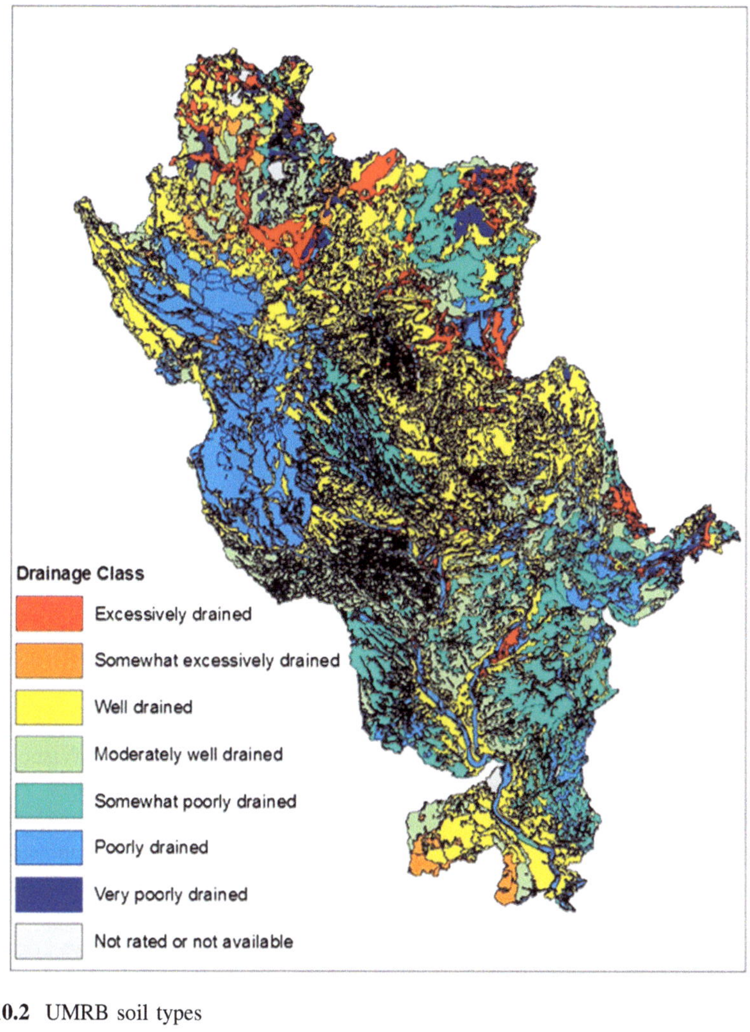

Fig. 10.2 UMRB soil types

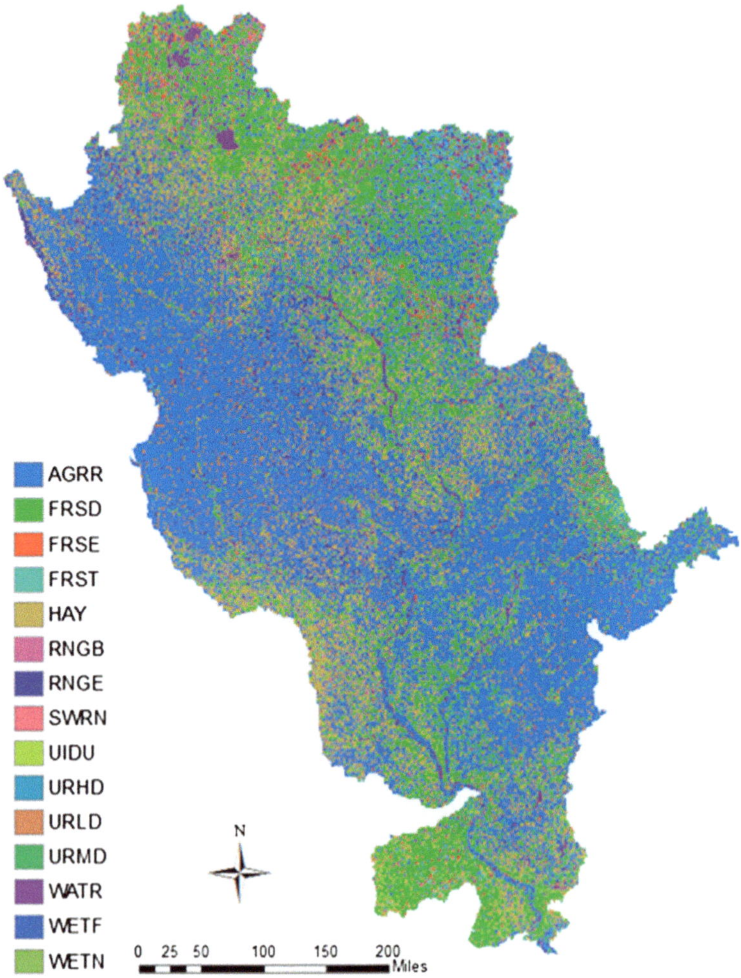

Fig. 10.3 UMRB land use and land cover

of nitrogen and phosphorous lost from the land to the stream in the UMRB is higher than in any other portion of the MRB (Fig. 10.3).

The climate of the UMRB is subhumid continental. The climate of the UMRB, which encompasses most of Illinois, Iowa, Minnesota, Missouri, and Wisconsin, is humid continental, with warm, moist summers and cold, dry winters. Average monthly temperatures vary significantly throughout the year, with maximums in July and minimums in January. The majority of the precipitation (approximately three-fourths) occurs between April and October. Average annual rainfall for the area is 91 cm. Monthly average temperatures range from −12 to 0 °C for January to 22–27 °C in July, with a year-round average of about 11 °C. Flows comprised of runoff from the UMRB's stream network support navigation and hydroelectric plants and fulfill municipal, industrial, and agricultural water requirements.

10.3 Integrated Watershed and Riverine Modeling System

10.3.1 SWAT Model

The SWAT is a lumped parameter watershed model developed and maintained by the U.S. Department of Agriculture (USDA) Agriculture Research Service (ARS). SWAT uses algorithms from a number of previous ARS models including CREAMS, GLEAMS, EPIC, and SWRRB (Arnold et al. 1998). The SWAT model was developed to assess the impact of land management and climate patterns on water, sediment, and agricultural chemical yields over long time periods in large watersheds. The watershed is partitioned into a number of subbasins. Each subbasin possesses a geographic position in the watershed and is spatially related to adjacent subbasins. Each subbasin is further divided into hydrological response units (HRU) based on topography, land use, and soil. HRUs are the smallest computational units in SWAT with unique land use, soil type and slope within a subbasin. Thus, SWAT can take two levels of the spatial heterogeneity into account. The first level (subbasin) supports the spatial heterogeneity associated with hydrology, and the second level (HRU) incorporates the spatial heterogeneity associated with land use, soil type and slope class. Within a subbasin, SWAT does not retain the spatial location of each HRU. The loss of spatial information within the subbasin introduces a measure of unrealism and requires caution in interpreting model results. In SWAT, hydrologic, soil, water quality and other processes are modeled within the subbasins through the use of HRUs. Flow generation, sediment yield, and pollutant loadings are summed across all HRUs in a subbasin, and the resulting flow and loads are then routed through channels, ponds, and/or reservoirs to the watershed outlet. SWAT typically produces daily results for every subbasin outlet, each of which can be summed to provide monthly and annual load estimates.

Major model components include climate, hydrology, nutrient cycle, pesticide, plant growth, and land management. For climate, SWAT uses the data from the station nearest to the centroid of each subbasin. The hydrology module simulates major hydrologic components and their interactions as simple responses using empirical relationships (Fig. 10.4). Precipitation in the form of either rainfall or snowfall is the major driving mechanism of the hydrologic cycle. SWAT calculates actual ET based on potential evapotranspiration (PET) from soils and plants separately. PET can be estimated by three methods: Priestley-Taylor (Priestley and Taylor 1972), Hargreaves (Hargreaves and Samani 1985), and Penman-Monteith (Allen et al. 1989). Surface runoff volume and infiltration are computed with the modified curve number method or Green and Ampt equation. The peak rate component uses Manning's formula to determine the watershed time of concentration and considers both overland and channel flow. The soil profile is subdivided into multiple layers that support soil water processes including infiltration, evaporation, plant uptake, lateral flow, and percolation to lower layer. A storage routing technique is used to calculate redistribution of water between layers in the soil profile. Lateral subsurface flow in the soil profile is calculated simultaneously with

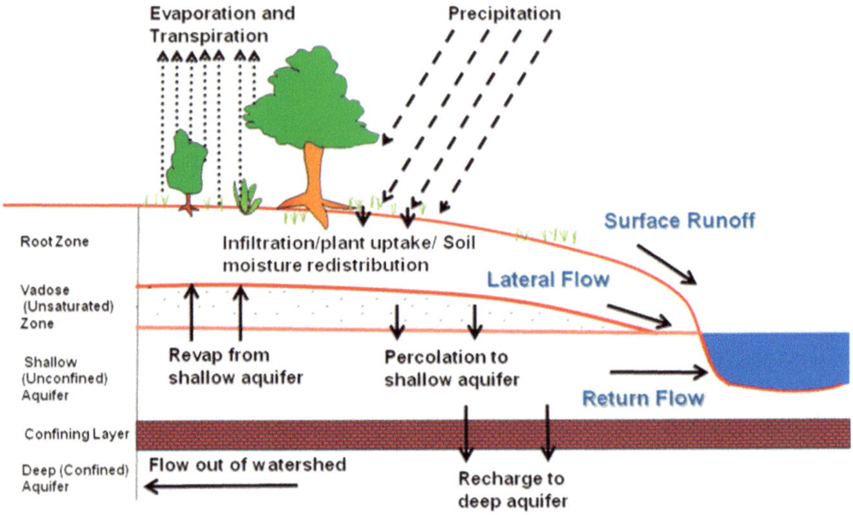

Fig. 10.4 SWAT hydrologic processes and interactions

percolation. Groundwater flow contribution to total streamflow is simulated by routing a shallow aquifer storage component to the stream (Arnold and Allen 1996). Channel routing is simulated using either the variable-storage method or the Muskingum method; both methods being variations of the kinematic wave model.

Erosion and sediment yield are estimated for each HRU with the Modified Universal Soil Loss Equation (Williams and Berndt 1977). The channel sediment routing uses a modification of Bagnold's sediment transport equation (Bagnold 1977) that estimates the transport concentration capacity as a function of velocity. The model either deposits excess sediment or re-entrains sediment through channel erosion depending on the sediment load entering the channel. The delivery ratio is estimated for each particle size as a linear function of fall velocity, travel time, and flow depth.

SWAT simulates the complete soil nutrient cycle for nitrogen and phosphorus. The soil nitrogen cycle is simulated using five different pools; two are inorganic forms (ammonium and nitrate) while the other three are organic forms (fresh, stable, and active). Similarly, SWAT simulates six different pools of phosphorus in soil; three are inorganic forms and the rest are organic forms. Primary biochemical transformations of nitrogen and phosphorus are simulated. Nitrate export with runoff, lateral flow, and percolation are estimated as products of the volume of water and the average concentration of nitrate in the soil layer. Organic nitrogen and organic phosphorus transport with sediment is calculated with a loading function developed by McElroy et al. (1976) and modified by Williams and Hann (1978) for application to individual runoff events. The loading function estimates daily organic nitrogen and phosphorus runoff loss based on the concentrations of constituents in the top soil layer, the sediment yield, and an enrichment ratio. The amount of soluble phosphorus removed in runoff is predicted using labile phosphorus concentration in the top soil layer, the runoff volume and a phosphorus soil partitioning coefficient. In-stream nutrient dynamics in SWAT are simulated using the kinetic routines from the QUAL2E in-stream water quality model (Brown and Barnwell 1987).

SWAT allows detailed agricultural management practices to be simulated, tracking planting, tillage, and fertilization operations and calculating resultant plant growth with specific dates or with a heat unit scheduling approach during the year. The plant growth component of SWAT utilizes routines for phenological plant development based on plant-specific input parameters such as energy and biomass conversion, temperature, water and nutrient constraints, canopy height and root depth, and shape of the growth curve. A single plant growth module is used in SWAT for simulating all crops and assessing removal of water and nutrients from the root zone, transpiration, and biomass/yield production.

10.3.2 HEC-RAS Model

Few models provide the ability to couple river flow quantity with sediment and water quality, and those that do are proprietary products that are both expensive as well as difficult, if not impossible, to modify to suit local conditions. HEC-RAS is a public domain model developed by the U.S. Army Corp of Engineers (USACE) (http://www.hec.usace.army.mil) and is widely used and accepted by the engineering community and many regulatory agencies. The HEC-RAS model contains 1D river analysis components for: (1) hydraulic simulation; (2) movable boundary sediment transport simulation; and (3) water quality analysis. A key element is that all three components use a common geometric data representation and common geometric and hydraulic computation routines. In addition to the three river analysis components, the model contains several hydraulic design features that can be invoked once the basic water surface profiles are computed, data storage and management capabilities, graphics and reporting facilities.

The hydraulic simulation is the key computation engine. It performs 1D steady and unsteady flow calculations on a network of natural or manmade open channels. Hydraulic calculations are performed at each cross section to compute water surface elevation, critical depth, energy grade elevation, and velocities. HEC-RAS is able to perform mixed flow regime (subcritical, supercritical, hydraulic jumps, and draw downs) calculations in the unsteady flow computations module. The hydraulic calculations for cross-sections, bridges, culverts, and other hydraulic structures that were developed for the steady flow component were incorporated into the unsteady flow module. The model can handle a full network of channels, a dendritic system, or a single river reach.

Sediment simulation in HEC-RAS utilizes one dimensional, cross-section averaged, hydraulic properties from RAS's hydraulic engines to compute sediment transport rates and update the channel geometry based on sediment continuity calculations. The sediment module runs in the quasi-unsteady mode, i.e. it computes the unsteady hydraulics as a series of steady state events. The sediment transport in HEC-RAS is based on shear stresses computed from these steady state events. However, the interaction of the bed profile with the entrainment and transport equations is quasi-dynamic in that the bed is adjusted for erosion or deposition by the Exner equation. Sediment transport module simulates suspended load (fine sediments moving at the same speed as water) and bedload (sand and gravel moving at a slower rate along the bed). Seven different transport functions are currently available in RAS including Ackers and White, Englund-Hansen, Laursen, Myer-Peter-Muller, Toffaleti, Yang, and Wilcock. Currently HECRAS employs Exner 5, a "three layer" algorithm to compute bed sorting mechanisms. Exner 5 divides the active layer into two sublayers, simulating bed coarsening by removing fines initially from a thin cover layer. During each time step, the composition of this cover layer is evaluated and if, according to a rough empirical relationship, the bed is partially or fully armored, the amount of material available to satisfy excess capacity can be limited.

HEC-RAS nutrient water quality module includes a set of nutrient simulation modules (NSM) (Zhang and Johnson 2012). NSM I computes riverine algal biomass, organic and inorganic nitrogen and phosphorus species, CBOD and DO. NSM II computes multiple algal biomass, nitrogen, phosphorus, and carbon cycling, DO, COD, alkalinity, pH and pathogen, as well as numerous additional

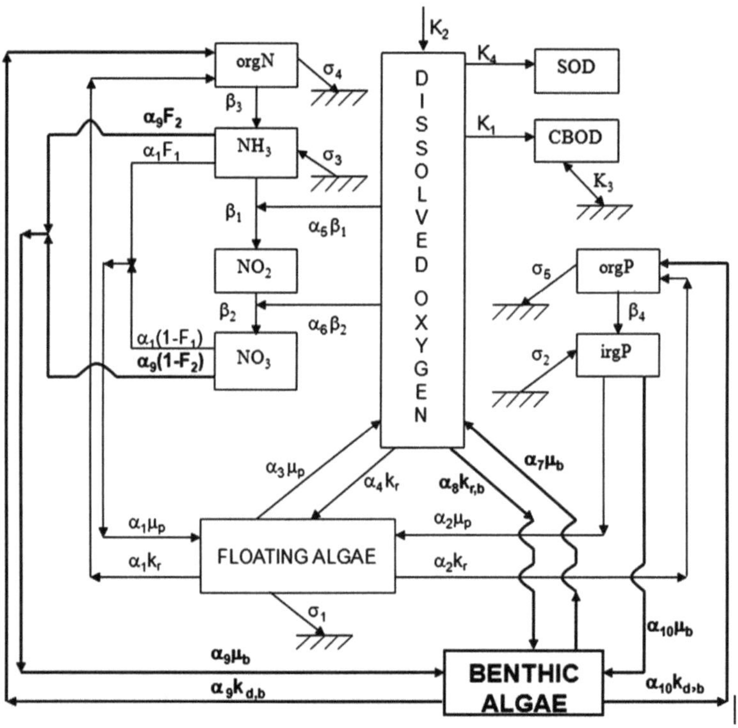

Fig. 10.5 Schematic representation of HEC-RAS-NSM I

composite constituents. In addition, NSM III incorporates a dynamic bed sediment diagenesis component, which simulates the chemical and biological processes undergone at the sediment–water interface after sediments are deposited. The schematic representation of HEC-RAS NSM I water quality processes is shown Fig. 10.5. The conservation-of-mass equation in HEC-RAS is solved using QUICKEST-ULTIMATE transport algorithm.

10.3.3 Integration of SWAT and HEC-RAS

Although SWAT includes riverine water hydraulic and water quality capabilities, the demands of tracking the dynamic transport and fate of nutrients and/or the specific characteristics of the receiving water system required additional capabilities not available within SWAT in order to address real-world water management concerns. Figure 10.6 shows a framework of integrated SWAT and HEC-RAS modeling system. Evaluation of the transport and fate of nutrients in the watershed was broken into the following steps:

1. Simulation of the 8 digit HUCs was conducted using a SWAT model
2. Extraction of time series SWAT outputs of runoff and water quality from each HUC watershed along the mainstem and used as boundary conditions to HEC-RAS.
3. Simulation of the Mainstem River Using a HEC-RAS Model
4. Assess the potential impacts of nutrient loading and land use changes in the basin.

As shown in Fig. 10.6, SWAT is linked to the HEC-RAS model providing tributary discharge and contaminant loads to the HEC-RAS model. Once the SWAT model was satisfactorily developed and calibrated the SWAT model results are imported into the Hydraulic Engineering Center's Data Storage System (HEC-DSS) through an interface (Fig. 10.7). For the HEC-RAS model, SWAT provides discharge and contaminant loads from the major streams and drainages tributary to

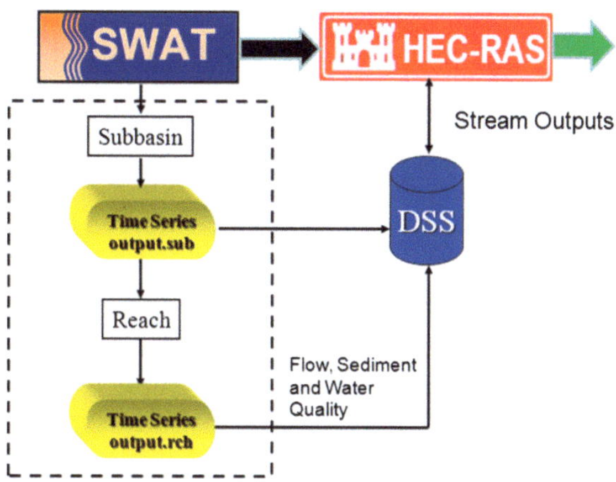

Fig. 10.6 Integated SWAT and HEC-RAS modeling system

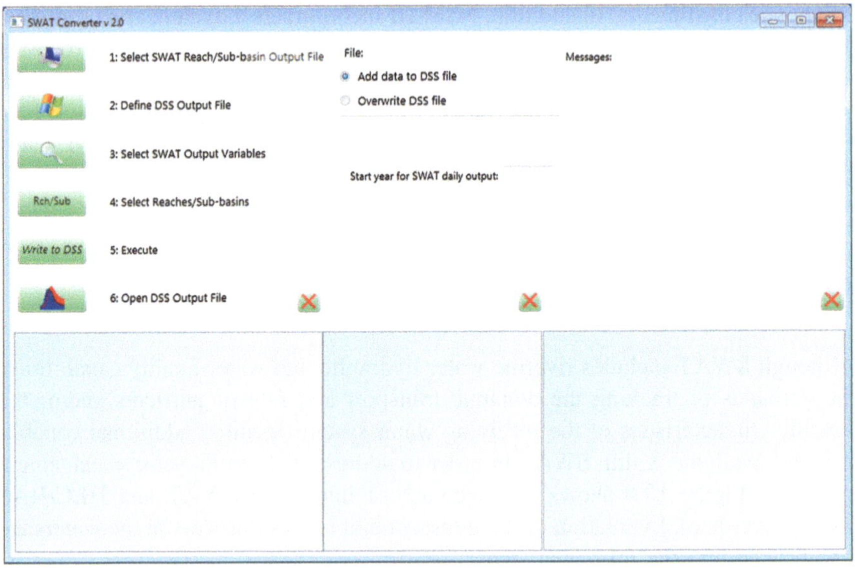

Fig. 10.7 HEC-DSS importer interface from SWAT outputs

the HEC-RAS model's domain as its boundary conditions. The HEC-RAS model is then calibrated based on measured in-stream flow, sediment and nutrient concentrations. The goal is not only to characterize each of those features but sequentially to link them together as part of a systematic evaluation of the large river basin based on a defined set of conditions.

Table 10.1 describes the state variables that SWAT passes to the HEC-RAS model. The state variables required for HEC-RAS are passed from SWAT using HEC-DSS. As shown in Table 10.1, SWAT provides daily time-series for flow and contaminant loads for HEC-RAS.

Table 10.1 SWAT state variables passed to HEC-RAS

SWAT source	HEC-RAS input	Unit
Flow	Flow	cms
Suspended sediment	Suspended sediment	ton/day
NH_4-N	NH_4-N	kg/day
NO_3-N	NO_3-N	kg/day
NO_2-N	NO_2-N	kg/day
Organic N	Organic N	kg/day
PO_4-P	PO_4-P	kg/day
Organic P	Organic P	kg/day
Algae	Algae	kg/day
DO	DO	kg/day
CBOD	CBOD	kg/day

10.4 Application of the Integrated Modeling System to the Upper Mississippi River Basin

The SWAT model inputs consist of topography, soil properties, land use/cover type, weather/climate data, and land management practices. The study watershed is divided into subbasins. The USGS divided the UMRB into 131 eight-digit-level subbasins with an average drainage area of 3,755 km^2 (Fig. 10.8). Each subbasin is further divided into hydrological response units (HRU) based on topography, land use, and soil.

We have calibrated and validated the SWAT outputs against observed data. Those results showed that the SWAT model properly reproduced observed flow, sediment, total nitrogen and phosphorus from the basin. We now have used those calibrated outputs from the SWAT model as input into the HEC-RAS model and calibrate the hydraulic and water quality processes in the UMR.

The dominant cause of nutrient flux to the UMR is agricultural activities in the UMRB. Corn and soybean fields are the largest source of N input (52 %) followed by atmospheric deposition (16 %). Among agricultural land uses, the largest P input was from pasture and range lands (animal manure; 37 %) followed by corn and soybeans (25 %). In contrast to N, urban sources (12 %) made a significant contribution of P to the UMR (Alexander et al. 2000). The SWAT model reveals a consistent trend of heavy nitrogen and phosphorus exports from areas of intensive agricultural production in the UMRB.

The SWAT model, in addition to runoff, outputs sediments and water quality variables, such as total suspended solids (TSS), nitrogen species (organic nitrogen, nitrate/nitrite, and ammonia/ammonium), phosphorus species (organic and mineral phosphorus), chlorophyll-a, dissolved oxygen (DO), and carbonaceous biochemical oxygen demand (CBOD). The HEC-RAS/NSM I model has the same water quality state variables as the SWAT model and accepts inputs in terms of one-to-one relationship. We used a separate interface tool that imports SWAT outputs into an appropriate time-series dataset within the HEC-DSS for HEC-RAS. The interface extracts daily SWAT outputs (runoff and its water quality constituents) at required reach or subbasin outlets into HEC-DSS file that is acceptable to HEC-RAS.

Principal tributaries of the UMR are the Minnesota, St. Croix, Wisconsin, Rock, Iowa, Des Moines, Illinois, and Missouri Rivers and several smaller rivers and streams. In the 1,076 km of river between the first lock, Upper St. Anthony Falls, and the last lock of the Channel Navigation Project, Lock 27, the UMR falls 128 m with an average slope of approximately 9.5 cm/km. Average flow of the UMR ranges from 280 $m^3\ s^{-1}$ at St. Paul, Minnesota, to 4,955 $m^3\ s^{-1}$ at St. Louis, Missouri. Data input for HEC-RAS included geometric data to represent river networks, channel cross-section data, and hydraulic structure data such as bridges and culvert data. Cross-section data includes station-elevation data, main channel bank stations, downstream reach lengths, roughness coefficients, and contraction and expansion coefficients. Hydrologic events are represented by flow data. Time-series predicted by SWAT for all of the required constituents and drainage sources

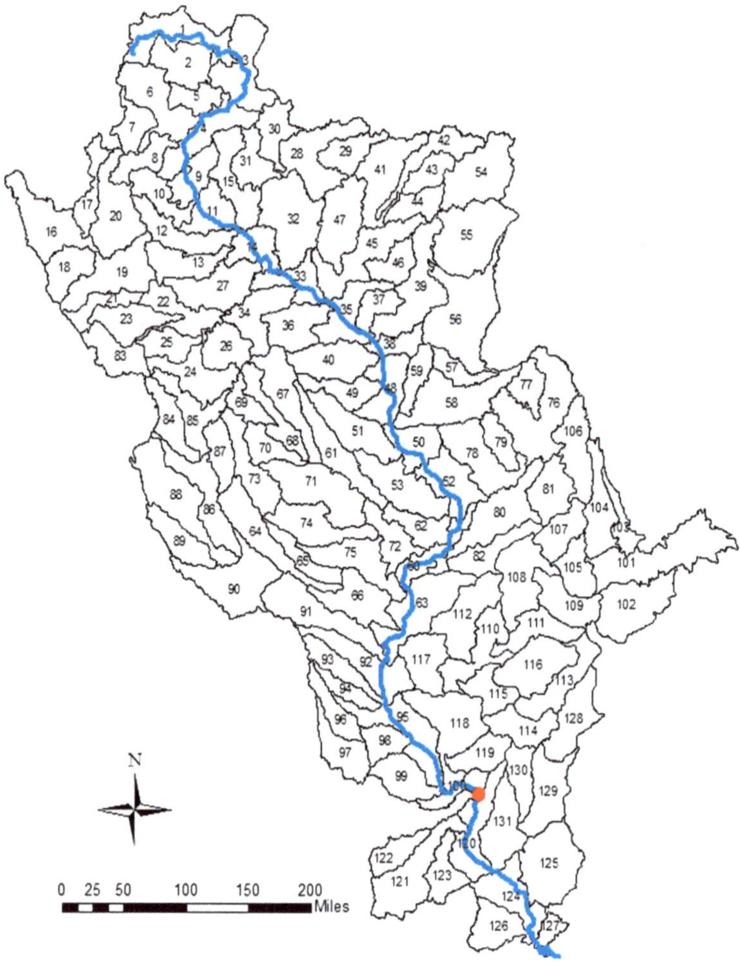

Fig. 10.8 SWAT UMRB subbasin distributions

to the mainstem river will provide the necessary boundary conditions to exercise HEC-RAS.

Modeled flow discharge, suspended sediment, total nitrogen and total phosphorus fluxes for a number of UMRB long-term sampling sites between 1990 through 2009 were compared with observed data. Figures 10.9, 10.10, 10.11 and 10.12 show the monthly modeled and observed comparison for the simulation period. Discharge and water quality data from the UMR were acquired from the US Geological Survey (USGS).

The difference between the measured and modeled monthly average stream flow for all four gages was less than 10 % for the simulation period. The hydrograph plot revealed a strong correlation between the measured and modeled flows as indicated by the R^2 and NSE values, which were 0.76 and 0.72, respectively at

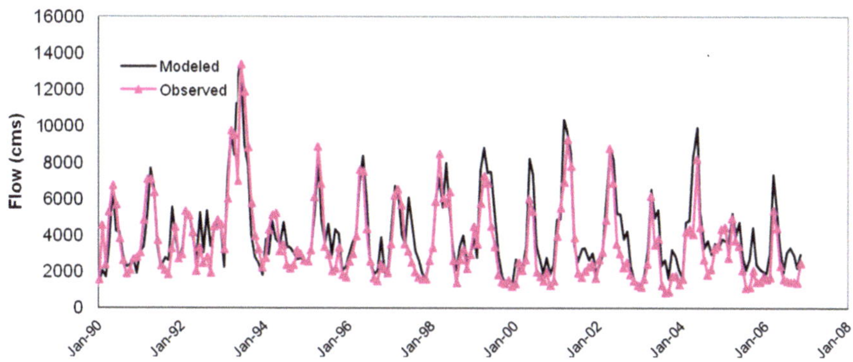

Fig. 10.9 Observed versus modeled monthly streamflow for the simulation period

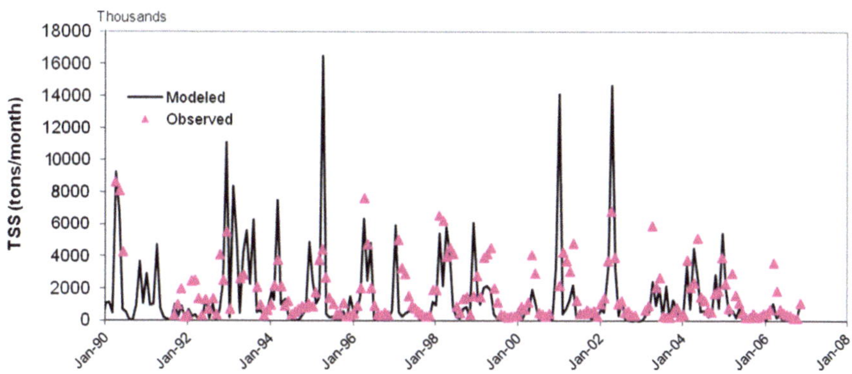

Fig. 10.10 Observed versus modeled monthly TSS for the simulation period

the watershed outlet. Compared with the streamflow results, the simulations on suspended sediment loads were less accurate. The modeled monthly TN and TP loads generally followed the month-to-month observed trends. The seasonal variation of the monthly streamflow TN and TP loads was fairly well reproduced in the calibration period. These results show a reasonable match between the modeled results and observed data, indicating the model's ability to adequately simulate both the watershed and in-stream processes involved within the UMRB.

Watershed characteristics, including tributary streams, point and non-point pollution sources, all influence mainstem river water quality. Variations in land use practices, cover types, and watershed area will determine the level and type of sediment, nutrient, and contaminant inputs into the UMR from their tributaries. Typical seasonal fluctuations in flow, as well as periodic extreme events, can have dramatic effects on river water quality. This modeling system is able to quantify the effects of agricultural production systems on nitrogen and phosphorus runoff to streams and the nutrient cycling in the waters delivered to the watershed oulet.

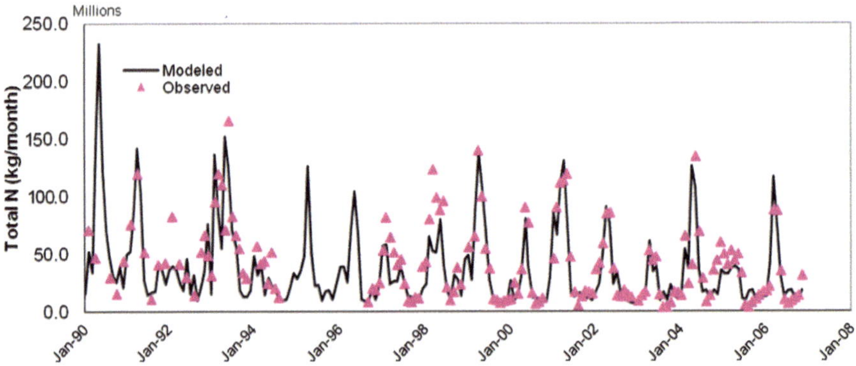

Fig. 10.11 Observed versus modeled monthly TN for the simulation period

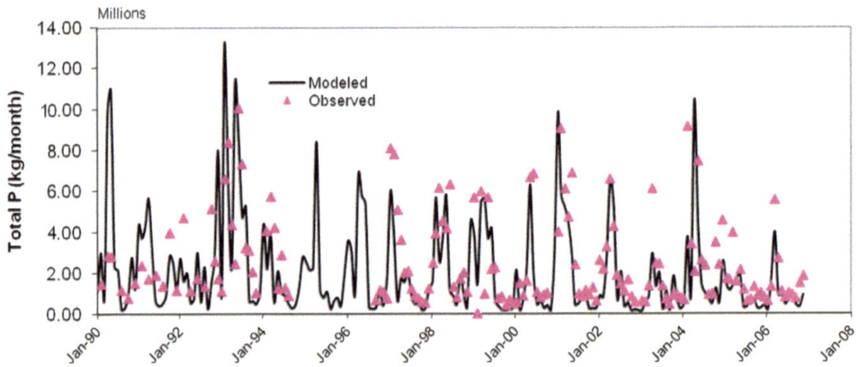

Fig. 10.12 Observed versus modeled monthly TP for the simulation period

Efficient reductions may also be achieved by targeting both nitrogen and phosphorus sources in close proximity to large rivers.

10.5 Summary

The purpose of this study was to develop an integrated watershed and riverine modeling system and use this integrated modeling system to study the transport and fate of nutrients in a large scale river basin. The SWAT model and the HEC-RAS model were loosely integrated so that output from the watershed model became input into the riverine model. The model integration considers spatial and temporal characteristics of the watershed systems, correspondence and transference of the state variables between two models, and file format specifics for proper communications between the models.

The SWAT model was successfully used to estimate flow and nutrient exports from each tributary in the UMRB. The HEC-RAS model was able to use daily SWAT outputs and model nutrient cycling and transport and fate of nutrients in the mainstem of the basin. Land use changes and different watershed management scenarios could be evaluated using the integrated modeling system to determine water quality impacts. The integrated watershed and riverine modeling system provides a systematic approach to evaluating nonpoint nutrient sources, transport, and fate in a large river basin. This approach will provide an effective management tool to investigate changes in the flow quantity and water quality in the Mississippi River.

References

Alexander RB, Smith RA, Schwarz GE. Effect of stream channel size on the delivery of nitrogen to the Gulf of Mexico. Nature. 2000;403:758–61.

Allen RG, Jensen ME, Wright JL, Burman RD. Operational estimates of evapotranspiration. Agron J. 1989;81:650–62.

Arnold JG, Allen PM. Estimating hydrologic budgets for three Illinois watersheds. J Hydrol. 1996;176:55–77.

Arnold JG, Srinivasan R, Muttiah RS, Williams JR. Large-area hydrologic modeling and assessment: part I. Model Dev J AWRA. 1998;34(1):73–89.

Bagnold RA. Bed-load transport by natural rivers. Water Resour Res. 1977;13:303–12.

Brown CL, Barnwell Jr. To the enhanced stream water quality models QUAL2E and QUAL2E-UNCAS documentation and user manual: USEPA. Environmental research laboratory. Athens. Georgia. 1987. EPA/600/3-85/040. p. 455.

Debele B, Srinivasan R, Parlange J-Y. Coupling upland watershed and downstream waterbody hydrodynamic and water quality models (SWAT and CE-QUAL-W2) for better water resources management in complex river basins. Environ Model Assess. 2008;13:135–53.

Devine J, Dorfman M, Rosselot KS. Missing Protection: Polluting the Mississippi river basin's small streams and wetlands. October: Natural resources defense council issue paper; 2008.

Diaz RJ, Rosenberg R. Spreading dead zones and consequences for marine ecosystems. Science. 2008;321:926–8.

Gassman PW, Reyes MR, Green CH, Arnold JG. The soil and water assessment tool: historical development, applications, and future research directions. Trans ASABE. 2007;50(4):1211–50.

Hargreaves GH, Samani ZA. Reference crop evapotranspiration from temperature. Appl Eng Agric. 1985;1:96–9.

Lian Y, Chan I-C, Singh J, Demissie M, Knapp V, Xie H. Coupling of hydrologic and hydraulic models for the Illinois river Basin. J Hydrol. 2007;344:210–22.

McElroy AD, Chiu SY, Nebgen JW, Aleti A, Bennett FW. Loading functions for assessment of water pollution from nonpoint sources. Environ Prot Tech Serv.1976; EPA 600/2-76-151.

Mueller DK, Spahr NE. Nutrients in streams and rivers across the Nation—1992–2001: U.S. geological survey scientific investigations report 2006–5107 (2006).

Priestley CHB, Taylor RJ. On the assessment of surface heat flux and evaporation using large-scale parameters. Mon Weather Rev. 1972;100:81–2.

USDOE (U.S. Department of Energy) U.S. Billion Ton Update: Biomass supply for a bioenergy and bioproducts industry. Perlack RD and Stokes BJ (Leads). ORNL/TM-2011/224. Oak Ridge National Laboratory, Oak Ridge, TN.2011, p. 227.

USEPA (U.S. Environmental Protection Agency). Hypoxia in the Northern Gulf of Mexico: an update by the EPA Science Advisory Board, 2007. EPA-SAB-08-003.

Wetzel RG. Limnology, lake and river ecosystems. 3rd ed. San Diego: Academic Press; 2001. p. 1006.
Williams JR, Berndt HD. Sediment yield prediction based on watershed hydrology. Trans ASAE. 1977;1100–04.
Williams JR, Hann RW. Optimal operation of large agricultural watersheds with water quality constraints. Texas Water Resources Institute, Texas A&M University, Technical Report. No. 96, 1978
Zhang Z, Johnson BE. Aquatic nutrient simulation modules (NSM). ERDC/EL-12-X, Draft, U.S. Army Engineer Research and Development Center, Vicksburg, MS. 2012.

Chapter 11
Remote Sensing and Geospatial Analysis for Landscape Pattern Characterization

Xiaojun Yang, Bojie Fu and Liding Chen

Abstract Landscape pattern characterization aims to map, quantify, and interpret landscape spatial patterns, and is therefore a fundamental pursuit in landscape ecology. The advances in remote sensing and geographic information systems (GIS) have greatly contributed to the development of quantitative methods for landscape pattern characterization. This chapter will review the utilities of remote sensing and GIS for the measurement, analysis, and interpretation of landscape spatial patterns. While remote sensing allows a direct observation of landscape patterns and processes at various scales, GIS provides a technical platform for data integration and synthesis in support of landscape pattern analysis and modeling. The chapter will begin with an overview on the research status identifying some gaps when landscape ecologists utilize remote sensing and GIS techniques in their research. Then, it will examine the utilities of remote sensing and landscape metrics for landscape pattern mapping and quantification, which will be followed by a discussion on GIS-based spatial analysis and modeling techniques for examining patterns, relationships, and emerging trends and for simulation and prediction. While the topics covered in this chapter span the entire spectrum in landscape pattern characterization, our emphasis is not on a comprehensive review but on some methodological issues highlighting caveats and cautions when using remote sensing and geospatial techniques. We believe the issues identified here can help landscape ecologists to better utilize remote sensing and GIS techniques in their specific applications.

X. Yang (✉)
Department of Geography, Florida State University,
Tallahassee, FL 32306, USA
e-mail: xyang@fsu.edu

B. Fu · L. Chen
Research Center for Eco-Environmental Science,
Chinese Academy of Sciences,
Beijing 100085, China

B. Fu and K. B. Jones (eds.), *Landscape Ecology for Sustainable Environment and Culture*, DOI: 10.1007/978-94-007-6530-6_11,

Keywords Landscape patterns · Remote sensing · GIS · Data quality · Thematic resolution · Spatial scale · Ecological processes

11.1 Introduction

Landscape pattern characterization aims to map, quantify, and interpret landscape spatial patterns, and is therefore critical to address the spatial interaction between landscape patterns and ecological processes (Turner 2005). An increasing number of indices or metrics have been developed to quantify various landscape aspects (e.g. McGarigal and Marks 1995; McGarigal et al. 2009). These metrics can be derived from categorical maps that are predominately produced through remote sensing (e.g. Yang and Liu 2005a; Wang and Yang 2012). And various landscape elements are normally represented as digital maps, and geographic information systems (GIS) can be used to relate these elements in search of their causal relationship (e.g. Serneels and Lambin 2001; Hietel et al. 2004). Furthermore, spatially explicit modeling techniques can be used to examine the underlying landscape dynamics being linked with various biophysical and socio-economic conditions (e.g. Yang and Lo 2003; Hepinstall et al. 2008; Paudel and Yuan 2011).

Although remote sensing, GIS, and spatial analysis have been widely used for landscape pattern characterization (*c.f.* Turner and Gardner 1991; Steiniger and Hay 2009), there have been some major limitations when landscape ecologists utilize these techniques in their specific applications. First, with an increasing use of remote sensing and GIS in landscape ecological studies, many do not address some key technical issues, such as the potential uncertainty or error relating to landscape pattern measure, analysis, and modeling. For example, according to a recent study conducted by Newton et al. (2009), among 438 research papers published in the journal *Landscape Ecology* for the years 2004–2008, more than one third of these studies explicitly mentioned remote sensing but there was a frequent lack of important technical details, with approximately three-quarter failing to provide any assessment of uncertainty or error in image classification and mapping. Without these critical technical details, the quality and credibility of the scientific research would be open to question.

On the other hand, landscape ecologists were among the earliest groups who benefited from the use of remote sensing and geospatial techniques, as attested by Carl Troll's pioneering work in African savannah landscape analysis through the use of aerial photographs (Troll 1939). However, few scholars in landscape ecology are fully aware of the latest development in these techniques. For example, existing landscape ecological studies that incorporated a remote sensing component have predominately used aerial photographs and Landsat imagery, with few targeting new types of data acquired in optical and microwave portions of the electromagnetic spectrum (Cohen and Goward 2004; Newton et al. 2009).

This chapter will provide an overview on the utilities of remote sensing and GIS techniques for landscape pattern characterization. While remote sensing allows a

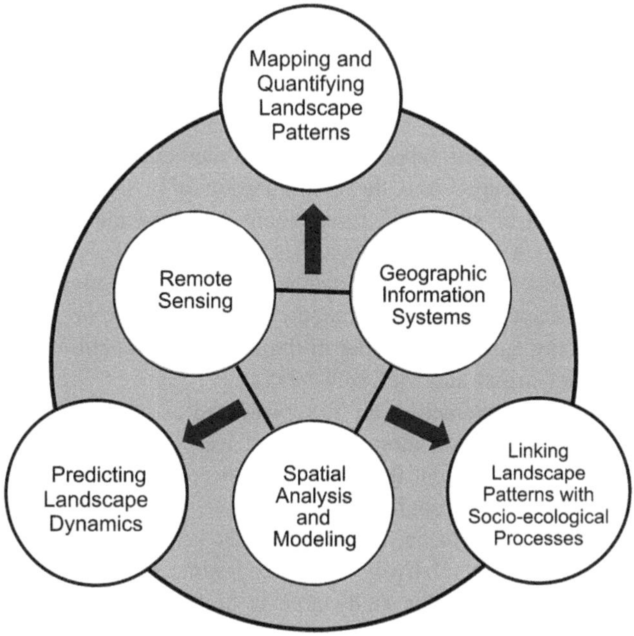

Fig. 11.1 A framework guiding the use of remote sensing and geospatial analysis for landscape pattern characterization. While remote sensing provides an indispensable source of data for landscape pattern mapping and quantification, GIS offers a platform for spatial data integration and synthesis that can help link observed landscape patterns with socio-ecological processes and for predicting landscape dynamics

direct observation of landscape pattern and process at various scales, GIS provides a platform for integration and synthesis of theories and technologies in support of landscape pattern analysis and modeling (Fig. 11.1). The chapter is organized into several major sections, beginning with a discussion of the research status identifying some gaps in the use of remote sensing and GIS techniques in landscape ecology. Section 11.2 will examine remote sensing and landscape metrics for landscape pattern mapping and quantification. Section 11.3 will discuss some GIS-based spatial statistical analysis and modeling techniques for examining patterns, relationships, and emerging trends and for simulation and prediction. The last section will summarize the major findings. While the topics covered in this chapter span the entire spectrum in landscape analysis, our emphasis is not on a comprehensive review but on some methodological issues highlighting caveats and cautions when using geospatial techniques in landscape ecology. We believe the issues identified here can help landscape ecologists to better utilize remote sensing and GIS techniques in their specific applications.

11.2 Landscape Pattern Mapping and Quantification

Landscape patterns consist of two components: composition and configuration (McGarigal and Marks 1995). The composition of a landscape refers to the number and occurrence of different types of landscape elements, which is a nonspatial measure of the landscape. And the configuration of a landscape refers to the physical distribution or structural arrangement of landscape elements, which is spatial by nature as it mainly deals with such an aspect as dimension, shape, or orientation of landscape elements. Together the spatial configuration and composition of landscape elements define the spatial pattern or heterogeneity of landscapes and play an important role in the ecological functionality and biological diversity (McGarigal and McComb 1995).

Although landscape elements can be represented as biotopes or habitats, they are commonly described in a more aggregated way, as land cover classes. Such land cover categories represent the interface between biophysical conditions and anthropogenic influences through time. Land cover types can be mapped through field surveys or remote sensing. The former can achieve an excellent mapping accuracy, especially for a relatively small geographic area. But this method can become less efficient when the study area is quite large or poorly accessible. Remote sensing, through sensors mounted on various aerospace platforms, can acquire photos or images over the visible, infrared and microwave portions of the electromagnetic spectrum within a short time period. They can revisit and acquire data over a specific study site, suitable for monitoring the dynamics of a landscape.

11.2.1 Land Cover Mapping

Critical to the entire landscape pattern analysis is the production of a land cover map by remote sensing, which relies upon the acquisition of remote sensor data and the identification of information extraction methods that are appropriate to the landscape characteristics under investigation (Yang and Lo 2002).

11.2.1.1 Remote Sensor Data Acquisition

Over the past several decades, data from various remote sensors have been used for land cover mapping. Earlier attempts were largely built upon the use of aerial photography. The acquisition of information on regional, national and global land cover has been the subject of numerous studies and evaluations since the early 1970s (e.g. Gaydos and Newland 1978; Jensen 1981; Haack et al. 1987; Yang and Lo 2002; Seto and Fragkias 2005; Lackner and Conway 2008; Bagan et al. 2012), which were largely stimulated by the launch of Earth Resources Technology Satellite-1 (ERTS-1; later renamed as Landsat) in 1972. Images acquired by the

US Landsat program and French SPOT satellites are the principal sources of data for land cover mapping. In addition, large volumes of valuable data have been acquired by the Indian remote sensing satellites (IRS), the NASA Terra satellite, the China-Brazil Earth resources satellites (CBERS), and several European, Canadian, and Japanese satellites carrying active remote sensing devices. Moreover, remote sensor data with high spatial resolution acquired by several commercial satellites have become available since the late 1990s, which allow a substantial proportion of the basic land cover units to be distinguished.

Although various types of remote sensor data are widely available, it can be quite difficult to identify an appropriate type of imagery in a land cover mapping application. To help ease this data acquisition process, one should carefully consider several important issues. Firstly, before actual data acquisition and processing, an appropriate land cover classification scheme should be identified or established considering specific research objectives and the characteristics of landscapes under investigation. There are many different land cover classification schemes available, and for the sake of data interoperability, one should consider those being widely used whenever possible, such as the one developed by United States Geological Survey (USGS) for use with remote sensor data (Anderson et al. 1976).

Secondly, when deliberating over a range of remote sensor data, one should consider different aspects of image characteristics, such as spatial, spectral, radiometric, and temporal resolutions. Traditionally, land cover mapping emphasizes the importance of image spatial resolution but recent studies suggest that the other three resolutions are also critical (Jensen 2005). The choice of spatial resolution should be linked with the thematic accuracy specified in the land cover classification scheme. In other words, images with a coarse spatial resolution can be only used for mapping broad land cover categories, while higher resolution images should be used for detailed land cover mapping. Images with higher spectral resolution can be quite useful for mapping different types of vegetation. And multi-temporal images can not only be used to analyze landscape changes but also map some land cover categories that are affected by seasonal or phonological variations.

Lastly, remote sensor data acquisition cost is another important consideration. It can vary greatly, from unaffordably expensive to virtually free. While most of the airborne products (e.g. AIRSAR, ATLAS, and LIDAR) and high-resolution satellite images are quite expensive, certain licenses could lead to the availability of these products to selected users at a substantially reduced price or even for free. For example, Terra Image USA, the master distributor for the U.S. market of satellite imagery from the SPOT constellation of high resolution Earth imaging satellites, and the University of Californian at Santa Barbara have formed a research partnership since 2005 that allows the students and faculty members at the campus to freely access the entire SPOT image dataset with spatial resolution comparable to aerial photography (http://www.ia.ucsb.edu/pa/display.aspx?pkey=1311). Images with coarse resolution (e.g. MODIS and AVHRR) have been freely available, and beginning in early 2009, the entire Landsat image archive collected by USGS EROS Data Center over nearly the past four decades can be downloaded via internet at no charge (http://landsat.usgs.gov). This moderate-resolution image archive is unmatched in quality, details, coverage, and length, and has been an invaluable resource for examining natural and anthropogenic changes on Earth's surface (Yang 2011a).

11.2.1.2 Information Extraction from Remote Sensor Data

Both visual interpretation and computer-based image classification techniques can be used to extract land cover information from remote sensor data. Through the combined use of various image elements with human intelligence, visual interpretation can achieve excellent mapping accuracy. And this technique can be implemented through on-screen digitizing in a GIS environment. However, it is manual by nature, and can be very much labor intensive for land over mapping over a large area. On the other hand, computer-based image classification can automate the entire land cover mapping process although some further research is still needed towards the operational use of this promising technique. In general, image classification is preferred over visual interpretation for land cover mapping over large areas (Jensen 2007).

Remote sensor image classification is largely based on the manipulation of statistical characteristics of one or more multispectral scenes. A variety of classification methods have been developed for remote sensing applications. By using specific criteria, these methods can be grouped in different ways: a parametric or a nonparametric classifier; a supervised or an unsupervised classifier; a "hard" or a "soft" classifier; a spectral or a spatial classifier; a sub-pixel, a pixel or an object-based classifier (Jensen 2005). Each of these methods has its own advantages and disadvantages, and there is no any single classifier that can be superior to another in all aspects (Duda et al. 2001).

Among all existing image classifiers, some are considered as advanced ones. However, most of the mapping applications have relied upon the use of a conventional classifier that largely manipulates a single image element (e.g. color or tone) in the multispectral pattern recognition (Jensen 2005). Conventional pattern recognition methods are largely based on the use of parametric statistics, which generally work well for medium-resolution scenes covering spectrally homogeneous areas, but not in heterogeneous regions or when scenes contain severe noises due to the increase of image spatial resolution. For years, substantial research efforts have been made to improve the performance of pattern classification for working with different types of remote sensor data and with spectrally complex landscapes. Some strategies have been developed as a result of such efforts: (1) the identification of various hybrid approaches that combine two or more classifiers, or incorporate pre- and post-classification image transformation and feature extraction techniques (e.g. Yang and Liu 2005b); (2) the development of 'soft' classifiers by introducing partial memberships for each pixel to accommodate the heterogeneous and imprecise nature of the real world (e.g. Shalan et al. 2003); (3) the decomposition of each pixel into independent endmembers or pure materials to conduct image classification at sub-pixel level (e.g. Verhoeye and Wulf 2002); (4) the incorporation of the spatial characteristics of the neighboring (contextual) pixels to develop object-oriented classification (e.g. Walker and Briggs 2007); (5) the fusion of multi-sensor, multi-temporal, or multi-source data for combining multiple spectral, spatial, and temporal features and ancillary information in the image classification (e.g. Tottrup 2004); and (6) the use of artificial intelligence technology, such as rule-based classifiers (e.g. Schmidt et al. 2004), artificial neural networks (e.g. Zhou and Yang 2011), and support vector machines (e.g. Yang 2011b), for pattern classification.

11.2.2 Landscape Pattern Quantification

Once a land cover map is available, landscape patterns can be quantified using landscape indices or metrics. These metrics are algorithms measuring the diversity, homogeneity or heterogeneity of a landscape. Although many earlier efforts have been made to identify various metrics that are meaningful for ecological functionality and biological diversity, it was the first primary release of the FRAGSTATS software package in 1995 that helped landscape ecologists to revolutionize the analysis of landscape structure (Kupfer 2012). FRAGSTATS defines eight major groups of structural or functional metrics: area/density/edge, shape, core area, isolation/proximity, contrast, contagion/interspersion, connectivity, and diversity (McGarigal and Marks 1995). Structural metrics measure the physical composition or configuration of a landscape without explicit reference to an ecological process, while functional metrics explicitly measure landscape pattern that is functionally relevant to the organism or process under consideration (McGarigal 2002). These metrics can be commonly measured at three levels: patch, class, and landscape. Patch-level metrics characterize the spatial context of patches, class-level metrics integrate over all patches of a given land cover type, and landscape-level metrics synthesize over the entire landscape. With the development of GIS software engineering, some metrics originally defined in FRAGSTATS have been incorporated into other software packages, such as Patch Analysis (http://www.cnfer.on.ca/SEP/patchanalyst/) and LANDISVIEW (http://kelab.tamu.edu/standard/restoration/restoration_tools.htm).

Landscape metrics offer an intuitive tool to measure landscape structure that can be linked with specific ecological processes. Because the input data (i.e. categorical land cover maps) can be derived from remote sensor imagery and the software toolkit is readily available, landscape metrics have been widely used for landscape pattern analysis. Nevertheless, there have been well documented concerns in the use of landscape metrics. Firstly, despite well-documented guidelines being given on the use of various metrics (e.g. McGarigal and Marks 1995; Haines-Young and Chopping 1996; McGarigal et al. 2002) and more than two decades of extensive research, interpreting landscape metrics beyond the simple quantitative description of landscape pattern remains difficult due to lack of generalizations concerning pattern-process relationships for many metrics (McGarigal 2002; Li and Wu 2004; Turner 2005; Fu et al. 2011).

Secondly, given the scale dependence of spatial heterogeneity, the statistic characteristics of landscape metrics can be affected by spatial extent and scale (e.g. Turner et al. 1989; Wu et al. 2002; Corry and Nassauer 2005). Turner et al. (1989) investigated the effects of changing grain size and extent of land cover data on spatial patterns. Landscape patterns were compared using metrics measuring diversity, dominance and contagion. They found that the diversity index decreased linearly as grain size increased, but dominance and contagion did not show such a linear relationship. While dominance and contagion increased with increasing spatial extent, diversity showed an erratic response. Corry and Nassauer (2005) found that the amount of linear habitat patches increased with increasing spatial resolution of land cover data. When many linear patches are present, metrics measuring landscape configuration may not be reliable; after resampling the land cover data into a coarse grain size, landscape configuration metrics became ecologically relevant. They found that composition metrics can be more useful in highly fragmented landscapes. A more comprehensive study was conducted by Wu et al. (2002) who examined how some common metrics responded to changing grain size and extent. Their found that the responses fell into three general categories: Predictable responses to changing scale with definable, simple scaling relationship; less predictable, staircase-like responses; and erratic responses without consistent scaling relationships.

Since landscape metrics are generally derived from categorical maps, the thematic resolution of land cover data and the classification scheme can also affect the statistical properties and behavior of landscape metrics. For example, Corry and Nassauer (2005) found that aggregation of land cover classes can reduce the number of patch types and thus increase the likelihood of contiguity. Huang et al. (2006) examined the sensitivity of two dozens of metrics to a number of land cover classes with different spatial patterns. They found that many metrics behaved predictably with increasing classification detail. At lower class numbers, metrics were quite sensitive to increasing classification detail. Their studies suggest the importance of land cover classification scheme in landscape pattern analysis.

Moreover, the quality of input data (i.e. land cover map) can affect the statistical characteristics of landscape metrics. For example, Wickham et al. (1997) tested the sensitivity of three common metrics (i.e. patch compaction, contagion, and fractal dimension) to land cover misclassification and differences in land cover composition. They found that differences in land cover composition need to be larger than the misclassification error in order to be confident that differences in landscape metrics are not due to misclassification. Corry and Nassauer (2005) noted that several data conversion procedures (e.g. vector to raster conversion, digitization of analogue data, and resampling) often introduce errors in land cover maps that can further affect the computed metric values. Langford et al. (2006) found that land cover classification error is not always a good predictor of errors in landscape metrics but maps with low misclassification rates can yield errors in landscape metrics with much larger magnitude and substantial variability. They also noted that certain image post-processing procedures such as smoothing might result in the underestimation of habitat fragmentation.

Lastly, with the development of landscape metrics over the past several decades, the choice of landscape metrics seems to be quite rich. But most higher-level metrics are derived from the same patch-level attributes, implying that many metrics can be partially or perfectly correlated with each other. This results in information redundancy. Moreover, a large number of metrics would be difficult to interpret and analyze. Practically, a small set of metrics that are not redundant but capture the major properties of a landscape are more desirable for a specific application. Ideally, these metrics should be consistent across spatial scale and time (e.g. Kelly et al. 2011). Recent studies suggest that landscape pattern can be characterized by using a small number of metrics but consensus has not reached on the choice of individual metrics (McGarigal 2002). For a specific application, one has to identify his/her own list of metrics to be used. This could be done through the use of landscape ecological principles or statistical methods or a combination of both. Many applications were based on the use of landscape ecological principles in selecting metrics (e.g. Zhang et al. 1997; Fuller 2001; Li et al. 2001; Baskent and Kadiogullari 2007). This approach may work well to reduce inherent redundancy but not empirical redundancy. On the other hand, statistical methods, such as principal component analysis, can help reduce data redundancy and select a parsimonious suite of independent metrics for landscape pattern analysis (e.g. Riitters et al. 1995; Herzog and Lausch 2001; Yang and Liu 2005a; Wang and Yang 2012).

Despite the above concerns, landscape metrics remain popular as they are seen by land managers and stakeholders as a simple tool for exploratory and descriptive analysis of spatio-temporal landscape pattern (Kupfer 2012). Encouragingly, recent advances in remote sensing and GIS software engineering allow more reliable land cover maps to be produced from multi-resolution imagery, and landscape metrics to be easily calculated through readily available software packages. The latest release of FRAGSTATS (4.0) extends the calculation of landscape metrics beyond categorical maps and into continuous maps through a moving window approach (McGarigal et al. 2012). This technical breakthrough can help minimize the variation of statistical properties of landscape metrics due to the modifiable area unit problem (MAUP). Moreover, the development of metrics rooted in graph, network, and circuit theory offers the promise of a more ecologically oriented approach to quantifying landscape pattern and process (Kupfer 2012).

11.3 Landscape Pattern-Process Analysis and Modeling

Landscape pattern characterization not only aims to map and quantify landscape spatial patterns, but also seeks to interpret them in relation to specific socio-ecological processes. Therefore, landscape pattern characterization serves as the centerpiece to address the spatial interaction between landscape patterns and ecological processes that can help understand the intrinsic causality and underlying landscape dynamics. On one hand, landscape patterns are strongly influenced by ecological processes that include all aspects of biological, chemical, physical, hydrological, and human-dimensional processes of the ecosystem. On the other hand, landscape patterns can significantly affect ecological processes and landscape dynamics across spatial and temporal scales. To assess the two ways of relationship, one needs to integrate the ecologically-oriented, vertical approach with the geographically-oriented, horizontal approach that incorporates aggregated, integrative environmental parameters (Bastian 2001). While a rich pool of landscape ecological literatures have discussed specific pattern-process relationships (e.g. Turner 1989; McGarigal and McComb 1995; Wu 2006), here we direct our attention on some generic methodological issues for integration and synthesis in a GIS environment.

11.3.1 Linking Landscape Patterns with Processes

Relating landscape spatial patterns with ecological processes involves the integration and synthesis of theories and technologies across spatial and temporal scales, which can be pursued through either the qualitative or quantitative approach. Given the scope of this chapter, we herewith limit our further discussion to some technical issues when using the quantitative approach, especially multivariate statistical analysis, to address the pattern-process linkage at the broad scale level. Multivariate statistical methods including linear and nonlinear multiple regression can be used to examine the pattern-process relationship. When constructing a multivariate regression model to assess how ecological processes would influence landscape patterns, each of the landscape metrics should be treated as a dependent variable, while various biological, chemical, physical, hydrological, and human-dimensional variables as independent variables (e.g. Lo and Yang 2002). On the other hand, when examining how landscape patterns would affect ecological processes, landscape metrics should be treated as independent variables, while each of specific ecological indicators as a dependent variable (e.g. Yang 2012). Several statistical methods, such as ordinary least squares (OLS) regression and logistic regression, can be used to determine the pattern-process relationship.

There are several issues one should pay close attention to when using the above empirical method to study the pattern-process relationship. Firstly, since all dependent and candidate independent variables are usually aggregated by areal units, how these units are defined in terms of scale and zoning systems can affect the results of parameter estimates in multivariate statistical analysis. This is actually the modifiable areal unit problem (MAUP) that has been extensively discussed in geography and spatial science literature (e.g. Openshaw 1984; Fotheringham and Wong 1991; Jelinski and Wu 1996; Dark and Bram 2007). The modifiable areal unit problem is considered as a fundamental problem inherent in the studies using spatially aggregated data because the results are always affected by the areal units used (Openshaw 1984). It can be essentially unpredictable in its intensity and effects in multivariate statistical analysis, and is therefore a much greater problem than in univariate or bivariate analysis (Fotheringham and Wong 1991). While the variations of statistical analysis due to the aggregation of smaller areal units into regions are generally well understood (e.g. Fotheringham and Wong 1991), the zoning problem is much less well understood (Jelinski and Wu 1996).

The number of observations or sample size is very important for multivariate statistical analysis. For geographically referenced data, we normally consider the number of spatial observation units being equivalent to sample size from a statistical perspective. In general, the number of observation units should be 5–10 times the number of candidate independent variables (Brace et al. 2012). For example, if a specific multivariate statistical model intends to include 5 independent variables, the number of observation units should be at least 25–50. Too many or too few units could lead statistical models to be over-fitting.

Because the total number of observation units is actually quite limited for many applications, one should be careful when selecting candidate independent variables to be included in multivariate statistical analysis. As mentioned before, many landscape metrics are perfectly or partially correlated with each other, which can cause information duplication. Therefore, when using landscape metrics as candidate independent variables to assess their impacts upon specific ecological processes, it is important to identify a small number of landscape metrics that are not duplicated but capture the major landscape properties.

A preprocessing procedure should be conducted for all dependent or independent variables. Because multivariate statistical analysis is sensitive to the variance of samples and data distribution, one should avoid using the raw data directly, particularly for those variables with a large statistical variance. For some environmental variables, such as water or air quality, one should use their average measurements by month, quarter or year. For landscape composition metrics, one should use the relative proportion rather than the total number. Before actual statistical analysis, raw data should be logarithmically transformed to improve their normality.

Before a statistical model is established, one should check the normality, multicollinearity, and spatial autocorrelation of independent variables. Data normality can be checked through the Kolmogorov-Smirnov test or the graphic approach using histograms and QQ plots. For some variables that do not show a clear normal distribution, one can transform the raw data logarithmically to improve the data normality. Any statistical models that show strong multicollinearity among the independent variables should be used with caution. The spatial autocorrelation can be computed by using Moran I or Geary C. If a strong spatial autocorrelation exists, one should use the strategies suggested by Legendre (1993) to reduce the spatial dependence.

When assessing the performance of different statistical models, one should pay attention on the number of independent variables included. In general, the explanatory power tends to be higher for a model that includes more independent variables. This suggests that any meaningful comparison of model performance should be based on the identical number of independent variables.

Finally, multiple regression analysis can make use of all or a subset of the sample in parameter estimation when building a statistical model. Although most of existing studies have built upon the use of all sample data in parameter estimations, recent development in spatial statistics suggests using a subset of the sample data can help reveal the variation of the cause-effect relationship across space (Fotheringham et al. 2002). This localized regression technique called geographically weighted regression has been included in some leading GIS software packages, such as ArcGIS 10. Although the software tool is readily available, it can be very much data demanding, particularly for some environmental data that can be only acquired through in situ measurements.

11.3.2 Modeling and Predicting Landscape Dynamics

Models developed to simulate and predict landscape dynamics as a physical process have become quite popular in recent years. This is necessary to understand the dynamics of complex ecosystems and to evaluate the consequences of landscape change on the environment (Yang and Lo 2003; Sutherland 2006). Ecologists were among the earliest groups who have demonstrated a strong interest in developing spatially explicit models to predict the impacts of different landscape configurations on plant and animal populations (Kareiva and Wennergren (1995). While many methods have been developed to predict the impacts at the species or population level (c.f. Sutherland 2006), here we direct our attention on the spatially explicit, dynamic models that are designed to work at the community or ecosystem level.

Over the past several decades, a variety of spatially explicit models have been developed by different communities, which can be either stochastic, such as the logit (e.g. Hu and Lo 2007), Markov (e.g. Myint and Wang 2006), cellular automata (e.g. Hagen-Zanker and Lajoie 2008), and agent-based models (e.g. Robinson et al. 2012), or processes based, such as dynamic ecosystem models (e.g. Euskirchen et al. 2006). Although these models are different by their underlying mechanism, they share many commonalities. The common approaches are the use of transition probabilities in a class transition matrix, multinominal logit methods, cellular or agent-based modeling, and GIS weighted overlay approach. These models consider different constraints by various biophysical, economical, and social parameters. Some of these parameters include land transition probabilities, topography, environmental protection, forest properties, transportation, population, economic indicators, human behavior, and policy. The role of remote sensing and GIS is indispensible in the entire model development process from model conceptualization to implementation that includes input data preparation, model calibration, and model validation. Comprehensive reviews on various spatial modeling approaches are beyond our scope in this chapter, and readers should refer to several other relevant publications (e.g. Agarwal et al. 2002; Parker et al. 2003; Verburg et al. 2004; An 2012).

While spatially explicit modeling approaches can help extend landscape analysis beyond quantitative pattern description and into the area of forecasting and prediction, there is a need to accept the limitations of prediction (Sutherland 2006). Most of the modeling efforts are technically driven, the justification and verification of ecological concepts and theories are not adequate. Some game-like simulators consider only a few untested factors, which often perform poorly when predicting for the future. On the other hand, some models tend to be too ambitious, considering too many variables, which are not easily parameterized. Many simulators do not contain components of model calibration and verification, and hence their results are generally not good enough for prediction and forecasting.

11.4 Concluding Remarks

In this chapter, we have reviewed the utilities of remote sensing and geospatial analysis for landscape pattern characterization, a fundamental pursuit in landscape ecology. While landscape ecologists were among the earliest groups benefiting from the use of remote sensing and GIS techniques, few of them are fully aware of the latest development in these technical areas. Essentially global coverage of remote sensor data with individual pixels ranging from sub-meters to a few kilometers can help make connections across various levels of landscape pattern analysis. The development of advanced image classification techniques and GIS software engineering has helped landscape ecologists to revolutionize the analysis of landscape structure by using pattern metrics. Nevertheless, the statistical properties and behavior of landscape metrics across a range of classification schemes and landscapes, as well as their sensitivity to changing landscape patterns, are still not fully understood. While GIS-based spatial analysis and modeling techniques can help examine patterns, relationships, emerging trends, and dynamics, landscape ecologists should also pay attention on some outstanding issues that we have discussed in this chapter. And landscape ecologists should fully understand both the strengths and weaknesses of remote sensing and geospatial techniques in order to better utilize these techniques in their specific applications. Finally, there is an increasing need to collaborate between the disciplines of landscape ecology and geospatial science that would not only lead to landscape ecology being taken more seriously but also help expand the inferential capabilities of geospatial research.

Acknowledgments This work has been partially supported by the CAS/SAFEA International Partnership Program for Creative Research Teams of "Ecosystem Processes and Services". The lead author would like to acknowledge the funding support from the U.S. Environmental Protection Agency Great Lakes and Estuarine research program and Florida State University for the time release in conducting the research.

References

Agarwal C, Green GM, Grove JM, Evans TP, Schweik CM. A review and assessment of land-use change models: dynamics of space, time, and human choice. Gen Tech Rep. NE-297. U.S. Department of Agriculture, Forest Service, Northeastern Research Station. 2002. p. 61. http://nrs.fs.fed.us/pubs/gtr/gtr_ne297.pdf. Last access on 28 July 2012.

An L. Modeling human decisions in coupled human and natural systems: review of agent-based models. Ecol Model. 2012;229:25–36.

Anderson JR, Hardy EE, Roach JT, Witmer RE. A land use and land cover classification system for use with remote sensor data. USGS professional paper 964. Online at: http://landcover.usgs.gov/pdf/anderson.pdf (1976). Last access on 4 July 2012.

Bagan H, Kinoshita T, Yamagata Y. Combination of AVNIR-2, PALSAR, and polarimetric parameters for land cover classification. IEEE T Geosci Remote. 2012;50(4):1318–28.

Baskent EZ, Kadiogullari AI. Spatial and temporal dynamics of land use pattern in Turkey: a case study in Inegol. Landscape Urban Plan. 2007;81:316–27.

Bastian O. Landscape ecology—towards a unified discipline? Landscape Ecol. 2001;16(8):757–66.

Brace N, Kemp R, Snelgar R. SPSS for psychologists. 5th ed. London: Routledge Academic; 2012.

Cohen WB, Goward SN. Landsat's role in ecological applications of remote sensing. Bioscience. 2004;54(6):535–45.

Corry RC, Nassauer JI. Limitations of using landscape pattern indices to evaluate the ecological consequences of alternative plans and designs. Landscape Urban Plan. 2005;72(4):265–80.

Dark SJ, Bram D. The modifiable areal unit problem (MAUP) in physical geography. Prog Phys Geog. 2007;31(5):471–9.

Duda RO, Hart PE, Stork DG. Pattern classification. New York: Wiley; 2001.

Euskirchen ES, McGuire AD, Kicklighter DW, Zhuang Q, Clein JS, Dargaville RJ, Dye DG, Kimball JS, McDonald KC, Melillo JM, Romanovsky VE, Smith NV. Importance of recent shifts in soil thermal dynamics on growing season length, productivity, and carbon sequestration in terrestrial high-latitude ecosystems. Glob Change Biol. 2006;12(4):731–50.

Fotheringham AS, Brunsdon C, Charlton M. Geographically weighted regression. New York: Wiley; 2002.

Fotheringham AS, Wong DWS. The modifiable areal unit problem in multivariate statistical analysis. Environ Plann A. 1991;23:1025–44.

Fu BJ, Liang D, Lu N. Landscape ecology: coupling of pattern, process, and scale. Chinese Geogr Sci. 2011;21(4):385–91.

Fuller DO. Forest fragmentation in Loudoun County, Virginia, USA evaluated with multitemporal Landsat imagery. Landscape Ecol. 2001;16:627–42.

Gaydos L, Newland WL. Inventory of land-use and land cover of Puget Sound region using Landsat digital data. J Res US Geol Surv. 1978;6:807–14.

Haack B, Bryant N, Adams S. An assessment of Landsat MSS and TM data for urban and near-urban land-cover digital classification. Remote Sens Environ. 1987;21:201–13.

Hagen-Zanker A, Lajoie G. Neutral models of landscape change as benchmarks in the assessment of model performance. Landscape Urban Plan. 2008;86(3–4):284–96.

Haines-Young R, Chopping M. Quantifying landscape structure: a review of landscape indices and their application to forested landscapes. Prog Phys Geog. 1996;20:418–45.

Hepinstall JA, Alberti M, Marzluff JM. Predicting land cover change and avian community responses in rapidly urbanizing environments. Landscape Ecol. 2008;23(10):1257–76.

Herzog F, Lausch A. Supplementing land-use statistics with landscape metrics: some methodological considerations. Environ Monit Assess. 2001;72:37–50.

Hietel E, Waldhardt R, Otte A. Analysing land-cover changes in relation to environmental variables in Hesse, Germany. Landscape Ecol. 2004;19(5):473–89.

Hu ZY, Lo CP. Modeling urban growth in Atlanta using logistic regression. Comput Environ Urban. 2007;31(6):667–88.
Huang C, Geiger E, Kupfer JA. Sensitivity of landscape metrics to classification scheme. Int J Remote Sens. 2006;27:2927–48.
Jelinski DE, Wu JG. The modifiable areal unit problem and implications for landscape ecology. Landscape Ecol. 1996;11(3):129–40.
Jensen JR. Urban change detection mapping using Landsat digital data. Am Cartographer. 1981;8:127–47.
Jensen JR. Introductory digital image processing: a remote sensing perspective. 3rd ed. New Jersey: Prentice Hall; 2005.
Jensen JR. Remote sensing of the environment: an Earth resource perspective. New Jersey: Prentice Hall; 2007.
Kareiva P, Wennergren U. Connecting landscape patterns to ecosystem and population processes. Nature. 1995;373(6512):299–302.
Kelly M, Tuxen KA, Stralberg D. Mapping changes to vegetation pattern in a restoring wetland: finding pattern metrics that are consistent across spatial scale and time. Ecol Indic. 2011;11(2):263–73.
Kupfer JA. Landscape ecology and biogeography: rethinking landscape metrics in a post-FRAGSTATS landscape. Prog Phys Geog 2012;36:400–420.
Lackner M, Conway TM. Determining land-use information from land cover through an object-oriented classification of IKONOS imagery. Can J Remote Sens. 2008;34(2):77–92.
Langford WT, Gergel SE, Dietterich TG, Cohen W. Map misclassification can cause large errors in landscape pattern indices: examples from habitat fragmentation. Ecosystems. 2006;9:474–88.
Legendre P. Spatial autocorrelation: trouble or new paradigm? Ecology. 1993;74:1659–73.
Li HB, Wu JG. Use and misuse of landscape indices. Landscape Ecol. 2004;19(4):389–99.
Li X, Lu L, Cheng G, Xiao H. Quantifying landscape structure of the Heihe River Basin, north-west China using FRAGSTATS. J Arid Environ. 2001;48:521–35.
Lo CP, Yang X. Drivers of land use/cover changes and dynamic modeling in Atlanta metropolitan region. Photogramm Eng Rem S. 2002;68:1073–82.
McGarigal K. Landscape pattern metrics. In: El-Shaarawi AH, Piegorsch WW, editors. Encyclopedia of environmetrics, vol. 2. Chichester: Wiley; 2002. p. 1135–42.
McGarigal K, Cushman SA, Neel MC, Ene E. FRAGSTATS v3: spatial pattern analysis program for categorical maps. Computer software program produced by the authors at the University of Massachusetts, Amherst. 2002. Available at the following web site: http://www.umass.edu/landeco/research/fragstats/fragstats.html. Last access on 10 July 2012.
McGarigal K, Cushman SA, Neel MC, Ene E. FRAGSTATS v4: spatial pattern analysis program for categorical and continuous maps. Computer software program produced by the authors at the University of Massachusetts, Amherst. 2012. Available at the following web site: http://www.umass.edu/landeco/research/fragstats/fragstats.html. Last access on 10 July 2012.
McGarigal K, Marks BJ. FRAGSTATS: spatial pattern analysis program for quantifying landscape structure. USDA forest service technical report PNW-GTR-351. 1995.
McGarigal K, McComb WC. Relationships between landscape structure and breeding birds in the Oregon coast range. Ecol Monogr. 1995;65:235–60.
McGarigal K, Tagil S, Cushman SA. Surface metrics: an alternative to patch metrics for the quantification of landscape structure. Landscape Ecol. 2009;24(3):433–50.
Myint SW, Wang L. Multicriteria decision approach for land use land cover change using Markov chain analysis and a cellular automata approach. Can J Remote Sens. 2006;32(6):390–404.
Newton AC, Hill RA, Echeverria C, Golicher D, Benayas JMR, Cayuela L, Hinsley SA. Remote sensing and the future of landscape ecology. Prog Phys Geog. 2009;33(4):528–46.
Openshaw S. The modifiable areal unit problem. CATMOG 38. GeoBooks. Norwich: England; 1984.
Parker DC, Manson SM, Janssen MA, Hoffmann MJ, Deadman P. Multi-agent systems for the simulation of land-use and land-cover change: a review. Ann Assoc Am Geogr. 2003;93(2): 314–37.

Paudel S, Yuan F. Assessing landscape changes and dynamics using patch analysis and GIS modeling. Int J Appl Earth Obs. 2011;16:66–76.
Riitters KH, O'neill RV, Hunsaker CT, Wickham JD, Yankee DH, Timmins SP, Jones KB, Jackson BL. A factor analysis of landscape pattern and structure metrics. Landscape Ecol. 1995;10:23–39.
Robinson DT, Murray-Rust D, Rieser V, Milicic V, Rounsevell M. Modelling the impacts of land system dynamics on human well-being: using an agent-based approach to cope with data limitations in Koper, Slovenia. Comput Environ Urban. 2012;36(2):164–76.
Schmidt KS, Skidmore AK, Kloosterman EH, Van Oosten H, Kumar L, Janssen JAM. Mapping coastal vegetation using an expert system and hyperspectral imagery. Photogramm Eng Rem S. 2004;70(7):703–15.
Serneels S, Lambin EF. Proximate causes of land-use change in Narok District, Kenya: a spatial statistical model. Agr Ecosyst Environ. 2001;85(1–3):65–81.
Seto KC, Fragkias M. Quantifying spatiotemporal patterns of urban land-use change in four cities of China with time series landscape metrics. Landscape Ecol. 2005;20:871–88.
Shalan MA, Arora MK, Ghosh SK. An evaluation of fuzzy classifications from IRS 1C LISS III imagery: a case study. Int J Remote Sens. 2003;24(15):3179–86.
Steiniger S, Hay GJ. Free and open source geographic information tools for landscape ecology. Ecol Inform. 2009;4(4):183–95.
Sutherland WJ. Predicting the ecological consequences of environmental change: a review of the methods. J Appl Ecol. 2006;43(4):599–616.
Tottrup C. Improving tropical forest mapping using multi-date Landsat TM data and pre-classification image smoothing. Int J Remote Sens. 2004;25:717–30.
Troll C. Luftbildplan und ökologische Bodenforschung (Aerial photography and ecological studies of the earth). Zeitschrift der Gesellschaft für Erdkunde. Berlin; 1939. p. 241–98.
Turner MG. Landscape ecology—the effect of pattern on process. Annu Rev Ecol Syst. 1989;20:171–97.
Turner MG. Landscape ecology: what is the state of the science? Annu Rev Ecol Evol S. 2005;36:319–44.
Turner MG, Gardner RH, editors. Quantitative methods in landscape ecology. New York: Springer; 1991.
Turner MG, O'Neill RV, Gardner RH, Milne BT. Effects of changing spatial scale on the analysis of landscape pattern. Landscape Ecol. 1989;3:153–62.
Verhoeye J, Wulf RD. Land cover mapping at sub-pixel scales using linear optimization techniques. Remote Sens Environ. 2002;79:96–104.
Verburg PH, Schot PP, Dijst MJ, Veldkamp A. Land use change modelling: current practice and research priorities. GeoJ. 2004;61:309–24.
Walker JS, Briggs JM. An object-oriented approach to urban forest mapping in Phoenix. Photogramm Eng Rem S. 2007;73(5):577–83.
Wang J, Yang X. A hierarchical approach to forest landscape pattern characterization. Environ Manage. 2012;49(1):64–81.
Wickham JD, Oneill RV, Riitters KH, Wade TG, Jones KB. Sensitivity of selected landscape pattern metrics to land-cover misclassification and differences in land-cover composition. Photogramm Eng Rem S. 1997;63(4):397–402.
Wu JG. Landscape ecology, cross-disciplinarity, and sustainability science. Landscape Ecol. 2006;21(1):1–4.
Wu JG, Shen WJ, Sun WZ, Tueller PT. Empirical patterns of the effects of changing scale on landscape metrics. Landscape Ecol. 2002;17(8):761–82.
Yang X. Use of archival Landsat data to monitor urban spatial growth. In: Yang X, editor. Urban remote sensing: monitoring, synthesis and modelling in the urban environment. Chichester: Wiley; 2011a. p. 15–34.
Yang X. Parameterizing support vector machines for land cover classification. Photogramm Eng Rem S. 2011b;77(1):27–37.

Yang X. An assessment of landscape characteristics affecting estuarine nitrogen loading in an urban watershed. J Environ Manage. 2012;90(1):50–60.

Yang X, Liu Z. Quantifying landscape pattern and its change in an estuarine watershed using satellite imagery and landscape metrics. Int J Remote Sens. 2005a;26:5297–323.

Yang X, Liu Z. Using satellite imagery and GIS to characterize land use and land cover changes in an estuarine watershed. Int J Remote Sens. 2005b;26:5275–96.

Yang X, Lo CP. Using a time series of satellite imagery to detect land use and land cover changes in the Atlanta, Georgia metropolitan area. Int J Remote Sens. 2002;23(9):1775–98.

Yang X, Lo CP. Modelling urban growth and landscape changes in the Atlanta metropolitan area. Int J Geogr Inf Sci. 2003;17(5):463–88.

Zhang D, Wallin DO, Hao Z. Rates and patterns of landscape change between 1972 and 1988 in the Changbai Mountain area of China and North Korea. Landscape Ecol. 1997;12:241–54.

Zhou L, Yang X. An assessment of internal neural network parameters affecting image classification accuracy. Photogramm Eng Rem S. 2011;77(12):1233–40.

Part III
Landscape Planning and Management

Chapter 12
Discursive Relationships Between Landscape Science, Policy and Management Practice: Concepts, Issues and Examples

Simon Swaffield, Jorgen Primdahl and Mark Hoversten

Abstract Different approaches have been proposed to help the science of landscape ecology achieve greater policy relevance. A common feature is the central role of landscape scientists as experts in solving 'place based problems' in effective ways. In practice however landscape ecologists have seldom had the impact they seek. This chapter uses concepts drawn from deliberative planning and case examples from the USA and Denmark to critically examine the science-practice interface between landscape ecology and landscape planning. It highlights the way that different roles, values, and interests interact at different stages in place based studies, and this may require a re-framing of landscape ecological science to become part of a multivalent discourse about landscape conditions and possibilities.

Keywords Deliberative planning · Landscape democracy · Place making · Spatial strategy

---

An erratum to this chapter is available at 10.1007/978-94-007-6530-6_17

S. Swaffield (✉)
Landscape Architecture, Lincoln University, Lincoln, New Zealand
e-mail: swaffies@lincoln.ac.nz

J. Primdahl
GeoScience and Nature Management, Copenhagen University, Copenhagen, Denmark
e-mail: jpr@life.ku.dk

M. Hoversten
Landscape Architecture, University of Idaho, Idaho, USA
e-mail: hoverstm@uidaho.edu

B. Fu and K. B. Jones (eds.), *Landscape Ecology for Sustainable Environment and Culture*, DOI: 10.1007/978-94-007-6530-6_12,

12.1 Introduction

Landscape ecology seeks greater practical and policy relevance (Wu and Hobbs 2002). As Hobbs (1997: 1) has noted, "The future of landscape ecology depends on whether landscape ecologists make the decision to take an active part in determining the future of landscapes". This realisation has led to calls to broaden the scope of the science to incorporate aspects of landscape planning and design. A variety of strategies have been proposed, including adoption of multi and trans-disciplinary research paradigms (Naveh 2005; Tress et al. 2003; Wu 2006), a change of focus from 'optimal patterns' to a search for the dynamic qualities of the landscape as defined by people (Haines-Young 2000), increased engagement with social science in a 'translational' approach to research and practice (Mussachio 2009a), participatory landscape ecology (Luz 2000), the use of a 'landscape services' framework (Termouzuien and Opdam 2009), and incorporation of 'design' as a complementary activity within science (Nassauer and Opdam 2008).

A common feature of these different strategies is the central role of landscape scientists as experts in solving 'place based problems' in an instrumentally rational way. Instrument rationality has been described and critiqued in the planning context by a number of authors, notably Friedmann (1987), and can be characterised along several dimensions. It works by identifying a desirable end state, and then logically considers and evaluates different means to achieve the desired ends. The emphasis of the approach is upon resolving choice and conflict as efficiently as possible, and maximising the utility of outcomes. It assumes that the future is sufficiently predictable to be able to make rational choices about how to proceed, and relies heavily upon expert knowledge, methods and skills to identify and realise solutions to place based problems (Alexander 2000; Allmendinger 2002; Mussachio 2009b; Amdam 2010).

However, experience from both rural land management (Duff et al. 2009) and urban planning (Flyvbjerg 2001) suggests that in order to achieve 'deep' social and policy relevance, it may be necessary to reconceptualise landscape science more fundamentally within a 'deliberative' paradigm of knowledge and action (Forrester 1999). The deliberative paradigm places emphasis upon argumentation (Fischer and Forester 1993), open discourse (Drysek 2000) and a combined 'internal and external perspective' on the planning process (Stein and Harper 2003). It is based upon what Flyvberg (2001) calls value rationality, where decisions are arrived at through open, discursive processes in which values, objectives and means are considered together. When expressed as communicative planning (Healey 1992), place making thus becomes understood as a locally situated collaborative social process with a significant learning dimension (Healey 1998) rather than technical problem solving at a local scale.

The difference can be illustrated by a hypothetical example. Consider a rural community faced with declining quality of life due to agricultural intensification and its effects on the landscape. An instrumentally rational approach might engage landscape scientists to measure public preferences for landscape, and to analyse and identify a technical change to the farming systems that could reduce the impacts of intensification upon those aspects of landscape that are identified as preferred by a majority of people. For example, it might implement a stock effluent management system to reduce nitrification of streams. In contrast, a deliberative approach using communicative and value rationality would engage the community, the farmers, and a range of experts in a series of workshops to identify and share understandings about their landscape. These might include collective consideration of the history of the landscape; the different values it represents for the people who live, work and visit; the dynamics and motivations that are driving change; and a vision or visions for how the landscape might support different aspirations for the future. Different possible ways to achieve agreed goals would be explored and debated, and a collaborative process established to implement change.

There are an increasing number of examples of this type of deliberative approach involving landscape scientists. Duff et al. (2009) reflect upon a decade of Australian experience of scientists working as facilitators with ranchers and indigenous communities. Austen (2011) reports upon a North American rural organisation which enrols science in support of cooperative and collaborative landscape actions. In New Zealand, Allen et al. (2011) describe a catchment based model of collaboration and deliberation involving land owners, communities, artists and scientists. A common feature of these examples is the engagement of science experts *within* a community based deliberative process.

This chapter explores how landscape science can engage with these notions of deliberative planning. We suggest that landscape ecology needs to do more than enrol social scientists in its problem solving teams. It needs to become engaged within collaborative, imaginative, and interactive forms of social process aimed at shaping future landscape pattern and character. In the next section we explore the limits of instrumental rationality in planning and place making, and introduce concepts from the 'deliberative' paradigm. We then examine decision making in two alternative landscape futures projects in the US, highlighting the way that different roles and interests interact discursively at different stages in place based studies. An example from Denmark then illustrates how experts can engage in a process of deliberation over the future of a rural community's own landscape. The chapter concludes by arguing that for landscape ecology to achieve the relevance it seeks, the objectivity and impartiality that is privileged within science needs to become reframed as one of several dimensions of value that are needed for decision making in a true landscape democracy (Arler 2008). We suggest that the role of scientists as experts must be expanded to include collaborators in a common and reflexive process of knowledge formation, and this raises both questions and challenges for the way that landscape ecology is practiced and validated.

12.2 Science, Rationality and Planning

12.2.1 Landscape Science as Rational Planning and its Limits

Modern science is widely characterised as an instrumental and solution driven endeavour, and this is reflected in the mainstream literature of landscape ecology. In reviewing the evolution of the discipline, Hobbs argued that landscape ecology had a unique role to play in "tackling todays major land use issues and in developing responses to the pressing problems arising as a result of human-induced global change" (1997: 1). The tools it has deployed for this applied programme have been drawn from both ecology and the geosciences (Weins 1992; Hobbs 1997). Debates over methodology have been framed within the science paradigm as a need to shift from established traditions of experimentation and falsification of formal hypotheses (Popper 1935, 1959) to investigative protocols better suited to the understanding and explanation of complex landscape systems (Pickett et al. 1994).

Landscape planning has also been largely characterised as a rational activity. Indeed, during the mid part of the 20th century, both planning and landscape theorists turned to science for their inspiration, and models of landscape planning processes privileged scientific understanding, technical analysis, and expert judgement (McHarg 1969; McAllister 1980). When the ecological science and rational planning traditions are drawn together, they create a trans-disciplinary research paradigm (Tress et al. 2003) of landscape ecological planning as an applied science (Ndubisi 2002), upon which contemporary proposals for increasing the relevance of landscape ecology draw directly. The process may involve a variety of modes of investigation, from empirical description and modelling (Opdam et al. 2002), and mediated and agent based modelling (Van der Belt 2004; Bakker and Doorn 2009), to an imaginative process of normative scenario building (Nassauer and Corry 2004), expressed recently as 'design in science' (Nassauer and Opdam 2008). Decision making processes are typically based on an assumption that different views can be reconciled and effectively integrated through rational examination and weighing up of options (Fry et al. 2007). Complex and frequently contested landscape dynamics are addressed by incorporating multiple scales of investigation (Mussachio 2009b).

This approach presumes well defined problems and clear decision making frameworks, in which values are a variable in the problem solving process (Termorshuizen and Opdam 2009). The role of experts is to lead the process (e.g. Steinitz et al. 2003). In practice, however, landscape ecologists have seldom had the impact they seek in place based problem solving (Stevens et al. 2007). A number of reasons have been suggested. These include the difficulty of re-scaling results and moving from the general to the particular (Stevens et al. 2007); insufficient engagement with the social sciences (Mussachio 2009a), and differences in the world view and culture of scientists on the one hand, and policy makers and managers on the other (Fischer 2009). Furthermore, the value frameworks of science and scientists are themselves subject to increasing scrutiny (Latour 2004). As we show below, even in rational, science based place-making processes (such as alternative futures planning) there are discursive moments—points at which the values of the experts involved shape the landscape outcomes by directing investigations down particular pathways. In short, the engagement of science, scientists, and scientific knowledge with planning and politics is now widely recognised as a major focus of tension more generally (Latour 2004), and a priority for investigation in landscape ecology in particular (Beunen and Opdam 2011).

The challenge of translating knowledge from the general to the particular is also a well-recognised problem in landscape planning. As Steinitz (1990) explains it in practical terms, what works well at one scale does not necessarily work well at another scale. A number of authors have addressed the problem. Nassauer and Opdam, for example, propose a stepwise process moving from science knowledge through generalizable pattern rules to place specific design solutions (2008: 642), and Theobald et al. (2005) propose the use of indicators to bridge between general knowledge and particular situations. Jensen et al. (2000) distinguish between the role of expertise in context independent knowledge—for example about genetic landscape processes- as opposed to context dependent knowledge about communities and their landscape practices, which is grounded in particular situations. Each of these may need different investigative strategies. However, the question remains of how to reconcile scientific credibility with problem salience, imagination, and local and political legitimacy (Cash et al. 2003).

The importance of legitimacy opens the issue of how best to understand and incorporate diverse social values. Opdam et al. (2002: 769) argued that "The future of landscape ecology lies in the understanding of how landscape pattern is related to the functioning of landscape systems, *placed in the context of (changing) social values* and land use" (our emphasis). This has led to a now widely accepted imperative to include social scientific expertise within the multidisciplinary teams undertaking applied landscape ecological projects (Mussachio 2009a). Nonetheless, introducing social science into landscape ecology per se does not necessarily achieve either practical results or legitimacy. There are a wide range of social science traditions and methodologies, and knowledge generated using methods aligned with the natural sciences may not adequately engage with ways of knowing about landscapes that are embedded in communities and practices. As Flyvberg (2001) demonstrated in an urban context, social sciences tend to be strongest where natural sciences are weakest, and vice versa- landscape ecology is strong on explanation and prediction, whereas social sciences overall may be most effective in interpretation and critique. It is for this reason that several authors have called for 'transdisciplinary' approaches (Tress et al. 2003; Mussachio 2009a), which can transcend particular methodologies.

However, drawing together knowledge from diverse sources is not a neutral process. Reflecting upon a decade of rural landscape ecological management in Northern Australia, Duff et al. (2009) note that attempts to 'integrate' across diverse interests and cultures seldom works because of power imbalances. Instead, they argue for collaborative 'working in combination', development of trust through embracing difference and developing shared understandings, brokering between interests, and investing heavily in communication to enhance adaptive learning. Flyvberg (2001: 154) reached a similar conclusion. Noting that "...power has a rationality that rationality does not know. Rationality, on the other hand, does not have a power that power does not know. The result is an unequal relationship between the two", he argued that to be effective in influencing urban policy and planning, social science had to set aside its ambition of adopting the instrumental rationality of the natural sciences, and turn instead to promoting greater rationality in expressing and debating values.

12.2.2 Deliberative Planning and Communicative Rationality

The deliberative paradigm (Forrester 1999) places emphasis upon processes of dialogue and argumentation, and upon communicative and value rationality. Forrester (1999) noted that societies construct their lived worlds through language, ideology and tradition, in which knowledge and power are intertwined, and this focuses attention upon the role of discourse in the planning process. A discourse is "a shared way of apprehending the world" (Dryzek 2005, p. 9). Discourses are thus descriptions of meaning, accounts, and stories (Foucault 1972) that reveal the worldviews that organize social life, including the planning processes themselves (Thompson et al. 1990). One can examine narratives about landscapes that are 'spoken' by individuals or groups, and particular storylines or narratives are inevitably associated with political power, in the sense that they can be used by individuals or groups to control the discussion, allow or not allow certain information to be used, persuade others, or get their way (Forrester 1989). Landscape ecological literature, for example, tends to privilege issues of biodiversity and ecological function over, say, spiritual or aesthetic values.

Deliberative planning draws in turn upon critical theory, a philosophical premise that seeks greater rationality in communication through which (ideally) all views and perspectives are given voice free of power bias (Habermas 1989; Leonard 1990; Dryzek 1987, 2000, 2005). Critical theorists argue that all communication is influenced by the point of view of the speaker, and hence any understanding of the world is based on individual biases and socially constructed understanding (Leonard 1990). Yet they believe that it is possible to be aware of one's own and other's biases so that mutual understanding is possible (Forrester 1989). Habermas (1989) proposed the idea of the 'public sphere' in which individuals consider what they are doing and determine how they will live together collectively (Keane 1984). An *authentic* public sphere is one in which the *ideal speech situation* exists, where those involved all have communicative competence, and can exchange views and understandings free from domination or deception (Dryzek 1987).

Habermas described solutions based on communicative rationality as *reasoned consensus* (Dryzek 1987). This does not require everyone to agree or even to like the eventual decision, but means that after consideration of all points of view, participants can live with a given course of action as the best option, given the situation. Of course, in practice, a planning discourse can seldom if ever take place in the ideal speech situation of communicative rationality. People express a diversity of interests to varying degrees and in varying ways, and reaching a reasoned consensus is difficult. Yet proponents of deliberative planning believe it is possible for people to change their position during the course of the planning process, at least to the extent needed to move forward towards a resolution of the issue at hand.

Closely linked to the idea of communicative rationality is the concept of value rationality. Initially developed by the social theorist Weber, value rationality is a process of deliberating openly upon the desired ends, rather than means. Dietz et al. (2005) identified three dimensions of environmental values that may be expressed in a community—usefulness, individual preference, and collective principles or morality. Value rationality is thus a process of determining desired outcomes in terms of how values might be realised, what individuals might prefer, and how to meet collective norms. This parallels the way Andrews (1979) conceptualised values in public decisions about landscape as intrinsic, preferences, and norms.

Flyvberg (2001) framed the application of value rationality in urban planning as a form of practical wisdom, and it is this melding of means and ends that characterises Duff et al. (2009) conclusions from their experience of collaborative landscape science and management in Australia. Similar combinations of modern science and practical wisdom are characteristic of best practice in co management of landscape resources in New Zealand (Wardle and Collins 2008) and reflect the emerging practice of collaborative landscape management in Denmark (Primdahl et al. 2010). As Demeritt put it, 'ultimately environmental narratives are not legitimated in the lofty heights of foundational epistemology but in the more approachable and more contested realm of public discourse (1994: 22).

In the next section of the chapter, we examine the implications of recognising and negotiating values in deliberation over landscape conditions and futures, in the context of the approach known as alternative ecological futures planning.

12.3 Alternative Futures as a Form of Deliberative Science

12.3.1 Alternative Futures Planning

Alternative futures (and scenario) planning provides useful insight into the consequences and challenges of a rational approach to planning through science. Development of scenarios and/or alternative futures has emerged as a powerful way to engage science with place, and projects typically use scientific knowledge to either predict landscape trajectories or to identify pathways towards desired future conditions. The advantage and appeal of identifying *alternative* pathways to the future, and different possible futures, rather than proposing a singular trajectory or outcome, is that it can accommodate a range of assumptions, where knowledge is uncertain, and enables comparative evaluation of alternative solutions. Most alternative futures and scenario projects are expert led (Hulse et al. 2002) and in many cases are entirely expert based (Steinitz et al. 2003). They are almost always interdisciplinary (Tress et al. 2003).

Studies that seek knowledge through projecting alternative futures have a history dating back to at least the 1950s, when Herman Kahn used the term 'scenario' to identify long range depictions of the future concentrating on "causal processes and decision points" (Kahn and Weiner 1967). In defining scenarios, Shearer (2005) identifies four common features- they are fictional descriptions of future change; they describe related situations; they describe what could happen as opposed to what will happen or even is likely to happen; and they organize knowledge within explicitly defined frameworks. In landscape ecological planning, scenarios are distinguished from alternative futures by their focus (Steinitz et al. 2003; Nassauer and Corry 2004; Shearer 2005). *Scenarios* describe different sets of assumptions that underlie potential change in landscape pattern (Hulse et al. 2002; Opdam et al. 2002; Nassauer and Corry 2004). *Normative* landscape scenarios describe futures that should exist or are preferable and can "inspire policy by providing images of landscapes that could meet societal goals" (Nassauer and Corry 2004, p. 344). They lead to processes of making alternative decisions and actions that could result in different courses of events. Therefore, they describe change that could, but not necessarily will, take place over time. Scenarios in turn result in *alternative futures,* which describe the functional consequences of scenarios (Nassauer and Corry 2004). Thus scenarios can be thought of as processes, while alternative futures can be seen as results of processes- the landscape outcomes.

From this perspective, alternative futures can be analyzed at many different times from the near future to very distant future. The alternative future at any given time is uniquely based on the scenarios (assumptions, decisions, actions, and events) that lead to it. Both scenarios and alternative futures are fictional in the sense that they have not yet occurred: actual decisions, actions, and events will lead to the concrete conditions of the future. Emmelin (1996) therefore proposed a methodology through which scenario studies and future landscapes can be used for landscape specific impact assessments of general policy proposals, such as changes of legislation and national/regional policies including agricultural policy.

The role of the scientists (such as landscape ecologists) in alternative futures is typically framed in terms of independent experts who investigate and present knowledge about alternatives and how they perform, from which the elected political decision makers can then choose a preferred policy. In some cases, there is involvement of stakeholders such as local communities in the development of alternative scenarios, and experts may be involved in identifying community preferences or values for different scenarios.

Nassauer and Corry (2004) and Nassauer and Opdam (2008) explicitly frame the alternative futures process as a scientific investigation, in which alternative normative outcomes are presented as hypotheses about how landscape *should* change, which can be tested under various assumptions about landscape dynamics. The results are then conveyed to political decision makers and citizens to act upon.

In this, the models follow Dryzek's (2005) argument that in order to ensure the critical integrity of the deliberative process, deliberation about what *should or could* be an outcome needs to be separated from consequential political decisions about what *will* be undertaken. The expert role is framed as a scientist or planner, not a decision maker.

The theoretical logic of separating the science deliberation from decision making is based upon a desire to ensure that analysis and deliberation is open, objective and unsullied by power imbalances. However, in expert led processes the practical effect can be quite the reverse of what is intended. Separation of stakeholders and decision makers from the process of investigating and analysing conditions and possibilities can lessen their commitment to the outcomes of this deliberation. This is exasperated in situations where office holders change during the process, and newcomers have little sense of 'ownership'.

The presumption of committed but independent scientists providing impartial advice to the decision-makers also fails to stand up to scrutiny when the evidence is considered. Analysis of several alternative futures cases suggests instead that quasi-political decisions are involved throughout the alternative futures modelling process. Alternative futures planning approaches in practice comprise a series of *discursive moments* that involve both deliberation *and* value based decision making. The decisions made at each moment impact all subsequent phases of the planning process, the science upon which it draws, and the eventual planning outcomes. Hence engagement of stakeholders and communities with the science is an essential requirement throughout the process, and this inevitably exposes scientists to the value rationality of decision making.

The two case studies upon which we base this argument took place in the US Mountain West in the latter part of the 1990s and early years of the 2000s (Fig. 12.1). The first case is the San Pedro project (Steinitz et al. 2003), located in the semi-arid region in southeast Arizona and northern Sonora, Mexico and includes the San Pedro Riparian National Conservation Area (SPRNCA). Research

Fig. 12.1 Map of western United States showing the location of the two projects in Oregon and Arizona. The Arizona project also included portions in Sonora, Mexico

was conducted by a multidisciplinary team assembled from Harvard University, regional based university departments and institutes, and the United States Army, and involved extensive landscape modeling using digital technologies. The San Pedro report identifies three major scenarios, with variations of each. They included current trajectories of change in development and water use, constrained scenarios, and open development orientated scenarios. San Pedro is one of a series of alternative futures projects undertaken by Harvard University for US federal agencies, and exemplifies the expert led approach to applied landscape science in alternative futures. The projects are tightly focused, technically sophisticated, and completed in relatively short time frames (typically 2 years or so).

The second case is the Willamette Valley, Oregon (Baker et al. 2004), which is bounded on the west by the Coastal Range and on the east by the Cascade Mountain Range. Two thirds of area is forested, primarily in upland areas, while much of the valley has been converted to agricultural use. Projected population growth is expected to place enormous demands on water and land resources. The study was funded by the U.S. Environmental Protection Agency (EPA) and completed by the Pacific Northwest Ecosystem Research Consortium (PNW-ERC) involving researchers at Oregon State University, the University of Oregon, the University of Washington, and the U.S. EPA, and again used sophisticated digital landscape models. Three visions of the future were created through to the year 2050—Plan Trend, Development and Conservation. The Willamette project exemplifies a strongly stakeholder based approach to alternative futures. Whilst also technically sophisticated, it is particularly notable for the institutional arrangements set up to engage a wide range of stakeholders and communities throughout the process and to assure that all scenarios would include plausible decisions and management practices as defined by stakeholders. The project ran for around a decade.

12.3.2 Discursive Moments in Alternative Futures

Analysis of the two contrasting cases has highlighted that irrespective of the style of engagement, both of these science based exercises involved a number of points at which decisions had to be made about similar questions, each of which would materially affect the project outcome. Each decision point- that we have termed 'discursive moments'—can be viewed as a fork in the road, a mix of deliberation and values based decision that determines future possibilities of both action and outcome. The moments are: identification of project scope; selection of the method and selection and assembly of the planning team; determination of the project design; data collection and management; development, selection and testing assumptions of scenarios; assessment of the effects of scenarios upon future landscapes; and selection of implementation outcomes and outputs.

1. *Identification of project scope*: This moment occurs before the project can begin. The institution(s) must become aware of a landscape management problem. It is likely to be motivated by the interests and concerns of key constituents, and previous studies might have defined underlying goals to be achieved. At a deeper level, questions about normative versus exploratory and deductive versus inductive approaches (Shearer 2005) will set the framework for the study. During this moment, questions about *what* and *why* may have lasting influence on the nature of communication throughout the project, and upon its possible outcomes.
2. *Selection and assembly of the planning team and planning method*: There is a wide variety in practice in the manner of selecting and assembling alternative futures planning teams, as well as the institutions represented. The inclusion or exclusion of particular disciplines or stakeholders will materially shape the scope and nature of *how* the science undertaken, *who* is involved, and its possible findings, as well as the way these findings might be translated into actions.
3. *Project design*: Although alternative futures projects share common characteristics (Baker et al. 2004) each focuses on unique ecological and social issues, incorporates distinctive approaches to stakeholder groups and public agencies, and utilizes its own data management system. Further, the fundamental rational for approaching scenarios and assumptions is defined during project design.
4. *Data selection and management*: Steinitz (1990) identified a range of fundamental questions about landscape that drive the landscape planning and modelling process. They include: How should the landscape be described? How does the landscape function? How does one know whether it works well or not? The responses shape the scope and character of the process.
5. *Selection and testing assumptions of scenarios:* Although there are an infinite number of possible scenarios, it is only feasible to pursue plausible ones. The makeup of those making these decisions and the process involved can determine the number of scenarios, the ease of modelling ecological and cultural systems, and the degree of political acceptance of the report.
6. *Assessing the effects of scenarios (futures)*: This phase uses science to predict outcomes, and implies a range of value judgements—from the most basic orientation of the process (is it testing hypotheses about normative futures, or evaluating impacts of alternative scenarios upon a given landscape), to detailed determination of criteria for evaluation.
7. *Selection of implementation outputs and outcomes*: This is perhaps the most difficult moment to examine, given the length of time required for political institutions to implement decisions, and the time required for implementation to make on the ground changes in landscape conditions. Nonetheless, implementation processes and plans are profoundly political, and hence express the values of the decision makers.

The implications of these moments for the nature of the landscape science and its relationship with wider planning processes are profound. According to Stein and Harper (2003) both a combination of 'internal' and 'external' perspectives is required to ensure effective, democratic and dialogical planning. An internal perspective means that the planner (or the landscape ecologist in our case) must participate in the planning process and relate to other participants as subjects (rather than objects), in order to fully understand the values behind the issues in question and to participate as a collaborator in the process of deliberation based upon value rationality—it thus provides social and political legitimacy.

An 'external' perspective analyses the planning from outside, as an object, using various theoretical 'lenses', and is needed in order to understand their relative effectiveness in achieving functional outcomes. Without this external perspective the participants will be unable to critically explain and evaluate the process and outcomes. An external perspective thus provides scientific credibility. However, without the insights achieved through (internal) participation the landscape ecologist will have no way to fully justify proposed planning solutions, apart from either individual interests or very general assumptions of 'right' and 'wrong'.

Traditional expert involvement places landscape science in an overtly 'external' perspective, although as we have shown above, in practice it still makes 'internal' decisions. A collaborative approach based upon value rationality involves the landscape scientists in the local 'internal' process and therefore enables them to "integrate and apply external knowledge into the internal framework." (Stein and Harper 2003, p. 132). In the next part of the chapter we present a case study of such expert-informed deliberation in place.

The practical effectiveness- or otherwise- of the two contrasting approaches in the case studies also deserves some comment. Outwardly, the strongly stakeholder focused Willamette project appears to have resulted in a more tangible outcome, in the form of conservation policies adopted and promoted by the EPA and local not-for-profit resource agencies. It could be inferred that the sense of ownership and engagement that resulted from the collaborative science process led to a commitment to act. In contrast, the San Pedro project did not appear to lead to a cohesive land planning response. However, there were consequences- and a decade later it is possible to identify significant changes in the water management regime within the military area. Hence the obvious planning outcome of a process may not be the only outcome, and a nuanced interpretation is needed. This is typical of alternative futures projects, and reflects another contrast between science as problem solving (outcome: problem solved, or not); and science as part of a deliberative process (with an outcome of improved understanding and collaboration, expressed in many ways). In the next section, we illustrate this more nuanced role for science through a Danish case study.

12.4 Deliberative Spatial Strategy Making in Place

12.4.1 Spatial Strategy

The distinguishing feature of landscape ecology is its concern for spatial relations in ecology, (Forman and Godron 1986) and how such spatial knowledge can be translated into practical land management and planning outcomes (Dramstad et al. 1996). In an ever more resource constrained world, knowing how best to act spatially—where to invest, where to protect, how to resolve competing demands on particular places, and how to build communities in place—is a critical role for landscape science. The recent growth of landscape ecology and its concern for relevance has paralleled the re-emergence of spatial strategy as a dimension of planning more generally.

Spatial plans were a key feature of town and regional planning as it developed in the mid 20th century, reflecting both the driving motivations- including management of land use conflicts, redevelopment of regions following wartime damage, and direction of new urban growth—and the practical implementation tools, particularly land use controls (Hall et al. 1973). Spatial relationships were also fundamental to the emergence of environmental planning in the 1960s, with its focus upon resource assessment and protection (McHarg 1969), and the development of spatial planning tools such as green belts and green ways (Ahern 2002).

The dominance of spatial thinking declined in many planning constituencies during the latter part of the 20th century as a result of two outwardly opposing dynamics- the emergence of participatory and advocacy planning (Davidoff 1965), and the ascendancy of more neoliberal planning paradigms that emphasised market processes (Friedmann 1987). However, several factors have now reversed this trend. They include: first, the realisation that participatory planning depends for much of its power and legitimacy upon the location of constituencies in particular places; second, the recognition that planning mechanisms based primarily upon non spatial market processes fail to deal with the cumulative consequences of development; third, that space is an increasingly scarce resource in urbanising regions; and fourth, that place itself is of great economic value—as technology overcomes the friction of distance for production, the quality of particular places becomes a major driver of economic success, as both workers and consumers seek out distinctive places to live, shop and work. Hence space has re—emerged as a key focus of strategic and place based planning.

Strategy has a number of interpretations. According to Shearer (2005), strategy can be summarised as having three possible dimensions: it may be a pre-active process, anticipating uncertain futures and establishing strategies to maintain profitability or viability of businesses or communities in the face of such uncertainty; it may be directive, guiding resources through strategic policy towards some given end; or it may be pro-active, actually making futures through strategic interventions. Strategy of all three kinds may also be seen as the combination of long term *visions* and short-termed *actions* of various kinds and nature (Albrechts 2004). Such visions must be shared by the groups, institutions and other stakeholders on whom the strategy depends.

Spatial strategies may integrate multiple dimensions- such as conservation of valued assets or resources, allocating investment or infrastructure to achieve particular purposes; and envisioning desirable future conditions to empower participants to act. Healey (2009) has analysed the spatial strategy making process with multiple stakeholders involved in the complex task of formulating clear and agreed 'directions' of spatial development. She argues that four dimensions of such a strategy making process are usually in play when such a process is unfolding: (1) *Mobilising attention* to the whole, that is creating a shared interest in the strategy, (2) *Capturing the situation*, thus clarifying the present context, its historic background, and the central goals of the strategy. (3) *Mobilising internal and external resources,* including knowledge. (4) *Generating a frame for strategy* (with a program over time and key projects). In a landscape context, the landscape ecologist obviously has much to contribute to the second and third knowledge focused dimensions, but it would have to be done within a practice context. The first and fourth dimensions require fundamental skills and knowledge in situated planning. In combination, the four dimensions of spatial strategy making expressed in this way are an example of deliberative planning rather than instrumental problem solving.

12.4.2 Place and Place Making

Place is a widely used concept in social science and spatial planning. It has varying definitions, but most express the three dimensions identified by Relph (1976), and conceive place as a nexus of distinctive biophysical characteristics, socio economic activities, and cultural significance- a concentration of form, practice and meaning in a defined locality (Hillier and Rookesby 2005). Place-making (Dovey et al. 1985; Schneekloth and Shibley 1995; Healey 1998) has been promoted by a range of disciplines as a process of active creation and cultivation of such qualities—through physically shaping places, empowering communities to collaborate in place building practices, and conserving, nurturing and projecting symbols of place.

Place is one level in a multilevel framework of phenomena, connecting geographic pattern with ecological and social process; and general knowledge with context dependent understanding. It has an uneasy relationship with landscape, and is frequently conflated, yet the two may also be conceived as fundamentally different. In his work on globalisation Castells (2000) distinguishes between the 'space of flows' as the way the material world is organized in interlinked networks to enable the fast growing flows of goods, information, energy, people and the 'space of place' in which people are living their daily life. He defines a 'place' as "…..a locale whose form, function, and meaning are self-contained within the boundaries of physical contiguity" (2000: 453). Spaces of place and of flows are very different, yet have to be seen together, like the external and internal perspectives described above: "The major danger in such a new historically spatial dichotomy is the breakdown of communication between power and people, between cities and citizen, and ultimately between a spatial technocratic instrumentalism and localistic fundamentalism" (Castells 1992: 75).

A physically bounded landscape may function as a place, a defined locale, but more typically an extensive landscape is a mosaic of contiguous places, just as it is a mosaic of ecosystems. The extent to which a landscape may be seen either as a defined area of space within which 'places' are located, or as a 'place' itself, as in the sense of a self-containing whole, was illustrated in a study of how Danish farmers in two different landscapes responded to the following question. The question was asked half an hour into a longer interview about how the farmers have experienced change in their landscape: "If you were talking on the phone with a remote relative who has not visited your area, and the relative asked you how it was where you live, what kind of landscape or place was it—how would you then reply?" The farmers (15 in each landscape) gave two kinds of answers, largely distinguished by the type of landscape in which they live and work. In one of the landscapes they all proudly referred to how it was a very nice area—located close to very nice (and for Danes well known) places. In the other landscape no one mentioned nearby attractions such as the spectacular dune systems on the North Sea Coast less than 10 km away. Instead they all referred to experiential features of the local landscape, such as the peacefulness (with no main roads), the flat landscape with the high sky (high 'ceilings'), and the new forests and the wildlife which came with them (Primdahl et al. 2010).

In the first landscape farmers talked about their landscape as a space relative to other locations, whereas in the second landscape they talked about their specific place within the landscape. One of the main differences between the two landscapes was that in the latter (place defined) landscape there has been a long tradition of co-operation on landscape issues, from heathland reclamation (in the 1950s) to afforestation (in the 1990s), as well as a shared and successful struggle against plans for locating a regional waste dump in their area, and common grazing of semi-natural salt marshes. These collective experiences may well have contributed to the strong sense of place in this area. In the former situation, landscape was an abstract concept, in the latter case it was lived—a distinction that has been widely recognised in the geographical literature, and identified in other similarly contrasting landscapes in very different countries (Primdahl and Swaffield 2004). Landscape as place becomes a focus of governance and spatial strategy making, framing attention and action in the way described by Healey (2009).

Landscape ecology has potential to contribute concepts and knowledge to both conceptions of landscape: landscape as a mosaic- which is the conventional focus for the discipline, or landscape as a concentration of meaning and experience, a locale. Most attention has been upon the former, with landscape ecology offering descriptive and explanatory knowledge about the relationships between landscape structure and function. In the latter case, of landscape as locale, perhaps the most valuable contribution of landscape ecology as a science is to inform communities and stakeholders about the landscape *context* in which they live, its characteristics and how it functions, and how this context shapes their everyday lives. In the next section we outline an example of place based spatial strategy making that draws upon such landscape ecological understanding.

12.4.3 Collaborative Local Planning in Denmark: The Lihme Project

Danish rural landscapes are farmed intensively by highly specialised pig, dairy, or cropping farmers producing commodities mainly for the world market. More than 90 % of all farm land is arable and affected by high concentrations of nutrients and pesticides. However, many of these rural landscapes are also relatively densely populated, and are currently affected by urbanisation processes, leading to significant in and out migrations of people. As a result, the vast majority of people living the rural regions are no longer commercial farmers or farm workers, and they are increasingly seeing the landscape and its values (or potential values) as a key resource for quality of life- thus attracting people to the area. These new rural populations are expressing interest in local landscape initiatives, and together with an administrative reform that has led to a decentralising of spatial planning, this has resulted in a growing interest in collaborative landscape planning. The focus of this section is a planning experiment in Lihme parish in central Jutland.

Lihme was one of five local areas included in an experimental planning project carried out by the municipality of Skive in close co-operation with researchers from University of Copenhagen. The project ran for 2 years with the aims to develop new forms of collaborative landscape planning and to develop new models for multifunctional rural landscape patterns. The key agent to drive the planning process forward was a working group in each of the five areas. Each group was established first by the municipality which contacted a few citizens in each area and asked them to form a group and appoint a leader. In Lihme the group varied over time between 7 and 10 representatives of the local community, including farmers as well as non-farmers.

The project goal was to create a strategy plan for the landscapes in the parish which had local 'ownership' and which could be incorporated in the legally binding municipal plan. The group worked closely together with a planner from the municipality, and this is critical to its success, with frequent contacts of different kinds with scientists and professional experts. The process started with a meeting for all five groups where the project objectives were outlined, and the first phase (app. 5 month) was to work out a broad analysis of strengths, weaknesses, opportunities and threats (SWOT) for the future socio-economic development of the parish.

Focus then shifted to the rural landscapes in Lihme, and this second phase started with a two day workshop for all five groups at which the SWOT analyses were presented and discussed with municipal planners, researchers and professionals. The first draft of landscape character maps carried out by a landscape architect was presented and given to the working groups to be discussed locally and modified to ensure it expressed local citizens perception of their 'own' landscapes. Excursions and various thematic lectures (including a presentation of a simple diagnostic tool for evaluation of the 'landscape conditions') were also part of the workshop. During the next four months the working group shared understanding of how the landscape functioned, and developed a landscape strategy plan for their parish. Regular meetings with municipal planners and workshops with researchers and professionals were included in this process to mobilise external as well as internal knowledge and ideas. The contribution of landscape ecologists can be a vital part of the diagnostic process- characterising the condition of the landscape as well as contributing to the preparation of a feasible strategy that recognises the possibilities, potentials and constraints of the landscape context.

The third phase of the strategy—the final design—started with the presentation of the strategy draft to a panel of 'landscape experts' (from university, consultancy, and public institutions including Skive municipality). After this presentation an invited expert panel (with an ecologist, a forester, a landscape historian, two landscape planners and a farm building architect) presented an alternative draft strategy worked out during a one day workshop. The 2 hour discussion following these two presentations functioned as a sort of 'confrontation dialogue' and turned out to be highly productive in shaping the final ideas for the strategy. During the next few months the final strategy was drawn up by one of the landscape planners participating in the panel in close contact with the working group. The strategy includes proposals for new green corridors linking the village to surrounding habitats, a new village forest, new recreational trails and new developments at the harbour in the village (Fig. 12.2).

Finally the whole strategy was presented and discussed at a public meeting in the parish. The community essentially took ownership of the strategy and parts of the strategy (including trail and corridors) are being implemented. Five thematic working groups in the parish are responsible for different aspects of the strategy, which has also been incorporated into the municipal plan.

The four dimensions of Healey's spatial strategy making process (see above) have been dealt in a number of different ways and at different stages in the Lihme process (Dias-Sardina et al. 2012). In this context, the different ways to mobilise and confront internal and external resources concerning knowledge, values and imaginations have been especially fruitful. However, more experience and more design proposals are needed before a more general culture of collaborative landscape planning can evolve. Systematically developed 'patterns' involving landscape ecologists, as proposed by Nassauer and Opdam (2008), would be highly beneficial, particularly in helping identify critical patterns and processes, and in helping prioritise where management interventions can be most effective and

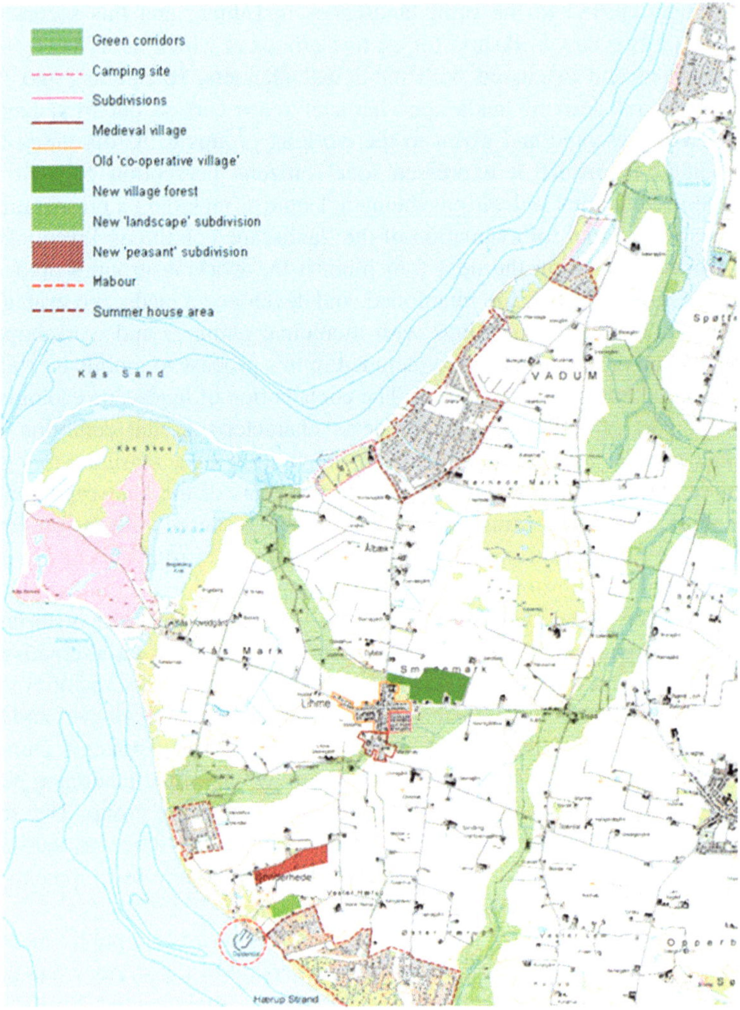

Fig. 12.2 Landscape strategy for the parish of Lihme. Key elements in the strategy are: new walking trails (not shown on the map), a new village forest, new 'rural subdivisions', new system of green corridor, new development plan for the habour. *Source* Primdahl et al. (2010)

efficient. However, such expert generated patterns cannot substitute or function as principal 'design solutions' for local rural landscapes. Local ownership of the design is essential if a landscape strategy plan is going to function as a frame for the innumerable decisions and actions taken by individuals and groups driving landscape change processes over time. One way to envisage this relationship between experts and locals in generating patterns is that the expert role is enabling, offering a spatial language and helping locals interweave the systematic knowledge of landscape ecology into their distinctive and evolving local landscape biography.

12.5 Landscape Ecology and Landscape Democracy

12.5.1 A True Landscape Democracy and Deliberative Landscape Science

What directions do these examples suggest about ways to reconcile the communicative rationality of deliberative landscape planning with the more technical and problem focused methods of landscape science? For this we need to return to the question of values. Responding to the imperatives of the European Landscape Convention, Arler (2008) has discussed his notion of 'a true landscape democracy' (an expression used in the explanatory report of the convention) that recognises three complementary types of values and decision making: self-determined, co-determined, and objective. Self-determined values express personal feelings and preferences, and express the dimension of landscape values that are most typically emphasised by economists and many social scientists, based on psychophysical or cognitive measures, and are widely used in landscape modelling. Co-determined values arise from informed and open deliberation over collective decisions- they are more than the aggregate of individual feelings, and express values arrived at socially. Objective values are based upon evidence and rational argument rather than power or rights, and correspond to the conventional 'truths' of science.

In recognising these different but complementary 'truths' of landscape, and the different ways in which they are shaped and identified, Arler then argues that they create a suite of possible and desirable roles for experts, as collaborators, brokers, mediators, and connoisseurs, as well as the source of conventional technocratic expertise. Involvement in landscape deliberations in what he describes as a true landscape democracy thus requires science experts to become participants in a conversation in which their knowledge is no more privileged than any other. Hobbs (1997) prefigured this shift, arguing that the future is made collaboratively, and Johnson and Campbell argued that implementing strategies to strengthen links between ecological science and public involvement will require 're-conceptualisation of the roles of both scientists and stakeholders so as to improve the integration of applied ecological science with democratic decision making' (Johnson and Campbell 1999: 502). Alternative futures planning based on collaborative institutions can provide one model, and other potential models that may help integrate science and collaboration include adaptive ecological management (Holling 1978; Williams and Brown 2012), and various forms of decision support, such as mediated modelling (Van der Belt 2004) and structured decision making (Gregory et al. 2001). The critical feature throughout, however, remains that which lies at the heart of the deliberative planning paradigm- the need to subsume the power of expertise within a situated process of collaborative deliberation.

12.5.2 Some Questions and Challenges

Re-conceptualising landscape ecological science within a collaborative and multivalent landscape planning paradigm thus destabilises the notion of science expertise as the 'given' role of landscape ecologists. As landscape scientists in post-colonial countries have found (Duff et al. 2009), engaging with collective forms of knowledge and practical wisdom requires development of a new humility and sensitivity to the possibility of multiple ways of knowing.

This raises a number of interesting questions for landscape ecology as it engages with planning and design. Landscape science is evolving towards a global discipline, and many of the drivers for knowledge are issues and problems that exist at a global scale. However, deliberative landscape science in the way we have described depends significantly upon the local public culture of decision making. Hence landscape ecology becomes far more context dependent that has been acknowledged to date, and this has profound implications for reporting and peer review. For example, how can reflective case studies on collaborative landscape projects be more widely and 'productively' be brought into the core journals of landscape ecology? How can scientists maintain credibility for their expertise while participating in values based deliberation (Cash et al. 2003)?

Nassauer and Opdam (2008) argue that design in science can fulfil this goal, but there is a risk that this continues to privilege science knowledge. A reframing of the process such that landscape ecological knowledge becomes one of several sources of knowledge that shapes landscape archetypes and design solutions can move values from being a sub set within the science endeavour, to become the framework within which wider deliberation occurs. The objective values of science thus become a participant in a conversation, rather than social values becoming a subset of science knowledge. The relationship is inverted.

One pathway may be to recognise the distinction noted earlier, between internal and external views. There are interesting precedents in social science reporting for the way that investigators can reframe their roles and findings to recognise that new knowledge may be co-produced with local participants. However, this raises questions for the editors and reviewers of science journals in landscape ecology, who need to balance demands for science legitimacy with the growing calls for relevance to place based landscape issues. Whilst there are multidisciplinary journals that specialise in such contextual science, if it remains marginalised from the mainstream journals then context sensitive science is unlikely to gain credibility in the discipline.

Finally, as Flyvberg (1998) argues, science knowledge is power. How will the discipline manage imbalances of social and economic power in landscape ecological projects? How can the increasingly global discipline of landscape ecology be accessible to the needs of different types of planning contexts and constituencies? Can such science contribute in an even handed way to the 'authentic public sphere' of deliberation proposed by Habermas?

Seeking greater relevance for landscape ecology is therefore a challenging pathway. Current models for enhanced engagement with planning and design tend to address social and cultural values by creating a subset of social science knowledge within instrumental landscape models. The insights of deliberative planning suggest that a more fundamental reorientation may be needed, by which landscape ecological knowledge becomes a subset of a wider framework of landscape values, and this raises challenges and opportunities for the science. Shifting from a focus upon technical knowledge to practical wisdom requires engagement with social processes as well as biophysical landscape conditions, and in a deliberative landscape democracy, neither is privileged.

Acknowledgments Observations upon alternative futures planning were drawn from work undertaken towards a PhD thesis by Mark Hoversten. We appreciate the invitation to submit the chapter by the Chinese Academy of Science and the insightful comments and suggestions from an anonymous reviewer. Michelle Collings helped prepare the manuscript.

References

Ahern J. Greenways as strategic landscape planning: theory and application. Wageningen NL: Wageningen University; 2002.

Albrechts L. Strategic (spatial) planning re-examined. Environ Plann B: Plann Des. 2004;31: 743–58.

Alexander LR. Rationality revisited: planning paradigms in a post-postmodernist perspective. J Plann Educ Res. 2000;19:242–56.

Allen W, Fenemor A, Kilvington M, Harmsworth G, Young RG, Deans N, Horn C, Phillips C, Montes de Oca O, Ataria J, Smith R. Building collaboration and learning in integrated catchment management: the importance of social process and multiple engagement approaches. NZ J Mar Freshwat Res. 2011;45(3):525–39.

Allmendinger P. Towards a Post–Positivist Typology of Planning Theory. Planning Theory. 2002;1(1):77–99.

Amdam R. Empowerment Planning in Regional Development. European Planning Studies. 2010;18(11):1805–19.

Andrews ANL. Landscape values in public policy. In: Elsner R and Smardon R, editors. Our National Landscapes: A conference on applied techniques for analysis and management of the visual landscape. USDA Forest Service, Berkeley Ca., Pacific SouthWest Forest and Range Experimental Station; 1979.

Arler F. A true landscape democracy. In: Arntzen S, Brady E, editors. Humans in the land: the ethics and aesthetics of the cultural landscape. Oslo: Oslo Academic Press; 2008. p. 75–99.

Austen DJ. Landscape conservation cooperatives: a science-based network in support of conservation. The Wildlife Professional, Fall. 2011;2011:32–7.

Baker JP, Hulse DW, Gregory SV, White D, Van Sickle J, Berger PA, et al. Alternative futures for the Willamette river basin. Ecol Appl. 2004;14(2):313–24.

Bakker MM, van Doorn AM. Farmer-specific relationships between land use change and landscape factors: Introducing agents in empirical land use modelling. Land Use Policy. 2009;26(3):809–17.

Beunen R, Opdam P. When landscape planning becomes landscape governance, what happens to the science? Landscape Urban Plann. 2011;100:324–6.
Cash DW, Clark WC, Alcock F, Dickson NM, Eckley N, Guston DH, Jager J, Mitchell RB. Knowledge systems for sustainable development. Proc Nat Acad Sci. 2003;100(14):8046–91.
Castells M. The world has changed: can planning change? Urban Landscape Plann. 1992;22: 73–8.
Castells M. The rise of the network society. 2nd ed. Oxford: Blackwell; 2000.
Davidoff P. Advocacy and pluralism in planning. J Am Plann Assoc. 1965;31(4):331–8.
Demeritt D. Ecology, objectivity and critique in writings on nature and human society. J Hist Geog. 1994;20(1):22–37.
Dias-Sardina I, Primdahl J, Craveiro D, Kristensen LS. Participatory processes for strategic planning in two rural regions: the case of Sáo Domingos Mine, Alentejo, Portugal and the case of Lihme, Jutland, Denmark. In: Paniagua Á, Bryant R, Kizos T, editors. The political ecology of depopulation: inequality, landscape, and people. Ceddar: Zaragoza; 2012. pp. 151–71.
Diertz T, Fitzgerald A, Shwom R. Environmental values. Annu Rev Environ Resour. 2005;30: 335–72.
Dovey K, Downton P, Missingham G. editors. Place and Placemaking. Melbourne; 1985.
Dramstad WE, Olson JD, Forman RTT. Landscape ecological principles in landscape architecture and land-use planning. Washington: Island Press; 1996.
Drysek J. Deliberative democracy and beyond: liberals, critics, Contestations. Oxford: Oxford University Press; 2000.
Dryzek JS. Disursive designs: critical theory and political institutions. Am J Polit Sci. 1987;31(3):656–79.
Dryzek JS. The politics of the Earth: environmental discourses. 2nd ed. Oxford: Oxford University Press; 2005.
Duff G, Garnett D, Jacklyn P, Landsberg J, Ludwig J, Morrison J, Novelly P, Walker D, Whitehead P. A collaborative design to adaptively manage for landscape sustainability in North Australia: lessons from a decade of cooperative research. Landscape Ecol. 2009;24:1135–43.
Emmelin L. Landscape impact analysis: a systematic approach to landscape impacts of policy. Landscape Res. 1996;21(1):13–34.
Fischer F. Democracy and expertise: reorientating policy inquiry. Oxford: Oxford University Press; 2009.
Fischer F, Forrester J. The argumentative turn in policy analysis and planning. Raleigh: Duke University Press; 1993.
Flyvberg B. Rationality and power: democracy in practice Chicago. Illinois: Chicago University Press; 1998.
Flyvberg B. Making social science matter: why social inquiry fails and how it can succeed again. Cambridge: Cambridge University Press; 2001.
Forman RTT, Godron M. Landscape ecology. New York: Wiley; 1986.
Forrester J. Planning in the face of power. Berkeley: University of California Press; 1989.
Forrester J. The deliberative practitioner: encouraging participatory planning process. Cambridge Mass: MIT Press; 1999.
Foucault M. The archaeology of knowledge and the discourse on language. New York: Pantheon Books; 1972.
Friedmann J. Planning in the public domain: from knowledge to action. Princeton: Princeton University Press; 1987.
Fry G, Tress B, Tress G. Integrative landscape research: facts and challenges. In: Wu J, Hobbs R, editors. Key topics in landscape ecology. Cambridge CUP; 2007. p. 246–68.
Gregory R, McDaniels T, Fields D. Decision aiding, not dispute resolution: creating insights through structured environmental decisions. J Policy, Anal Manage. 2001;20:415–32.
Habermas J The structural transformation of the public sphere. (Lawrence TB. Trans). Massechusetts Institute of Technology. Cambridge. 1989.

Haines-Young R. Sustainable development and sustainable landscapes: defining a new paradigm for landscape ecology. Fennia. 2000;178(1):7–14.
Hall P, Graley H, Drewitt R The containment of urban Britain. Allen and Unwin London. 1973.
Healey P. Planning through debate- the communicative turn in planning theory. Town Plann Rev. 1992;63(2):233–49.
Healey P. Collaborative planning in a stakeholder society. Town Plann Rev. 1998;69(1):1–21.
Healey P. In Search of the "strategic" in spatial strategy making. Plann Theory Practice. 2009;10(4):439–57.
Hillier J, Rookesby E, editors. Habitus: a sense of place. Aldershot: Ashgate; 2005.
Hobbs R. Future landscapes and the future of landscape ecology. Landscape Urban Plann. 1997;37:1–9.
Holling CS (1978) Adaptive environmental assessment and management. Caldwell NJ, John Wiley and Sons.
Hulse D, Gregory S, Baker J (2002) Willamette River Basin Planning Atlas. In: Hulse D, Gregory S, Baker J, editors. Corvallis. Oregon State University Press. Oregon. 2002.
Jensen MB, Person B, Guldager S, Reeh U, Nilsson K. Green structure and sustainability-developing a tool for planning. Landscape Urban Plann. 2000;52:117–33.
Johnson BR, Campbell R. Ecology and participation in landscape—based planning within the Pacific Northwest. Policy Stud J. 1999;27(3):502–29.
Kahn H, Weiner AJ. The year 2000: a framework for speculation on the next thirty years. New York: Macmillan; 1967.
Keane J. Public life and late capitalism. Cambridge: Cambridge University Press; 1984.
Latour B. Politics of nature: how to bring sciences into democracy. Cambridge: Harvard University Press; 2004.
Leonard ST. Critical theory in political practice. Princeton: Princeton University Press; 1990.
Luz F. Participatory landscape ecology-a basis for acceptance and implementation. Landscape Urban Plann. 2000;50:157–66.
McAllister DM. Evaluation in environmental planning. Cambridge: Mass: MIT Press; 1980.
McHarg I. Design with nature. NY: Doubleday; 1969.
Mussachio LR. The ecology and culture of landscape sustainability: emerging knowledge and innovation in landscape research and practice. Landscape Ecol. 2009a;24:989–92.
Mussachio LR. The scientific basis for the design of landscape sustainability: A conceptual framework for translational landscape research and practice of designed landscapes and the six Es of landscape sustainability. Landscape Ecol. 2009b;24:993–1013.
Nassauer JI, Corry RC. Using normative scenarios in landscape ecology. Landscape Ecol. 2004;19:343–56.
Nassauer JI, Opdam P. Design in science: extending the landscape ecology paradigm. Landscape Ecol. 2008;23:633–44.
Naveh Z (2005) Towards a transdisciplinary landscape science. In: Weins J, Moss M, editors. Issues and perspectives in landscape ecology. CUP. p. 346–54.
Ndubisi F. Ecological Planning: a historical and comparative synthesis. Baltimore: Johns Hopkins Press; 2002.
Opdam P, Foppen R, Vos C. Bridging the gap between ecology and spatial planning in landscape ecology. Landscape Ecol. 2002;16:767–79.
Pickett STA, Kolasa J, Jones CG Ecological understanding: the nature of theory and the theory of nature. Academic Press. New York. 1994.
Popper K Logik der Forschung Verlag von Julius Springer, Vienna, Austria. First published in English as Popper K 1959 The Logic of Scientific Discovery Hutchinson and Sons. London. 1935.
Primdahl J, Kristensen LS, Busck AG, Vejre H. Functional and structural changes of agricultural landscapes: how changes are conceived by local farmers in two Danish rural communities. Landscape Res. 2010;35(6):633–53.

Primdahl J, Swaffield S. Segregation and multifunctionality in New Zealand landscapes. In: Brouwer Floor, editor. Sustaining agriculture and the rural environment: governance, policy and multifunctionality. Cheltenham: Edward Elgar; 2004. p. 266–85.
Primdahl, J, Handgaard H, Kristensen LK, Stahlschmidt P, Rask A, Willadsen H Det åbne land i kommuneplanprocessen. Byplan 1. 2010. pp. 32–9.
Relph E Place and placelessness. London: Pion. 1976.
Shearer AW. Approaching scenario-based studies: three perceptions about the future and considerations for landscape planning. Environ Plann B: Plann Des. 2005;32:67–87.
Schneekloth LH, Shibley CG. Placemaking: the art and practice of building communities. New York: Wiley; 1995.
Stein SM, Harper TL. Power, trust, and planning. J Plann Edu Res. 2003;23:125–39.
Steinitz C. A framework for theory application to the education of landscape architects (and other environmental design professionals). Landscape J. 1990;9:136–43.
Steinitz C, Arias H, Bassett S, Flaxman M, Goode T, Maddock IIIT, Mouat D, Peiser R, Shearer A. (2003) Alternative futures for changing landscapes: the Upper San Pedro river basin in Arizona and Sonora. Washington: Island Press; 2003.
Stevens CJ, Fraser I, Mitchley J, Thomas MB. Making ecological science policy-relevant: Issues of scale and disciplinary integration. Landscape Ecol. 2007;22:799–809.
Termorshuizen J, Opdam P. Landscape services as a bridge between landscape ecology and sustainable development. Landscape Ecol. 2009;24:1037–52.
Theobald DM, Spies T, Kane J, Maxwell B, Hobbs NT, Dale VH. Ecological support for rural land use planning. Ecol Appl. 2005;15(6):1906–14.
Thompson M, Ellis R, Wildavsky A. *Cultural Theory*. Westview Press: Boulder; 1990.
Tress B, Tress G, van der Valle A, Fry G Interdisciplinary and transdisciplinary landscape studies: potentials and limitations. Delta Series 2. Wageningen. 2003.
Van der Belt M Mediated modeling: a systems dynamics approach to environmental consensus building. Washington: Island Press; 2004.
Wardle P, Collins H, editors. Beyond reasonable drought: adapting dryland farming to climate change. Nelson. New Zealand Land Care Trust; 2008.
Williams BK, Brown ED Adaptive management: The U.S. department of the interior applications guide. Adaptive management working group, U.S. department of the interior. Washington DC. http://www.usgs.gov/sdc/publications.html. 2012.
Weins JA. What is landscape ecology, really? Landscape Ecol. 1992;7:149–50.
Wu J. Landscape ecology, cross-disciplinarity and sustainability science. Landscape Ecol. 2006;21:1–4.
Wu J, Hobbs R. Key issues and research priorities in landscape ecology: an idiosyncratic synthesis. Landscape Ecol. 2002;17:355–65.

Chapter 13
Planning Multifunctional Measures for Efficient Landscape Management: Quantifying and Comparing the Added Value of Integrated and Segregated Management Concepts

Carolin Galler, Christina von Haaren and Christian Albert

Abstract Scientists often argue that landscape and environmental planning should aim for developing multifunctional landscapes in order to enhance implementation effectiveness and public spending efficiency. In planning and decision-making practice however, multifunctional effects are usually neither quantitatively assessed nor explicitly and transparently considered. In this research, a procedure is developed for quantitative assessment of multifunctional effects and trade-offs of conservation measures on landscape functions. The procedure is applied on local scale and uses available data. The method is tested for a core set of landscape functions in a case study region in Germany. The results provide empirical confirmation that integrative management strategies can be considerably more effective and efficient than sectoral ones. However, the added value of integrative environmental measures highly depends on their spatial allocation within areas of overlapping requirements for multiple landscape functions. The results of the analysis can help directing implementation resources towards areas and management measures that maximize attainable benefits. The analysis thus provides very useful support for planning decisions.

Keywords Multifunctionality · Multifunctional measures · Landscape functions · Landscape management · Landscape planning · Environmental planning · Funding efficiency

---

C. Galler (✉) · C. von Haaren · C. Albert
Institute of Environmental Planning, Leibniz Universität Hannover,
Herrenhäuser Str. 2 30419 Hannover, Germany
e-mail: galler@umwelt.uni-hannover.de
URL: www.umwelt.uni-hannover.de

C. von Haaren
e-mail: haaren@umwelt.uni-hannover.de

C. Albert
e-mail: albert@umwelt.uni-hannover.de

B. Fu and K. B. Jones (eds.), *Landscape Ecology for Sustainable Environment and Culture*, DOI: 10.1007/978-94-007-6530-6_13,

13.1 Introduction

Environmental planning and funding schemes are organized primarily by sectoral administrations that have different constituencies and diverse competencies. This situation produces uncoordinated approaches to environmental management that focus on separate ecosystem compartments, such as water, biodiversity or soil. River basin management, habitat networks planning and concepts for soil protection and restoration exemplify such approaches. Even in comprehensive environmental planning such as landscape planning (according to the European Landscape Convention (Council of Europe 2000) and German landscape plans), multifunctional measures are not deduced systematically and their added value is not assessed. Instead, landscape planners employ experience and intuition in combination with stakeholder input in order to identify multifunctional areas and management measures (von Haaren and Galler 2013).

Clearly, a better scientific basis and method for selecting multifunctional measures and assessing their added value would be most helpful for applying an integrative concept in practice. Furthermore, such findings could support integrative approaches in environmental management that could promote more efficient use of public money. More specifically, the consideration of cost-benefit ratios of measures or the ratio between benefit and the amount of land required for implementation force planners to focus on the efficiency of their plans. Planners must consider the amount of land required for improvement measures and how exclusively it is dedicated to environmental purposes, because land is a scarce resource. A very relevant application for such optimization concepts is, for example, the design of agri-environmental measures in the Common Agricultural Policy.

Current scientific research employs different definitions of multifunctionality. In agricultural policy, multifunctionality is used to describe societal and non-commodity functions that are provided by farmers in addition to agricultural products. Increasingly, the EU interprets multifunctionality of agriculture as a justification for continued financial support of farmers to renumerate them for the provision of non-commodity outputs (Marsden and Sonnino 2008). In landscape sciences, multifunctionality is understood more broadly as the capacities of landscapes (and not only agricultural land uses) to simultaneously provide several ecological and socio-economic functions (cf. e.g. Helming and Wiggering 2003; Mander et al. 2007). Here, multifunctionality has been interpreted as a general objective for landscape development (cf. Brandt et al. 2000).

This research focusses on the multifunctionality of landscape management measures. This is understood as the positive effects of measures on the provision of multiple landscape functions. Landscape functions are defined as the capacity of a landscape and its subspaces to sustainably fulfill basic, lasting and socially legitimised material or immaterial human demands (von Haaren and Albert 2011). This definition stresses the normative character of the functions which we examine. As such, our definition of landscape functions differs from a prevailing understanding of ecosystem functions in the ecosystem services literature, which often restricts functions to 'operations' and processes in ecosystems (Fisher et al. 2009; Costanza et al. 1997).

Multifunctional effects of measures can be assessed with respect to the effectiveness and the efficiency of measures. The effectiveness of a measure describes the degree to which it contributes to improving one or multiple landscape functions. Efficiency means the ratio of achieved effects in relation to the costs of implementation.

Landscape multifunctionality as well as multifunctional effects on landscape functions must be considered in landscape planning and management in order to sustain the capacity of landscape to provide ecosystem services (OECD 2001; Willemen et al. 2008; De Groot and Hein 2007). Research has been conducted about landscape functions and landscape multifunctionality (e.g. Willemen et al. 2008; Potschin and Haines-Young 2006; Dijst et al. 2005; Helming and Wiggering 2003; Mander et al. 2007). Furthermore, the need to quantitatively investigate multifunctional effects has been identified (Osterburg and Runge 2007). However, until now very little empirical evidence exists about: (1) The dimension of multifunctional effects in different landscapes. For example, Rüter (2008) quantified multifunctional effects of landscape structures on species connectivity and water retention. (2) The preconditions of structural synergies among different landscape functions (e.g. von Haaren et al. 2011), or (3) methods for systematically generating and allocating multifunctional measures.

13.2 Research Objectives

The objective of this research is to develop and test a method for investigating and comparing the effectiveness and efficiency of sectoral and integrative landscape management strategies. The hypothesis is that integrative management strategies, are more effective and efficient than sectoral strategies. A case study is used to test a procedure for an integrative planning process that systematically identifies options for multifunctional measures, optimizes multifunctional effects and quantifies the added value.

The results of the analysis are expected to provide support for the consideration of multifunctionality in planning and implementation practice. Furthermore, they should form the basis for further research about approaches for integrating multifunctional landscape management into environmental and land use planning as well as in the design of Agri-Environmental Funding schemes and 'Payments for Ecosystem-Services' approaches.

13.3 Procedure and Methodology

13.3.1 Outline of the Methodological Steps

Based on an existing set of methods for landscape functional analysis (von Haaren 2004), a new procedure has been developed for systematically deriving multi-functional environmental measures in a transparent manner and for measuring their added value. The methodology focuses on a core set of landscape functions that represent different human demands on the ecosystem and demonstrate the benefits and shortcomings of an integrative approach. The method has been applied and tested in a case study.

The research consists of four main steps, which are summarized below and described in detail in the following sections (Fig. 13.1):

- First, the specific need and opportunities for safeguarding, enhancing or restoring a core set of landscape functions were identified for the entire case study area. The assessment considered each function's capacity for providing respective services, the specific sensitivities and impacts. Furthermore, environmental quality objectives at the local scale were derived for each landscape function in the case study area.
- Second, appropriate implementation measures were compiled and their general effects on achieving the objectives were estimated.
- Third, two scenarios were generated:
  - *Scenario of uncoordinated sectoral measures* (*scenario* 1). It combines four separately developed sectoral scenarios that were each optimized for one landscape function. In particular, the sectoral scenarios focus on safeguarding biodiversity (scenario B), climate change mitigation (scenario C), water quality conservation (scenario W) and erosion prevention (scenario E). Each scenario consisted of a map that illustrated the spatial allocation of appropriate measures for attaining the above described environmental quality objectives. The scenario of uncoordinated sectoral measures was created by overlaying the individual sectoral scenarios. Areas were identified, in which identical management measures were proposed in order to attain the environmental quality objectives (EQO) of different landscape functions (unintentional multifunctionality of measures). Furthermore, additional benefits of sectoral measures were assessed. The implementation costs were calculated.
  - *Scenario of planned integrative measures* (*scenario* 2). The integrative scenario is generated from the analysis of possible synergies and conflicts in the sectoral scenarios. In the integrative scenario, measures were combined in ways most beneficial for enhancing all considered landscape functions.

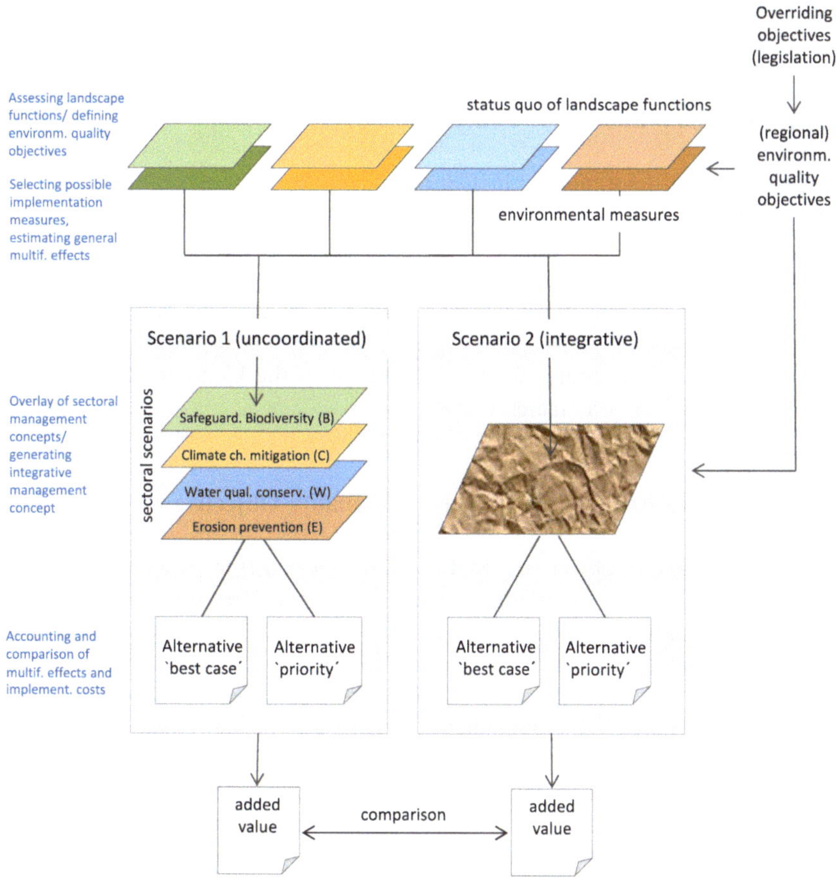

Fig. 13.1 Flowchart of the working steps

- Fourth, the effectiveness of the uncoordinated and integrative scenarios for attaining environmental quality objectives was assessed and the respective implementation costs were calculated. The uncoordinated scenario and the integrative scenario were then compared to determine their degree of effectiveness for enhancing the provision of several landscape functions and the efficiency of the resources that were spent. In this way, we could compare the unintentional multifunctional effects that resulted from the uncoordinated scenario with the multifunctional effects resulting from the planned multifunctionality of the integrative scenario. The comparison provides insights on how integrative management increases or decreases implementation costs in comparison to sectoral management.

13.3.2 *Assessing Landscape Functions and Deriving Regional Environmental Quality Objectives*

13.3.2.1 Assessing Landscape Functions

The assessment of the status quo includes mapping the capacities of the case study landscape to provide ecosystem services as well as identifying sensitivities, impacts and underlying pressures. This analysis focuses on four landscape functions: natural yields, water resources, climate change mitigation and safeguarding biodiversity. The assessment relies on existing data, namely the landscape master plan of the case study area as well as specific environmental information systems for soil and water. Procedures for the assessment of the landscape functions are described in detail in von Haaren (2004).

13.3.2.2 Deriving Regional Environmental Quality Objectives

Environmental quality objectives (EQO) can be derived from relevant legislation and directives as well as superordinate planning targets. They serve as operational objectives for landscape management at the local scale and form the basis for developing a spatially explicit implementation concept. Furthermore, environmental quality objectives can be used as reference values for monitoring and quantifying the effects of environmental measures. The environmental quality objectives used in the case study are listed in Table 13.1. They were adapted from regional environmental planning, as much as possible, or directly derived from legislation. Some objectives were defined as maximum or minimum (threshold) limits (e.g. maximum N-concentration in percolate water in order to reach a good chemical groundwater status). They provide a precise quantitative target and make it possible to quantify the difference between the status quo and the intended final development status. Here, the assessment indicators refer directly to the environmental quality objectives (e.g. a reduction of agricultural nitrogen input). They are suitable for application in other European regions. Until now existing standards for environmental quality or acceptable pollution levels do not address the whole range of landscape functional aspects. Furthermore, some objectives in the EU Directives or national legislation are not precisely formulated for every conceivable landscape context, e.g. the demand to stop the loss of biodiversity. Such superordinate general objectives must be operationalized for a specific landscape situation. In other cases, it may be possible to establish clear and quantifiable standards for national or European legislation in the future.

The progress made in achieving the quality objectives is assessed by using indicators (Table 13.1). They are described as follows:

- Habitat value points (VP) express the difference between status quo and targeted status of biotopes. The term ‘biotopes’ comprises landscape areas with specific

Table 13.1 Overview of specific overriding objectives and standards for the landscape functions, quality objectives (for local scale) and indicators for assessing the effects of environmental measures

Overriding objectives (legally binding)	Environmental quality objectives for the district scale derived from overriding objectives and planning targets	Generally applicable indicators for assessing environmental quality and the effects of environmental measures
Safeguarding biodiversity		
CBD: conservation of biological diversity, sustainable use of its components, fair and equitable sharing of the benefits; **Natura 2000/Habitats directive**: conservation status of natural habitats and species of community interest; **Federal Nature Conservation Act**: three primary objectives include safeguarding biodiversity, protection and improvement of landscape functions, preserving and development of aesthetic landscape functions	**Biotope concept of the landscape master plan**; explanation: the landscape master plan concretizes overriding objectives within a spatially explicit development concept by allocating targeted biotope types Furthermore, the landscape master plan provides information about areas of importance for preserving diversity of species and biotopes that can be used to establish priorities for implementation. Especially the improvement of areas with international, national and state wide importance (e.g. areas for species protection, Natura 2000 network, Nature protection areas) is of high priority	**Habitate value points**; Difference in value of the biotope types—status quo and targeted status after implementation of the measure (in reference to official list of biotope types in Lower Saxony (NLWKN 2012)), multiplied by size of area (similar to procedures for assessing the need for compensation measures within German impact mitigation regulation) Additionally, specific effects for species protection are assessed on an ordinal scale and assigned to biotopes classified as having special importance for species protection
Natural yield		
Soil protection directive, soil thematic strategy, German Federal Soil Protection Act, German Federal Nature Conservation Act: maintain natural yields capacity and soil fertility by avoiding soil erosion, requirements for good agricultural practice	**Erosion prevention on all erosion prone sites**; explanation: Erosion of soil caused by wind, water and floods is considered a main factor for damage to the natural yields function. The landscape master plan points out areas of high erosion risk (priority areas to implement measures for protection against erosion)	**Area with measures for protection against erosion within sensitive areas** (three level ordinal scale: little, periodical effects in lowering/reduction of soil erosion (+), medium effects (++), high, continuous effects in total prevention of erosion (+++)

(continued)

Table 13.1 (continued)

Overriding objectives (legally binding)	Environmental quality objectives for the district scale derived from overriding objectives and planning targets	Generally applicable indicators for assessing environmental quality and the effects of environmental measures
Water resources		
WFD: good status of groundwater bodies until 2027 (particularly good chemical status), Directives, especially EU Nitrates Directive, Drinking Water directive, Groundwater Directive	For Groundwater: **50 mg NO3/l maximum concentration in percolate water below all agricultural land**, for surface water: **3 mg/l maximum N-concentration of influent water (in case of good status of groundwater)**; explanation: the environmental quality objective refers to the limit of 50 ml/l of nitrogen in groundwater according to the Groundwater Directive (Annex 1). For a given area, the amount of N-reduction (kg/ha*a) on agricultural land that is required to achieve good groundwater status can be calculated.	**Reduction of agricultural N-input (in kg N/ha*a)** The quantified effects of measures on reduction of N-input are adapted from Osterburg and Runge (2007)
Climate protection		
United nations framework convention on climate change/Kyoto protocol on climate change, National strategy on climate protection: reduction of GHG-emissions and safeguarding of ecosystems as carbon sinks and storages	The overriding objectives for lowering GHG-emissions are not operationalized for regional or local scale (no defined emission limits for GHG). However, it is possible to estimate the potential for GHG retention and emission of landscapes, especially CO_2-retention of (natural) soils and potential CO_2-emissions caused by agricultural land use (Flessa et al. 2012; Saathoff et al. 2012). The reduction of CO_2-emissions that can be achieved by implementing measures within a given area is used as a quality objective	Effects for CO_2-balance/ reduction of emissions caused by agricultural land use and increase in CO_2 retention potential of soils according to Saathoff et al. (2012)

characteristics such as abiotic conditions, land use and part of the biocenosis (vegetation) that typically occur together. Biotopes are classified into types of defined, homogenous ecological conditions (Drachenfels 2013; von Haaren et al. 2012). Habitat value points were used to express the effects of measures for biodiversity. The increase in VPs depend on whether, for example, a measure is implemented on cropland (value I–II on value scale) or grassland (value II–V), as well as on site-specific characteristics. Benchmarks and criteria for valuing each biotope type are given in NLWKN (2012) and Drachenfels (2013). The habitat value of the status quo was taken from the landscape master plan, whereas the value of the target habitat types were referred to the general habitat values found in NLWKN (2012). The upgrading is measured in VP. They were multiplied by the size of the area covered by the particular biotope types in order to assess the total effect for the biotope function within a given area. This procedure was adapted from the assessment of need for compensation measures found in the German impact mitigation regulation (von Haaren et al. 2012).

- The landscape master plan identifies areas that are susceptible to soil erosion, as a result of a model based analysis according to the Universal Soil Loss Equation (e.g. von Haaren 2004). Measures have potential effects for soil erosion prevention, when they: (1) extent the period of time when the ground is covered by vegetation, (2) help shorten the length of steep slopes, (3) lower wind velocity. Furthermore, effective soil protection in floodplains requires a permanent ground cover. However, the soil protection effects could not be quantified based on the given information. They were assessed according to the size of measures on erosion sensitive sites and to an additional ordinal scaling of the effects (low (+), medium (++), high (+++)) that considers the time periods of protection (periodical or permanent effects).
- The reduction of N-input on agricultural land was used as an indicator for assessing water quality conservation effects of environmental measures. The accounting refers to the total amount of N-input and is adapted from Osterburg and Runge (2007). Doing so, the assessment of effects on water resource quality was restricted to nitrogen input in groundwater and surface waters. Possible inputs of phosphate were not considered in the case study.
- The effects of environmental measures on mitigating Green House Gases were calculated for CO_2 according to the CO_2-retention potentials of soil types and the potential CO_2-emissions that are caused by agricultural land use as stated in Saathoff et al. (2012). For fens and bogs (groundwater $\sim$10 cm under surface), they calculated a retention potential of 1,700–2,600 t CO_2/ha. This is assumed to be the potential reduction of CO_2-emissions when agricultural land on organic soils is rewetted and converted into near-natural biotopes. A CO_2-retention of 70–160 t CO_2/ha was calculated for hydromophic soils that are used as grassland (>5 years duration), such as Pseudogley, Fluvisols, marsch soils, Gleysole as well as Pozol.

13.3.3 Selecting Appropriate Implementation Measures and Estimating Their General Effects on Achieving Environmental Quality Objectives

13.3.3.1 Selecting Appropriate Measures

A range of suitable implementation measures with different effects and average implementation costs have been identified. The measures were selected from a catalog of possible measures that best conserve the respective landscape functions (Table 13.2). They can be grouped into classes according to their requirements for land use: (1) Measures that may be integrated into current land uses. These measures allow for a continuation of the current land use type or require only slight adaptation of existing practices; (2) measures that require abandoning, reducing or radically changing the land use e.g. to initiate natural dynamics or conservation measures that cannot be integrated into existing land uses and instead, require public land ownership for their implementation; (3) conservation measures for maintaining the status quo of a biotope where land use is already strongly restricted (biotope maintenance measures, especially for cultural biotopes).

The selection of alternative measures followed general decision guidelines. For soil erosion prevention these are: (1) If possible, employ a combination of land use-integrated measures (e.g. catch crops or undersown, non-plough tillage) and linear measures that affect slope length, wind velocity and surface runoff (e.g. hedges, field margins). (2) Aim for designating measures that involve as little extra management effort as possible. (3) Recommend a change of cropland areas within floodplains into grassland or forest because these areas are prone to erosion during floods. Decision guidelines for water quality conservation are: (1) Selecting measures that yield high N-reduction (minimum limit 20 kg N/ha*a) and high cost-effectiveness (maximum limit 3.50 €/kg). They were selected from the catalog of suitable water quality conservation measures provided by Osterburg and Runge (2007). (2) Prefer choosing measures that allow for the continuation of current land uses (e.g. continuation as cropland/as grassland) in order to exert only minimal effects on cultivation. (3) Enabling for combining measures on one site to increase effects, e.g. through targeting different aspects of crop rotation, or through addressing different phases of cultivation (e.g. implementing measures during the vegetation period or in winter). However, the measures selected for climate change mitigation and safeguarding biodiversity must not be understood as alternative measures. Instead, these measures listed in Table 13.2 are recommended for specific sites.

13.3.3.2 Estimating General Multifunctional Effects and Conflicts of Measures

The general effects of measures are estimated using the indicators described in Sect. 13.3.2.2. An overview of the estimates is given in Table 13.2.

Table 13.2 General effects of measures for landscape functions

Used in sectoral concept				Land use (status quo) cropland (C), grassland (G)	Type of restoration/improvement measure	Effectiveness for achieving objectives				Cost €/ha*a, T: in total
B	W	K	E			B: upgrade in habitat value points, additional effects for species protection (+)[h]	C: decrease of CO_2-emissions in t/a	W: reduction of nitrogen input (kg N/ha*a)	E: effectiveness of measures for erosion prevention (on ordinal scale 0 +, ++, +++)	
	X		X	C	M1: catch crops/undersown	0	0	20	++	70[i]
	X			C/G	M2: groundwater compatible techniques for the distribution of slurry and digestate	0	0	25	0	25[i]
	X			C	M3: reduced nitrogen fertilizer (limit for amount of fertilizer 10–20 % below ideal value, no delayed N application for grains)	0	0	30	0	80[i]
	X			C	M4: no distribution of manure after the harvest (no fertilizing in phase of straw decomposition)	0	0	30	0	20[i]
	X			C/G	M5: extension of time period with no manure application (1 Okt. to 15 Feb.)	0	0	20	0	25[i]
	X			G/pasture	M6: extensive grassland cultivation (no application of synthetic fertilizer)	GI(II)-GE(III) 1	0	30	0	100[i]
	X			G/pasture	M7: extensive pasture farming	GI(II)-GE(III) 1	0	40	0	77[i]

(continued)

Table 13.2 (continued)

Used in sectoral concept				Land use (status quo) cropland (C), grassland (G)	Type of restoration/improvement measure	Effectiveness for achieving objectives				Cost €/ha*a, T: in total
B	W	K	E			B: upgrade in habitat value points, additional effects for species protection (+)[h]	C: decrease of CO_2-emissions in t/ha	W: reduction of nitrogen input (kg N/ ha*a)	E: effectiveness of measures for erosion prevention (on ordinal scale 0 +, ++, +++)	
			X	C/suscept.	M8: non-plough tillage	0	[g]	0	+	40[i]
			X	C/suscept.	M9: hedgerows, field margins (on 10 % of the cropland)	A(I)-HN(IV) 3	70–160[b]	60[a,c]	0–+++	T: 125,984[j]
		X	X	C/suscept.	M10: land use type with permanent ground cover (extensive grassland, forest)	A(I)-G?(II-III)/ W?(III) 1–2	70–160[b]	50	+++	400[i], 162[j] (on cropl., on other sites)
		X		C/G	M11: bog/fen regeneration	A(I)-M?(V), GI(II)- M?(V) 3–4	1,700–2,600[b]	60/30[a,c]	+++	T: 27,236[j]
X				C/G	M12: small groups of shrubs and trees within agricultural landscape (hedgerows, copses, rows of trees, fruit trees, pollarded willows (for selected areas), on 10 % of the agricultural landscape	A(I)/GI(II)/??- HN(IV) 0–3	0	60/0[a,d]	0–+++	T: 125,984[j]
X				C/G	M13: extensive grassland cultivation (when necessary, species specific requirements)	GI(II)/G?(III-V)- G?(V) 1–3+	0	30	+++ (on cropl.), 0 (on grassl.)	162 (on grassl.)[j], 400 (on cropl.)[i]

(continued)

Table 13.2 (continued)

Used in sectoral concept				Land use (status quo) cropland (C), grassland (G)	Type of restoration/improvement measure	Effectiveness for achieving objectives				Cost €/ha*a, T: in total
B	W	K	E			B: upgrade in habitat value points, additional effects for species protection (+)[h]	C: decrease of CO_2-emissions in t/ha	W: reduction of nitrogen input (kg N/ha*a)	E: effectiveness of measures for erosion prevention (on ordinal scale 0 +, ++, +++)	
X				Unspec.	M14: development of near-natural forests	??-W?(V) 1–4	1,700–2,600[b], 70–160[b]	0–60[a,e]	+++ (on cropl.), 0 (on grassl.)	T: 13,348[j] (on cropl.; also used for impl. in forests)
X				Unspec.	M15: renaturation of water bodies	F?/S?(II)-F?/S?(V) 3+	(indirect effects)	0	0	T: 22,676/100 m[j]
X				Unspec.	M16: development (including maintenance) of highly valuable biotops (bog regeneration)	??(V) 2–4+[f]	1,700–2,600[b]	0–60[a,e]	0, +++ (on cropl.)	T: 27,236[j]
X				Unspec.	M17: development (including maintenance) of highly valuable biotopes (establishing and maintaining heathland/oligotrophic grassland)	RS/RH/RN (V) 2–4+[f]	0	0–60[a,e]	0, +++ (on cropl.)	T: 68,021[j]

(continued)

Table 13.2 (continued)

Used in sectoral concept				Land use (status quo) cropland (C), grassland (G)	Type of restoration/improvement measure	Effectiveness for achieving objectives				Cost €/ha*a, T: in total
B	W	K	E			B: upgrade in habitat value points, additional effects for species protection (+)[h]	C: decrease of CO_2-emissions in t/ha	W: reduction of nitrogen input (kg N/ha*a)	E: effectiveness of measures for erosion prevention (on ordinal scale 0 +, ++, +++)	
X				G	M18: maintenance of (existing) special grassland forms, when required special species protection measures (e.g. wet grassland, oligotrophic grassland)	G?(IV-V) 2–4+[f]	0	30	0	162[j]

[a] Permanent total reduction of N-input

[b] When measure is allocated on sites with specific soil conditions (fens/bogs or hydromorphic soils)

[c] For measures located on cropland a N-reduction of 60 kg N/ha*a is calculated (according to reduction rate of fallow fields), for measures on grassland 30 kg N/ha*a. It is assumed that hedgerows/herbal strips cover about 10 % of the cropland; therefore, the N-reduction potential is, related to the entire field area, relatively low

[d] For hedgerows/groups of shrubs on cropland a N-reduction of 60 kg N/ha*a is calculated, according to reduction rate of fallow fields. Single trees (e.g. fruit trees on grassland) have no N-reduction effects

[e] N-reduction depends on status quo. For measures on cropland 60 kg N/ha*a is calculated (according to reduction rate of fallow fields), for measures on grassland 30 kg N/ha*a is calculated, measures on other land use types have no N-reduction effects

[f] Value points for upgrading or (in case of maintenance measures) for prevention of degradation

[g] The effects on reduction of CO_2-emissions are not assumed as considerably high

[h] Shortcuts of biotope types and habitat value (I–V) according to NLWKN (2012) e. g. A (cropland), G (grassland), I (intensively used), E (extensively used), F (rivers, small streams), and average upgrading of biotope value (0–4)

[i] *Source* Osterburg and Runge (2007)

[j] *Source* Thuringian Ministry for Agriculture, Forestry, Environment and Nature Conservation (2003); calculated are costs for implementation of the measure as well as maintenance; costs for renaturation of water bodies are calculated per 100 m of Fließgewässer and include a buffer strip of 5 m

The compilation of measures and their effects illustrates that certain measures serve multiple landscape functions (Fig. 13.2). Some measures (M1, M10) are used in different sectoral scenarios. These are potential multifunctional measures. Furthermore, some measures have synergizing effects or additional benefits for multiple landscape functions. Measures for minimizing CO_2 emissions and for safeguarding biodiversity (e.g. M11, M13, M14, M16, M17) show the highest general multifunctionality.

In the overview of Table 13.2, there are no generally conflicting measures. However, when the measures are applied to the site, conflicts may occur (von Haaren et al. 2011). Also the extent to which the (general) effects actually occur depends on the allocation of measures and respective site-conditions.

13.3.3.3 Efficiency of Measures

The efficiency analysis of the different management measures was based on implementation costs that were derived from two sources. First, from Osterburg and Runge (2007) who give an overview of average implementation costs for management measures that are included in agri-environmental (AEM) programs. The provided costs can be used to calculate land use integrated measures. Second, the Thuringian Ministry for Agriculture, Forestry, Environment and Nature Conservation (2003) provides average cost rates for compensation measures from the German Impact Regulation. These rates were used to calculate the costs of implementing nature conservation measures that are not land use integrated or allow only minor additional revenues from the land where the measure is taken. In both cases theses average cost rates neglect site conditions or farm specifics which may influence the real costs needed for implementation. Still, the applied costs were considered sufficiently reliable for a comparison of the different scenarios and an assessment of the differences in costs.

13.3.4 Developing an Uncoordinated and an Integrative Management Scenario

13.3.4.1 Scenario of Uncoordinated Sectoral Measures

The scenario of uncoordinated sectoral measures (scenario 1) is build up of a combination of four separate sectoral scenarios: erosion prevention (sectoral scenario E), safeguarding biodiversity (sectoral scenario B), climate change mitigation (sectoral scenario C) and water quality conservation (sectoral scanrio W). In each sectoral scenario, measures were selected and allocated to safeguard and enhance the particular landscape function. Two alternatives were calculated: In the first alternative, termed 'best case', measures were chosen that completely fulfilled

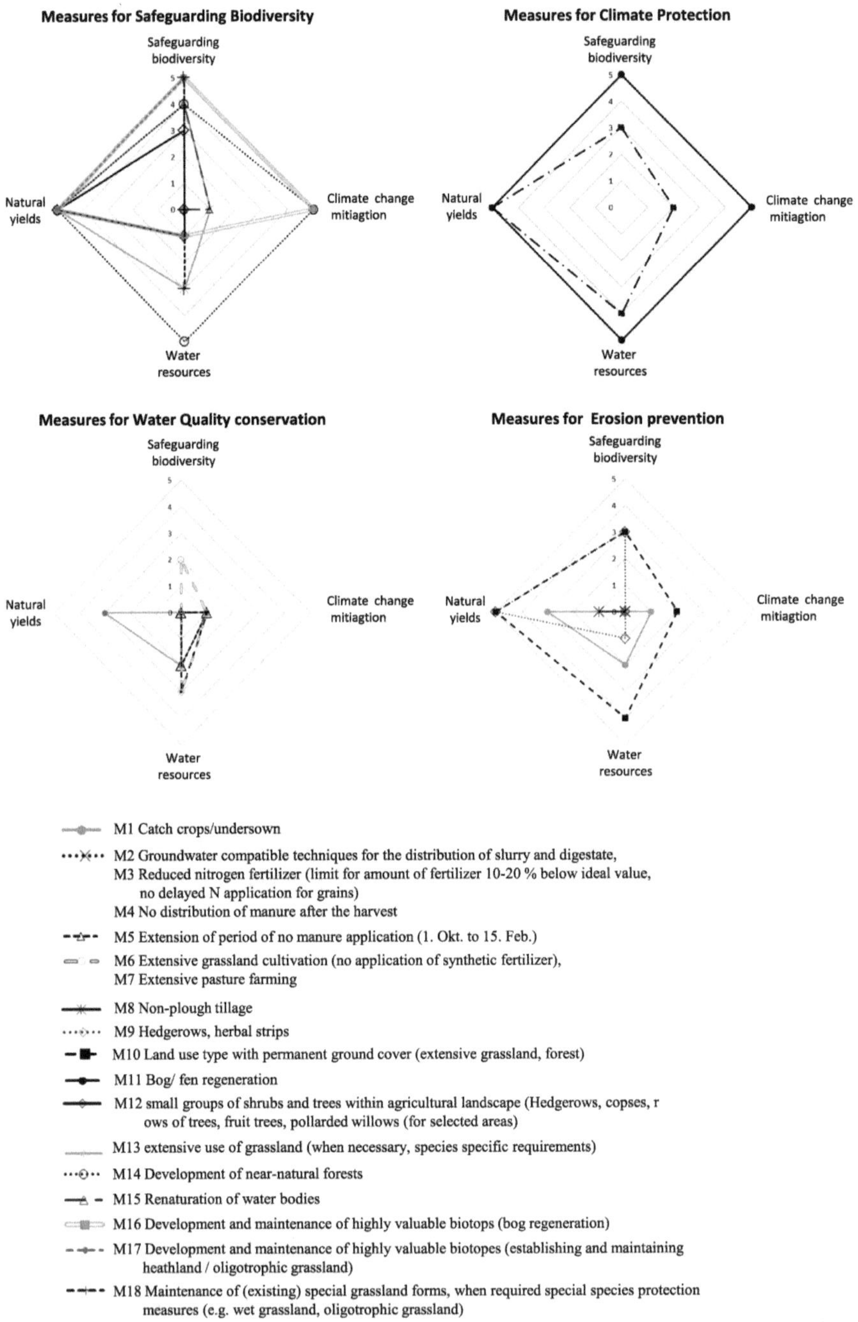

Fig. 13.2 General potentials of measures to enhance the provision of four core landscape functions (on a six-step lickert scale from 0 to 5, derived from the quantitative estimate of effects of measures in Table 13.2), grouped according to their implementation within sectoral scenarios

the sectoral environmental quality objectives. The second alternative, termed 'priority', is based on limited funds of 10 Mio. € for implementation.

When alternative measures can be chosen (e.g. for water quality conservation M1–5 on cropland, M2, 5–7 on grassland), then they are used in equal proportion in the sectoral scenarios.

All sectoral scenarios were overlaid in order to identify areas where measures for multiple functions overlap. Then the unintentional multifunctional effects were analysed. Unintentional multifunctional effects can occur in areas where different sectoral scenarios allocate similar or synergetic measures. They can also arise when proposed measures for two or more different functions are not similar. For example, the measure M10 (allocated for climate protection purpose) also has unintended benefits for the biotope function (increase of biotope value) regardless if the management concept for safeguarding biodiversity recommends measures for this site.

13.3.4.2 Scenario of Planned Integrative Measures

The scenario of planned integrative measures (scenario 2) attempts to achieve the environmental quality objectives of all the landscape functions simultaneously. The area where measures are allocated ('area for action'), includes all areas for measures of the sectoral scenarios. The areas for action were classified according to the number and combination of measures that were identified by the overlay of the sectoral scenarios. The number of measures that overlap built different categories, i.e. levels, of multifunctionality in the areas of action. For example, multifunctional Level-2 is when two measures overlap. In these areas (Level-2 to Level-4) synergies as well as possible conflicts must be addressed. It was analyzed whether the multiple (sectoral) objectives could be achieved by implementing either one specific multifunctional measure or different overlapping measures. Decision guidelines were documented to make decisions between alternative or conflicting development objectives transparent.

Corresponding to the procedure applied in scenario 1, two alternatives are calculated. The "best case" alternative refers to the same sectoral environmental quality objectives as the sectoral scenarios. The "priority" alternative is calculated for limited funds of 40 Mio. €, corresponding to the total amount spent for the four sectoral scenarios within scenario 1.

13.3.5 Assessing and Comparing Environmental Effects

The quantitative assessment of the scenarios' effectiveness for achieving the different environmental quality objectives used the indicators defined in Table 13.1. For each sectoral scenario within scenario 1 the intended effects for the targeted landscape function as well as unintentional multifunctional effects for other landscape functions were calculated. It was assessed and compared to what degree the regional environmental quality objectives were fulfilled by scenario 1 and by scenario 2. The costs for achieving specific sectoral and multifunctional effects were compared between the sectoral and integrative scenarios.

13.3.6 Case Study Area and Data Basis

The method for investigating the impacts of different mono- and multifunctional management strategies on the provision of ecosystem services was tested in the case study region of Verden County, Germany. Verden covers an area of approximately 79,000 ha in the state of Lower Saxony, near the city of Bremen. The landscape is dominated by agricultural land with a high proportion of grassland along the Aller river floodplain and dominating cropland on the sandy soils of the moraine area. In addition, moors, fens and marshes are characteristics of the landscape.

The assessment of landscape functions (status quo) as well as the generation of the scenarios is based on the landscape master plan of the County of Verden (2008) and additional environmental information systems from the state government and accessible www map servers (State Authority for Mining, Energy and Geology; River Basin Commission Weser). The management concepts do not include the settlement areas.

13.4 Case Study Results

13.4.1 Effectiveness and Efficiency of the Uncoordinated Scenario

The sectoral scenarios describe how each of the landscape functions can be individually optimized. The documentation of the outcomes of the sectoral scenarios includes: (1) which environmental quality objectives have been achieved, and to what extent (2) which measures, i.e. types of measures, are included in the management concept, (3) the total area where measures are implemented, as well as (4) implementation costs for a 10 year period (Table 13.3). The allocation of the measures is shown in maps (exemplarily Fig. 13.3).

13.4.1.1 Sectoral Scenario for Soil Erosion Prevention

The sectoral scenario for erosion prevention is based on the following underlying assumptions: (1) The allocation of one type of measure within an area is sufficient to avoid soil erosion to its full extent. (2) erosion protection measures are required

Table 13.3 Effects of the sectoral management strategies (erosion prevention (E), ...), including intended effects on the secoral EQO, additional unintended effects for other sectoral EQO and unintended multifunctional measures

	Area for measures (ha)	Percentage of case study area	Implementation costs (10 years)	Benefits for sectoral objectives				Multifunct. measures[a]
				E (extention of measures on erosive sites), objective fulfillment (%)	W (N-reduction in kg/ha*a), objective fulfillment (%)	C (reduction of CO_2-emissions in t/ha), objective fulfillment (%)	B (upgrading of Habitat value points), objective fulfillment (%)	
EQO (targeted in sectoral scenario)				19,506	1,500,000	10,604,116	63,995	
E (sum)	15,610	22.4	84,129,484	**19,506**	403,777	419,415	1,328	5,950
(percentage)				(100 %)	(26.9 %)	4.0	2.1	
M1, M8, M9	9,089	13.0	59,289,526	12,985	103,877	0	866	
M10	523	0.7	847,945	523	0	0	462	
M10 (within floodpl.)	5,998	8.6	23,992,013	5,998	299,900	419,415	0	
W (sum)	52,667	75.4	27,712,817	2,597	**1,498,423**	0	5,328	(unclear)
(percentage)				(13.3 %)	(99.9 %)		8.3	
M1-5	30,868	44.2	15,342,122	2,597	871,711	0	0	
M2, M5, M6, M7	21,799	31.2	12,370, 695	0	626,711	0	5,328	
C (sum)	32,869	47.1	249,817,669	17,217	1,565,823	**10,604,116**	38,858	6,383
(percentage)				(88.3 %)	(104.4 %)	(100 %)	60.7	
M10	27,775	39.8	111,098,557	16,408	1,388,732	1,944,225	27,775	
M11	5,094	7.3	138,719,112	809	177,091	8,659,891	11,083	
B (sum)	24,725	35.4	289,377,641	7,037	803,485	2.612,819	**63,995**	6,783
(percentage)				(36.1 %)	(53.6 %)	(24.6 %)	(100 %)	
M12	565	0.8	71,255,793	860	30,099	0	1,126	
M13	12,874	18.4	30,889,856	3,885	470,536	290,018	32,504	
M14	7,832	11.2	104,541,324	1,272	134,997	108,969	22,332	
M15	11	0.0	598,346	0	0	0	141	

(continued)

Table 13.3 (continued)

	Area for measures (ha)	Percentage of case study area	Implementation costs (10 years)	Benefits for sectoral objectives				Multifunct. measures[a]
				E (extention of measures on erosive sites), objective fulfillment (%)	W (N-reduction in kg/ha*a), objective fulfillment (%)	C (reduction of CO_2-emissions in t/ha), objective fulfillment (%)	B (upgrading of Habitat value points), objective fulfillment (%)	
M16	1,829	2.6	49,827,464	401	51,358	1,981,670	4,168	
M17	221	0.3	15,052,503	212	11,065	14,815	664	
M18	284	0.4	459,618	0	0	0	567	
M13/M12	27	0.0	221,003	27	746	942	67	
M14/M12	609	0.9	7,896,557	326	98,237	16,172	1,522	
M16/M13	60	0.1	902,677	30	2,573	51,089	178	
M16/M13/M14	108	0.2	1,538,305	21	3,874	149,146	241	
M16/M14	305	0.4	6,194,196	3	0	0	485	

[a] unintentional spatial overlap of similar or synergizing measures (sum (ha))

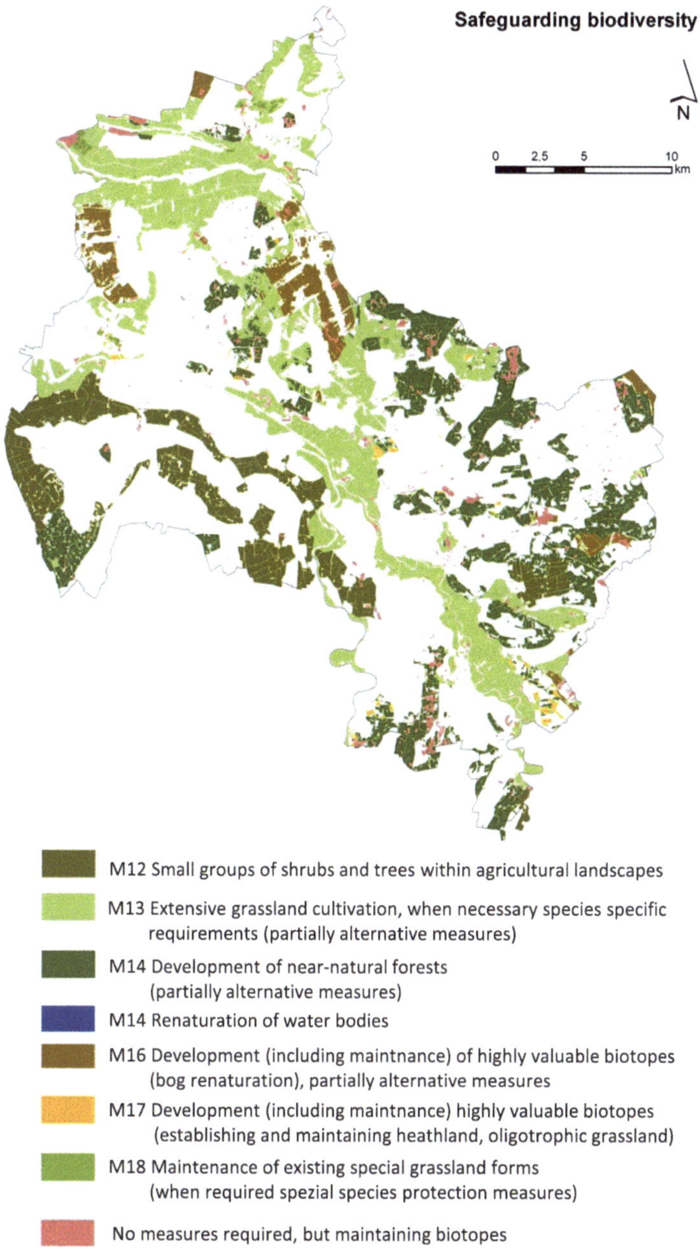

Fig. 13.3 Allocation of measures within the sectoral scenario for safeguarding biodiversity ('best case' alternative)

in all sites where cropland lies on erosion susceptible soils, even when linear biotope structures already exist. (3) The calculation of the effects and costs for hedgerows or field margins assumes that the implementation of the measure requires 10 % of the cropland.

Measures are implemented as listed in Table 13.3. Measures that can be integrated in cropland use (M1, M8) are allocated on 8,656 ha. Linear hedgerows or field margins (M9) are implemented on 433 ha. A change of land use (e.g. cropland within floodplains) into grassland or forest is required on 5,998 ha. The costs for implementing all measures total 84.13 Mio. € over the 10 year period. With a limited budget, e.g. 10 Mio. € (1 Mio. € per year), erosion preventions would only be guaranteed on 11.9 % of the susceptible areas.

13.4.1.2 Sectoral Scenario for Conservation of Water Quality

In order to achieve the mandatory groundwater quality objective in the case study, some areas require a N-reduction for agricultural land (UAA) of >10–20 kg/ha*a others as much as >20–40 kg/ha per year (Kreins et al. 2009). To reach the good status of surface waters an additional reduction of >5–10 kg/ha *a (UAA) is required in particular subunits of the area, in other subunits even >10–15 kg/ha*a (Kreins et al. 2009). This constitutes a total required N-reduction of about 1.500 t N per year.

To achieve these objectives, it is necessary to implement measures on each agricultural site (in total 52,667 ha). In fact, two different N-reduction measures are necessary on approximately 4.000 ha (combination of two measures on one site). The implementation of the water quality protection strategy costs about 27.71 Mio. €/10 years. With a limited budget of 10 Mio. €, more than one third (36.1 %) of the targets can be met.

13.4.1.3 Sectoral Scenario for Climate Change Mitigation

The implementation of climate change mitigation measures within the case study area can reduce CO_2-emissions on 10,604,116 t. This theoretical potential for the reduction of CO_2-emissions is considered the environmental quality objective for this landscape function.

Appropriate measures for the reduction of CO_2-emissions include M11 and M10. Reaching the CO_2 emission reduction target for the case study area requires implementing measures on 32,869 ha (5,094 ha M11, 27,775 ha M10). These measures cost 249.82 Mio. €. With a limited budget of 10 Mio. € (about 4 % of the total amount of spending required for reaching the environmental quality objective), 5.9 % of the targeted reduction of CO_2-emissions can be attained, if the most effective measure M11 is implemented.

13.4.1.4 Sectoral Scenario for Safeguarding Biodiversity

To meet the objectives of the landscape master plan's biotope concept, targets and measures were assigned to biotopes that are being impaired by current land use or whose quality could be improved. If the landscape master plan proposed alternative objectives, we allocated the alternative target biotope types in equal area proportions.

In order to meet the total biodiversity quality objectives, measures must be implemented on 24,725 ha (35.4 % of the case study area, Table 13.3). The benefits of these measures are expressed in habitat value points, in this case 63,995 VP. The implementation of all the measures would cost approximately 289.38 Mio. € for a 10 year period. For a limited budget, measures should be implemented in high priority areas. For the case study this means that a budget of 10 Mio. € can achieve about 2,212 VP (3.5 % of the target).

13.4.1.5 Unintentional Multifunctional Effects of the Sectoral Scenarios

Table 13.3 gives an overview of effects of the sectoral management strategies. This includes (1) the sectoral (intended) effects, (2) additional benefits for other landscape functions (global effects, that occur even if the measures are not quite similar), (3) effects of unintentional multifunctional measures (when similar or synergizing measures spatially overlap).

When the unintentional side-effects (additional benefits) of the different sectoral scenarios are compared, it becomes clear that the scenario for climate change mitigation has by far the highest additional effects on other landscape functions. The implementation of this sectoral scenario achieves more than 100 % of the target for water quality protection. Furthermore, 88.3 % of erosion prone sites are improved by this management strategy, and it achieves half of the optimized gain in habitat value of the sectoral biodiversity scenario. In addition to the climate protection scenario, the biodiversity strategy produces unintentional benefits for other landscape functions that are significantly higher than those of water and erosion protection scenarios. The implementation of the sectoral biodiversity scenario can satisfy ~25 to ~54 % of other sectoral objectives. In contrast, the optimized sectoral scenarios for erosion and water quality conservation do not have comparable benefits for other sectoral scenarios. They have only little effects on the objectives for climate change mitigation (4 % within erosion scenario) and safeguarding biodiversity (~8 % within water scenario). However, the erosion prevention scenario does fulfill the water conservation objectives to ~27 %, and the water quality scenario contributes the regional objectives for erosion prevention to ~13 %. It should be noted that the area of measures in the sectoral scenarios differs from 15,610 ha (for erosion prevention) to 52,667 ha (for water quality conservation).

The analysis of the actual location of measures in the case study shows that there are areas where measures overlap that are per se multifunctional and where they support the objectives of other sectoral strategies. Areas with multifunctional measures cover 38 % of the total area for measures within the sectoral strategy for erosion prevention, 22.8 % of the area for safeguarding biodiversity and 19.4 % within the climate protection concept. However, the spatial overlap of multifunctional measures is unclear for the sectoral scenarios for water quality protection and in part for erosion prevention. For these scenarios, the measures are not necessarily site specific and their spatial allocation is not determined. Therefore, they are difficult to include in the spatial analysis.

Furthermore, some measures are not compatible and present difficulties for the combined implementation of the sectoral management scenarios. Some combinations of measures cannot be implemented (e.g. M6 and M14 cannot be implemented simultaneously).

13.4.2 Scenario of Planned Integrative Measures (Scenario 2)

13.4.2.1 Multifunctionality Levels of Areas

Areas of multifunctionality levels 2, 3 and 4 occur in different combinations and extent (Table 13.4). Level-4 areas are always located on cropland, because a need for protection against soil erosion exists exclusively on cropland and on small and rare sites with uncovered soil. They extent on 20.6 % of the cropland (6,312 ha), which is 9 % of the study area. In these areas, measures are implemented that contribute to the optimization of all landscape functions. Level-3 areas are located on agricultural land, including cropland and grassland. Here, three sectoral measures overlap in different combinations and extent (694–10,905 ha). A coincidence of improvement requirements for two landscape functions (Level-2 areas) occurs on almost half of the area for action (18,154 ha, about a quarter of the case study area), in four different combinations. Areas with improvement requirements for only a single landscape function (monofunctional areas) are predominantly restricted to areas with water protection objectives (8,905 ha). In fact, on these sites (about 1/5 of the areas in the sectoral scenario of water quality conservation measures) water quality conservation cannot be implemented integratively with other measures for other landscape functions. Also on 5,922 ha (especially in forests), biodiversity measures cannot be integrated with other objectives because no respective need for water, soil or climate restoration has been identified in these areas. On about 10,727 ha (15.4 % of the case study area), no measures are required, but the status quo needs, in part, to be safeguarded.

The integrative scenario includes measures for all areas for action (59,113 ha), regardless their multifunctional level and contains monofunctional areas. The planning guidelines on how integrative measures are generated and allocated within the categories of the area for action are listed briefly in Table 13.4.

Table 13.4 Multifunctional effects of the integrative scenario (erosion prevention (E), water quality preservation (W), climate change mitigation (C), safeguarding biodiversity (B))

Degree multi-funct.	Overlaying needs/focus areas				Rules/principles for generating an integrative management concept, examples of allocated integrative measures	Area for measures (ha)	Percentage of case study area	Implem. costs (10 years)	Percentage of (theoret.) costs for implem. all measures	Erosion prevent. on erosive sites, objective fulfillment (%)	W: N-reduction in kg/*a, objective fulfillment (%)	C: reduction of CO_2-emissions in t objective fulfillment (%)	B: upgrading of Habitat value points, objective fulfillment (%)
	E	W	C	B									
1 EF				X	Measures of the sectoral scenario are adapted	5,922	8.5	87,263,275	22.0	0	0	0	15,965 (21.5 %)
		X			Measures of the sectoral scenario are adapted	8,905	12.8	4,999,434	1.3	0	254,422 (17.0 %)	0	2,785 (3.8 %)
	X				Measures of the sectoral scenario are adapted	493	0.7	798,142	0.2	493 (2.5 %)	0	0	431 (0.6 %)
2 EF	X			X	Lead: B (M13, M14)	31	0.0	816,220	0.2	31 (0.2 %)	0	0	123 (0.2 %)
		X		X	Lead: B (M13, M14)	9,129	13.1	45,296,858	11.4	0	232,824 (15.5 %)	0	21,392 (28.9 %)
		X	X		Lead: C (M10, M11)	7,923	11.3	52,380,440	13.2	0	378,349 (25.2 %)	2,005,876 (18.9 %)	8,727 (11.8 %)
	X	X			M10 (where allocated), selection of M1 where alternative measures recomm. (M1, M8, M9)	1,071	1.5	767,097	0.2	1,071 (5.5 %)	21,583 (1.4 %)	0	0

(continued)

Table 13.4 (continued)

Degree multi-funct.	Overlaying needs/focus areas				Rules/principles for generating an integrative management concept, examples of allocated integrative measures	Area for measures (ha)	Percentage of case study area	Implem. costs (10 years)	Percentage of (theoret.) costs for implem. all measures	Erosion prevent. on erosive sites, objective fulfillment (%)	W: N-reduction in kg/*a, objective fulfillment (%)	C: reduction of CO_2-emissions in t objective fulfillment (%)	B: upgrading of Habitat value points, objective fulfillment (%)
	E	W	C	B									
3 EF		X	X	X	Lead: B, additional requirements for rewetting	7,728	11.1	82,824,156	20.9	0	165,701 (11.0 %)	1,753,987 (16.5 %)	16,672 (22.5 %)
	X	X		X	Lead: B, add. requirm. on allocation of M12 to achieve effects for erosion prevent.; when M12 also M1	694	1.0	9,123,834	2.3	694 (3.6 %)	62,423 (4.2 %)	0	1,825 (2.5 %)
	X	X	X		Lead: C (M10, M11)	10,905	15.6	45,696,543	11.5	10,905 (55.9 %)	546,150 (36.4 %)	908,990 (8.6 %)	11,084 (15.0 %)
4 EF	X	X	X	X	Lead: B, add. requirm. for rewetting (M16), possibly implem. of M12 & M10 within the same area	6,312	9.0	66,694,725	16.8	6,312 (32.4 %)	327,479 (21.8 %)	908,064 (8.6 %)	18,790 (25.3 %)
						59,113	84.6	396,660,724	100	19,506 (100 %)	1,988,931 (132.6 %)	5,576,918 (52.6 %)	97,794 (132.1 %)

13.4.2.2 Effects of Integratively Planned Measures

Total of regional objective fulfillment

The integrative planning results in achieving or even exceeding the sectoral EQO. Erosion prevention as well as the targeted CO_2-retention are fully achieved, whereas nitrogene-input reduction goes beyond the sectoral target (~133 % of the EQO). The improvement of biotopes leads to a habitat value that is about 32 % higher than the habitat value achieved in the sectoral strategy for safeguarding biodiversity. This is to be expected, because the spatial extent of the measures in the integrative scenario is larger than any single sectoral concept. However, the area where measures are allocated can be reduced while still fulfilling the regional objectives. For example, monofunctional areas for water quality conservation measures on cropland (8,905 ha) as well as Level-2 areas for water quality conservation and the improvement of the biotope function (9,129 ha), could be excluded. Herewith, the remaining area where measures are necessary is reduced to 41,079 ha.

Proportional regional objective fulfillment in areas of action with different categories of multifunctionality.

From a cross-sectoral perspective, it appears to be more effective to allocate measures in areas where it is possible to enhance multiple landscape functions. If integrative measures are implemented in areas where all landscape functions need improvement (Level-4 multifunctionality), which is 9 % of the case study area, each sectoral EQO (quantified for the whole case study area) can be achieved between a minimum of 8.6 % (for climate protection) and a maximum of ~32 % (for erosion prevention). If, additional, measures are implemented on Level-3 sites (improvement for three functions), ~ 34 % of the sectoral objective for climate protection can be achieved, ~ 65 % of the sectoral objective for biodiversity, ~ 74 % for water and ~ 92 % for soil protection. These sum up to 264 % that (in sum of all four sectoral objectives) can be fulfilled, which is about two third of the regional objectives (in total 400 %) (Fig. 13.4). These results can be achieved on only ~37 % of the case study area and 43 % of the total focus area. Areas with multifunctionality Level-1 or -2 cover 48 % of the case study area. Here, 67 % of the sectoral objectives for biodiversity can be reached, ~59 % for water, ~19 % for climate and ~8 % for erosion prevention. The assessment also takes into account assumed additional benefits for biodiversity in areas where sectoral biodiversity measures are not officially required.

Proportional fulfillment of regional objectives related to a reference area unit.

Within a reference area unit of 1,000 ha, the sectoral objectives (accept EQO for erosion prevention) can be fulfilled to different proportions. In sum, the regional objective fulfillment varies from 2.3 % in monofunctional areas for water quality conservation to 14.6 % in Level-3 (EWB) areas (Fig. 13.5). The sum of all sectoral objective fulfillments within the reference area unit of 1,000 ha is in Level-4 areas and in specific Level-3 areas (EWC) up to six times higher than in

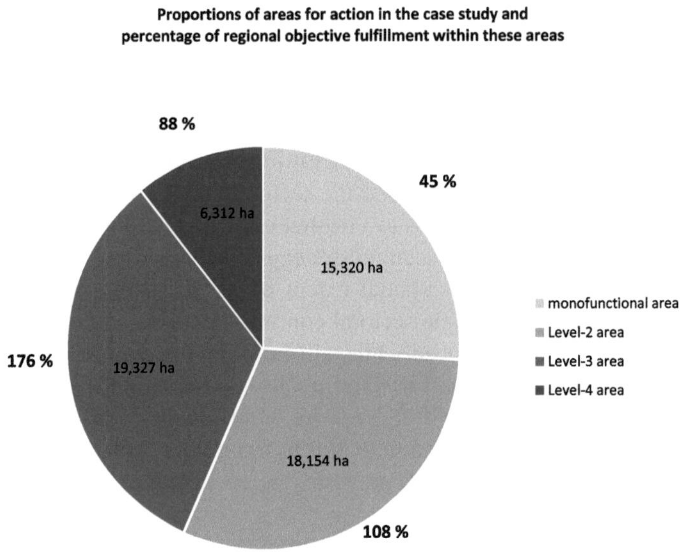

Fig. 13.4 Proportions of areas for action in the case study and percentage of regional objective fulfillment within these areas

monofunctional areas and half times to three times higher than in Level-2 areas. Thus, measures in specific high level multifunctionality areas are about six times to half times more efficient in terms of the extent of area needed for measures than in monofunctional or Level-2 areas. In comparison to sectoral scenarios, the average effects (in terms of total objective fulfillment) that can be reached by measures within a given area unit is more than six times higher in Level-4 areas than in the sectoral scenario for water quality conservation (with the lowest effects per area unit) and about half times higher than in the sectoral scenario for climate change mitigation (with highest average effects per area unit).

Implementation Costs

The costs for implementing measures in all focus areas (areas with mono- or multifunctional need for any conservation management) amount to 396.66 Mio. €. For the sectoral scenario, the theoretical sum of implementation costs for the different sectoral concepts of measures amounts to 651.04 Mio. €, disregarding the fact, that a complete implementation of all sectoral concepts of measures is not possible, because they are spatially overlapping and conflicting. Thus, the integrative scenario undercuts the theoretical costs of the sectoral scenario by almost 40 %.

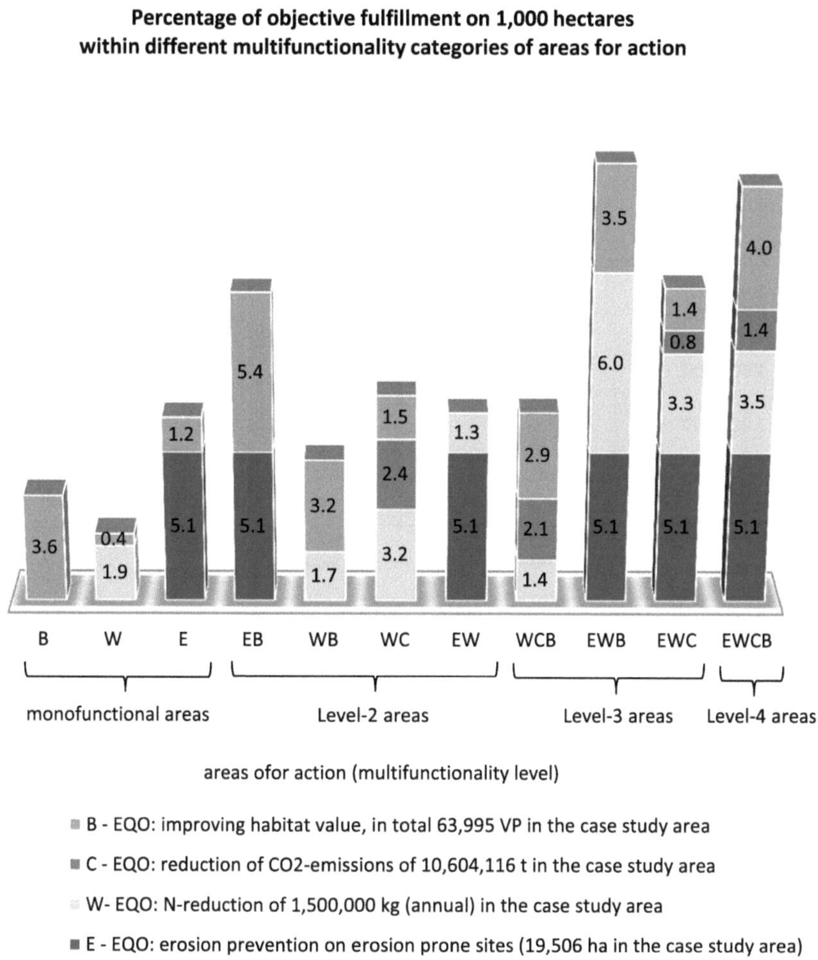

Fig. 13.5 Comparison of regional objective fulfillment in areas for action with specific multifunctionality related to the reference area unit of 1,000 ha (erosion prevention (E), water quality conservation (W), climate protection (C), safeguarding biodiversity (B))

Cost Efficiency Related to Area for Action

Figure 13.6 shows the average implementation costs for measures per 1,000 ha in the integrative scenario and the sectoral scenarios. Related to 1,000 ha, the average implementation costs for measures within the integrative scenario amount to 6.71 Mio. €, whereas implementing sectoral measures for safeguarding biodiversity costs about 11.7 Mio. €/1,000 ha, for climate change mitigation 7.60 Mio. €. In contrast, the costs for water quality conservation measures (0.53 Mio. €) and erosion prevention (5.39 Mio. €) are lower than the costs for the implementation of integrative measures related to the same area unit of 1,000 ha.

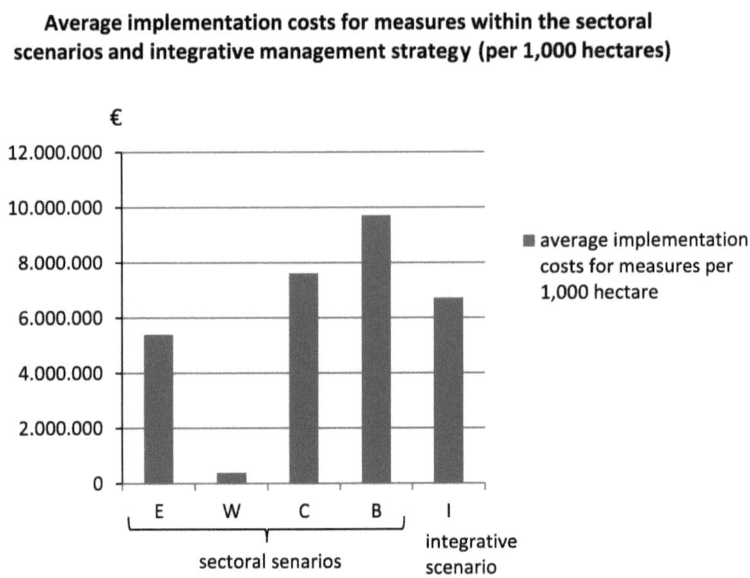

Fig. 13.6 Average costs for implementing measures on 1.000 ha (sectoral management strategies: erosion prevention (E), water quality conservation (W), climate change mitigation (C), safeguarding biodiversity (B); integrative management strategy (I))

It has to be noticed that the average effect on N-reduction of the selected sectoral water quality conservation measures (28.5 kg N/ha*a) is lower than the average effect on N-reduction of the integrative measures (34.5 kg N/ha*a). Nevertheless, related to their effects the costs for integrative measures are still higher than costs for the selected set of most efficient water quality conservation measures.

Cost Efficiency Related to Regional Objective fulfillment

Figure 13.7 shows costs for implementing (integrative) measures allocated in different areas of action to fulfill 1 % of the regional quality objectives. It illustrates, that the costs for measures needed to fulfill 1 % of the regional objectives decrease when multifunctional effects are achieved. The figure shows that multifunctional measures are not per se more efficient. A specific measure becomes more efficient (i.e. the implementation costs related to 1 % regional objective fulfillment), the more multifunctional effects are generated. However, the implementation costs in monofunctional areas for nitrogen reduction or erosion prevention (W, E) as well as in Level-2 areas with needs for both, nitrogen-reduction and erosion prevention (EW), are lower, compared to measures with multifunctional effects on biodiversity or climate protection (e.g. EB, WB, WC). This is due to the fact, that costs for implementing (integrative) biotope conservation measures

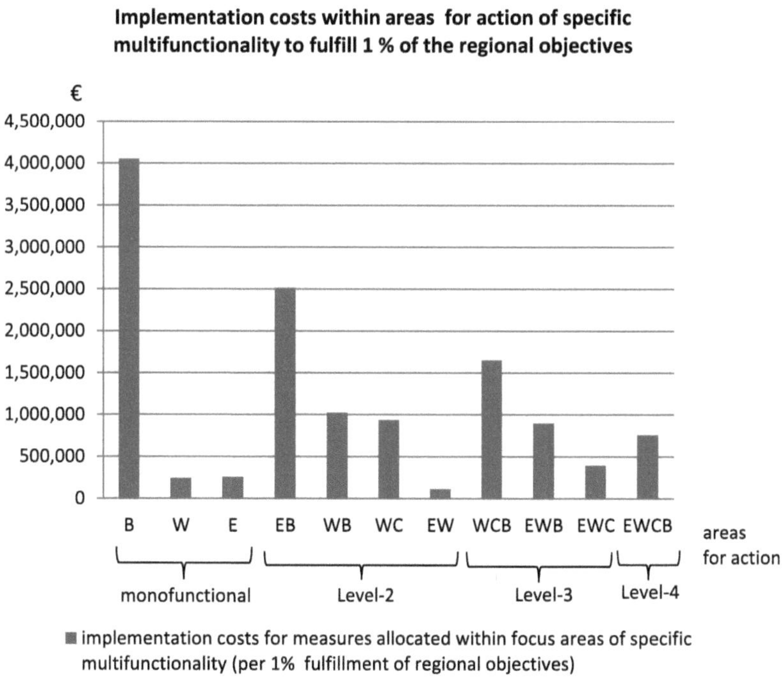

Fig. 13.7 Implementation costs for integrative measures within different focus areas of specific multifunctionaity to fulfill 1 % of the objectives (sum of sectoral objectives)

as well as restoration of fens and bogs for climate change mitigation are higher than costs for water quality conservation or erosion prevention (Table 13.2). In sum, costs for 100 % fulfillment of regional objectives for safeguarding biodiversity in scenario 1 are about 14 times higher than for reaching the objectives of water quality conservation (Table 13.3). Nevertheless the financial resources are used more efficiently, when measures are allocated in areas where multifunctional benefits can be maximized. The efficiency of measures for safeguarding biodiversity is considerably lower in areas where monofunctional restoration is needed because no multifunctional effects or added value for other functions can be generated.

Regional Objective Fulfillment with Limited Funds

With a limited budget of 40 Mio. € for 10 years, the regional environmental quality objectives can be fulfilled to 42.1% (sum of sectoral objective fulfillment) when measures are allocated proportionally in all areas of action regardless of their level of multifunctionality. The sum of all sectoral objectives that can be fulfilled spending 40 Mio. € can reach 52.8 %, when measures are allocated in areas where

Table 13.5 Effects (in percentage of regional objective fulfillment) of sectoral and integrative scenario when the budget is limited to 40 Mio. €

Effects on	Sectoral scenario (objective fulfillment in %)[a]					Integrative scenario (objective fulfillment in %)[b]
	E (10 Mio. €)	W (10 Mio. €)	C (10 Mio. €)	B (10 Mio. €)	Sum of effects (40 Mio. €)	40 Mio. €
Erosion prevention (E)	11.9	4.8	3.5	1.3	21.5	19.4
Water quality conservation (W)	3.2	36.1	4.2	1.9	45.4	13.1
Climate change mitigation (C)	0.5	0	4.0	0.9	5.4	5.1
Safeg. biodiversity (B)	0.2	3.0	2.4	3.5	9.1	15.2
Sum	15.8	43.9	14.1	7.6	81.4	52.8

[a] optimized selection and allocation of measures for achieving maximum effects in each sectoral management strategy
[b] maximization of effects by allocating measures in Level-4 areas

maximum multifunctional effects can be reached (areas with 3- and 4-level multifunctional needs) (Table 13.5). However, this implies a trade-off for the fulfillment of sectoral objectives which can be achieved at relatively low costs: If we would give each sector an equal share of 10 Mio. € for implementation, in particular the water objective could be achieved at a rate of 36.1 %. In the integrative scenario only 13.4 % of the sectoral objective for water quality conservation are fulfilled spending 40 Mio. € proportionally in all areas of action. Looking at the specific effects for landscape functions, climate change mitigation as well as safeguarding biodiversity would profit from an integrative management, whereas the benefits for water quality conservation are lower, compared to the effects achieved within the sectoral scenario.

13.5 Discussion

A method has been developed and tested for investigating and comparing the effectiveness and efficiency of sectoral and integrative landscape management strategies. In a case study the effects and the efficiency of the strategies were quantified. The results showed that water quality conservation measures have few additional benefits for other landscape functions. In contrast, measures for climate change mitigation and safeguarding biodiversity are generally multifunctional. However, these measures very often require land use restrictions or even the abandonment of land use. In addition, their implementation costs are higher than for integrated land use measures that preserve water quality. In fact, achieving the sectoral biodiversity-objectives (in the case study) costs about 10 times more than achieving water quality objectives.

We confirmed the hypothesis that integrative management strategies are considerably more effective and efficient than sectoral ones. However, the added value of an integrated strategy requires a model of integrated environmental politics. The implementation of multifunctional measures causes a redistribution of resources among the different environmental policy sectors. The shift mainly enhances safeguarding biodiversity and climate change mitigation, because the measures that improve these landscape functions are usually multifunctional and expensive. In contrast, when funds are distributed equally to each sector, water quality conservation measures can be implemented to a greater extent. This is due to the fact that sectoral measures for water quality conservation are relatively inexpensive, compared to integrative measures.

Whether or not the potential multifunctional effects can actually be achieved depends greatly on the allocation of measures. Multifunctional effects can be optimized within an integrative environmental management strategy by implementing potential multifunctional or synergizing measures on sites that require an improvement of multiple landscape functions. This leads to efficient landscape management. However, the implementation of measures that only benefit one ecosystem function may also be worthwhile, although their cost efficiency related to the sum of regional objective fulfillment is generally lower. For example, biodiversity measures that are located in forests cannot be combined with other objectives because there is generally no need for water, soil or climate restoration in forests. Furthermore, monofunctional measures are not per se inefficient. Especially water quality conservation can be efficiently implemented by sectoral measures that are integrated into land uses.

The case study confirmed that measures for safeguarding biodiversity as well as measures for climate change mitigation can be used as leading measures for optimizing effects for all landscape functions. Although no conflicting measures have been identified in the analysis on the landscape scale, it cannot be ruled out that conflicts may occur in a more detailed, site-specific analysis. Nevertheless, not all types of measures for safeguarding biodiversity (that are used as leading measures within the integrative scenario) generate the maximum effects for climate change mitigation (e.g. for M16 maximum CO_2-retention can be asumed, whereas the implementation of M14 or M13 on cropland contribute reducing CO_2-emissions, but do not necessarily lead to an optimal CO_2-retention. However, the biodiversity priority areas that can be defined with respect to their international, national or regional importance, cannot be used as a sole indicator for effectively and efficiency guiding implementation measures (allocation and time scheduling). These priority areas include areas of all multifunctionality levels, also those of low effectiveness and efficiency (monofunctional areas and areas with only two overlaying improvement requirements). Therefore, measures within areas of high importance for safeguarding biodiversity are effective and efficient in a cross-sectoral sense, if they are allocated on sites that have a high level of multifunctionality.

The investigation showed, that the method can be applied using the data of a regional landscape plan. However, shortcomings of the cost calculation should be addressed in the future with respect to the used data base. The estimation of costs for measures that are land use integrated and of those that address biotope maintenance and re-establishment is based on lump sums. For the land use integrated measures, the data used in this study did not include the specific site conditions, which would have differentiated the loss of revenue. In addition, the cost for compensating farmers for a decrease in revenue per hectare, were not included in the calculation for measures that require distinct land use changes. Therefore, in a real implementation situation, the costs for some of the measures, especially for biodiversity and climate mitigation, may be even higher than we presumed, particularly if the public must purchase the land. We consider this shortcoming to be acceptable in this study because the cost of compensation depends greatly on the legal conditions in a real situation. For example there may be no need for compensation at all if (1) a legal ordinance is in place, (2) the EU Cross Compliance regulation would demand that a part of the land remains unused, or (3) if public land for example along rural tracks and farm roads is illicitly used as farmland. Also the maintenance costs that we used do not include the revenue which could be generated, also for a maintenance situation.

The method is suitable for an initial estimate of multifunctional effects that includes possible impacts of policy objectives and their implementation instruments. This can help to prevent unexpected negative side-effects. A prominent example of this is the funding of energy crop cultivation that has the overriding objective of contributing to climate change mitigation. The extensive financial support accepts trade-offs for landscape functions (the provision of ecosystem services) such as drinking water, biodiversity and even CO_2-retention function of soils and ecosystems. Greiff et al. (2010) emphasize the need for including multifunctional effects in the design of funding programs.

The analysis results on multifunctionality are not indended to be taken as a planning guideline. However, the developed method is well suitable for the analysis of multifunctional effects and trade-offs between landscape functions and can be used for decision support purpose within planning and implementation processes.

13.6 Conclusions

The research results show that aiming for integrative, multifunctional landscape management can indeed provide added value for several landscape functions as argued for in the context of an integrated model of environmental politics that is propagated by the EU. Environmental impact assessment, e.g. strategic environmental assessment (SEA) or European Directives such as WFD, already follow this integrative approach for assessing multiple environmental issues as well as interaction and cumulative effects. In an integrated context it is clearly beneficial to place more emphasis on considering multifunctional effects in planning and implementation processes. However, in environmental management practice, planning as well as funding instruments are generally sector-oriented. From a broader environmental perspective, measures that have been identified (as) efficient for achieving a sectoral objective (e.g. funding of energy crop cultivation), may be less effective and efficient if multifunctional aspects are included in the assessment. When the presented procedure is applied, the effectiveness and efficiency of measures (in a cross-sectoral sense) can be assessed for each plot of land within a defined area. This information may help to design funding programs that maximize the effects of environmental measures while having a limited budget. Implementing such a multifunctionality strategy requires integrative planning systems and funding programs. Both could probably be fostered by a cross-sectoral environmental administration system.

References

Brandt J, Tress B, Tress G (editors). Multifunctional landscapes: interdisciplinary approaches to landscape research and management. Conf mater. 2000. http://forskningsbasen.deff.dk/Share.external?sp=S1befc68d-4d52-4b60-b377-02af62a25193&sp=Sruc

Costanza R, d'Arge R, de Groot RS, Farber S, Grasso M, Hannon B, Limburg K, Naeem S, O'Neill RV, Paruelo J, Raskin RG, Sutton P, van den Belt M. The value of the world's ecosystem services and natural capital. Nature. 1997;387:253–260.

De Groot RS, Hein L. Concept and valuation of landscape functions at different scales. In: Mander Ü, Wiggering H, Helming K, editors. Multifunctional land use. Meeting future demands for landscape goods and services. Berlin: Springer; 2007.

Dijst M, Elbersen B, Willis K. The challenges of multi-functional land use in rural areas. J Environ Plan Manag. 2005; 48(1):3–6

von Drachenfels O. Kartierschlüssel für biotoptypen in niedersachsen. Naturschutz und landschaftspflege in niedersachsen A/4; Niedersächsischer Landesbetrieb für Wasserwirtschaft, Küsten- und Naturschutz (NLWKN) (editor). 2013. http://www.nlwkn.niedersachsen.de/portal//search.php?_psmand=26&q=Kartierschl%C3%BCssel

Fisher B, Turner RK, Morling P. Defining and classifying ecosystem services for decision making. Ecol Econ 2009;68(3):643–653.

Flessa H, Müller D, Plassmann K, Osterburg B, Techen A, Nitsch H, Nieberg H, Sanders J, Meyer zu Hartlage O, Beckmann E, Anspach V. Studie zur Vorbereitung einer effizienten und gut abgestimmten Klimaschutzpolitik für den Agrarsektor. Landbauforschung vTI Agriculture and Forestry Research Special Issue 2012;361:1–472.

Greiff K, Weber-Blaschke G, Faulstich M, Haaren C. Förderung eines umweltschonenden Energiepflanzenanbaus. Naturschutz und Landschaftsplanung. 2010;42:101–7.

Helming K, Wiggering H (editors). Sustainable Development of Multifunctional Landscapes. Springer; 2003.

Kreins P, Behrendt, Gömann H, Heidecke C, Hirt U, Kunkel R, Seidel K, Tetzlaff B, Wendland F. Analyse von Agrar- und Umweltmaßnahmen im Bereich des landwirtschaftlichen Gewässerschutzes vor dem Hintergrund der EG-Wasserrahmenrichtlinie in der Flussgebietseinheit Weser. Landbauforschung vTI Agriculture and Forestry Research Special Issue 2009;336: 1–308.

Mander Ü, Wiggering H, Helming K (editors). Multifunctional land use: meeting future demands for landscape goods and services. Landscape Tomorrow. Springer; 2007.

Marsden T, Sonnino R. Rural development and the regional state: denying multifunctional agriculture in the UK. J Rural Stud. 2008;24:422–31.
Niedersächsischer Landesbetrieb für Wasserwirtschaft, Küsten- und Naturschutz (NLWKN). Liste der Biotoptypen in Niedersachsen, korrigierte Fassung 20. August 2012; Informationsdienst Naturschutz Niedersachsen. 2012;32(1)
Organisation for Economic Co-Operation and Development (OECD). Multifunctionality. Towards an analytical framework. 2001. www.oecd.org. Accessed 12 June 2009.
Osterburg B, Runge T. Maßnahmen zur Reduzierung von Stickstoffeinträgen in Gewässer – eine wasserschutzorientierte Landwirtschaft zur Umsetzung der Wasserrahmenrichtlinie. Landbauforschung Völkenrode FAL Agricultural Research Special Issue 2007;307:1–302.
Potschin M, Haines-Young R. "Rio+10", sustainability science and Landscape Ecology. Landscape Urban Plan. 2006;75:162–174.
Rüter S. Biotopverbund und Abflussretention in der Agrarlandschaft. Modellanalytische Untersuchungen am Beispiel des sächsischen Lösshügellandes. Dissertation, Gottfried Wilhelm Leibniz Universität Hannover. 2008.
Saathoff W, von Haaren C, Dechow R, Lovett A. Farm-level assessment of CO_2 and N_2O emissions in lower saxony and comparison of implementation potentials for mitigation measures in Germany and England. Reg Environ Change. 2012. doi:10.1007/s10113-012-0364-8.
Thuringian Ministry for Agriculture, Forestry, Environment and Nature Conservation. Die Eingriffsregelung in Thüringen – Kostendatei für Ersatzmaßnahmen. 2003. www.thueringen.de/imperia/md/content/tmlnu/104.pdf. Accessed 15 Oct 2012.
von Haaren C. Landschaftsplanung. 528 S. Ulmer; 2004.
von Haaren C, Albert C. Integrating ecosystem services and environmental planning: limitations and synergies. Int J Biodivers Sci, Ecosyst Serv Manage. 2011;7:150–67.
von Haaren C, Galler C. Landschaftsplanung. Grundlage vorsorgenden Handelns. Federal Agency of Nature Conservation, editors. (2013).
von Haaren C, Kempa D, Vogel K, Rüter S. Assessing biodiversity on the farm scale as basis for ecosystem service payments. J Environ Manage. 2012, http://www.sciencedirect.com/science/article/pii/S0301479712004021.
von Haaren C, Saathoff W, Galler C. Integrating climate protection and mitigation functions with other landscape functions in rural areas: a landscape planning approach. J Environ Plann Manage. 2011;1–18. doi:10.1080/09640568.2011.580558.
Willemen L, Verburg P, Hein L, van Mensvoort M. Spatial characterization of landscape functions. Landscape Urban Plan. 2008;88:34–43. doi:10.1016/j.landurbplan.2008.08.004.

Cited Legislation, Conventions and Plans

Council Regulation (EC) No 1698/2005 of 20 September 2005 on support for rural development by the European Agricultural Fund for Rural Development (EAFRD).
Council of Europe 2000: European Landscape Convention (ELC). http://conventions.coe.int/Treaty/en/Treaties/Html/176.htm.
County of Verden. Landscape master plan of the County of Verden. 2008, http://www.landkreis-verden.de/internet/page.php?typ=2&site=901000228&rubrik=901000021.
Directive 2006/118/EC of the European parliament and of the council of 12 December 2006 on the protection of groundwater against pollution and deterioration (Groundwater Directive).
River Basin Commission Weser. Map server http://www.fgg-weser.de/agrum_ergebnis/agrum_web/Karten_LN.htm. Accessed 11 Sept 2012.
State Authority for Mining, Energy and Geology Map server. http://nibis.lbeg.de/cardomap3/. Accessed 21 Dec 2011.

Chapter 14
The Structure and Dynamics of Agricultural Landscapes as Drivers of Biodiversity

Francoise Burel, Stephanie Aviron, Jacques Baudry, Violette Le Féon and Chloe Vasseur

Abstract The study of the relationships between agriculture and biodiversity is important to sustain biodiversity for the future. The landscape level has an influence, which has been until now mainly related to the importance of semi-natural elements. But in agricultural areas crop land is often dominant and acts on biodiversity by the resources it provides and the effects of disturbances induced by agricultural practices. The mosaic of crops is ephemeral and highly dynamic in space and time according to farming practices and crop rotations. The aim of this chapter is to assess the role of agricultural landscape heterogeneity on biodiversity. Landscape heterogeneity may be measured from different perspectives, considering non-cropped areas versus crop ones, or taking into consideration the dynamics of the mosaic of crops and agricultural practices. From studies on a long term ecological research site in Brittany, France, we present how these different approaches of landscape heterogeneity allow a better understanding of the diversity of processes driving biodiversity in agricultural landscapes. Most of all we underline the necessity to include knowledge of farming systems and farming practices in the analyses.

Keywords Agricultural landscape · Biodiversity · Heterogeneity · Crop mosaic · Dynamics

F. Burel (✉)
UMR Ecobio, CNRS Université de Rennes 1, bât 14,
campus de Beaulieu 35042 Rennes cedex, France
e-mail: Francoise.burel@univ-rennes1.fr

S. Aviron · J. Baudry · C. Vasseur
INRA SAD Paysage, 65 rue de Saint Brieuc,
35642 Rennes cedex, France

V. Le Féon
INRA, UR406 Abeilles et Environnement, Site Agroparc,
Domaine Saint-Paul 84914 Avignon cedex 9, France

B. Fu and K. B. Jones (eds.), *Landscape Ecology for Sustainable Environment and Culture*, DOI: 10.1007/978-94-007-6530-6_14,

14.1 Introduction

Agriculture and biodiversity are two faces of the same coin. On the one hand, biodiversity is the basis of agriculture: plants that are grown, animals that are raised are species shaped over centuries for the benefit of people; they all depend on other living organisms. On the other hand, the fate of biodiversity in many landscapes of the world depends on its use and management by rural/farming communities. Some agricultural landscapes may be more diverse than "natural landscapes" and thus are a source of biodiversity, but many farming practices at field and landscape levels are a threat for biodiversity.

Between the 1960s and the 1990s the negative impacts of agriculture on biodiversity have become increasingly documented. Among the main drivers of biodiversity decline are agricultural practices and changes in landscape structure (Robinson and Sutherland 2002). For a long period, public policies neglected the problem but then, under pressure from environmentalists and the general public, regulations were established to protect rare or emblematic species. Nature reserves were created and were extended to Natura 2000 zones in the European Union as an ecological network of protected areas, designated to protect habitats and species present on red lists in negotiations between NGOs and policy makers. Even though farming techniques such as haying or grazing were used as integrative management techniques for those protected areas, the general concept remained to segregate the agricultural, productive areas from the nature protection areas (Fisher 2008). The growing insight regarding the benefits of ecosystem services led to an interest in species providing those services and to the idea that biodiversity should be managed and protected everywhere because it is of use everywhere. Nowadays the maintenance of beneficial insects and birds, pollination, water purification etc. are services that farmers must be aware of in their crop production and management of land.

The objective of this chapter is to tackle these different issues. We utilize results from the different projects carried out on a Long Term Ecological Research site, the "Zone Atelier Armorique". Landscape ecology is the conceptual framework we use. That is to say we consider landscape patterns, their heterogeneity and connectivity, as major drivers of plant and animal population dynamics. Heterogeneity and connectivity are key concepts for biodiversity conservation and management (Burel and Baudry 2003) that need to be defined as specific metrics for the different questions and biodiversity groups we studied.

Landscape heterogeneity has many expressions. In the binary segmentation between semi-natural and cropping areas, heterogeneity increases if the share of the two components approaches 50 % of the area. The heterogeneity of the cultivated mosaic is also an important expression. This mosaic can be highly heterogeneous in space and time, as a result of the diversity of agricultural practices, and their spatial and temporal organizations by farmers (Vasseur et al. in press). The diversity of agricultural practices (cultivated species and varieties, rotations, technical operations) that can be observed at the landscape level, is due to farmers' decisions (Joannon et al. 2008). Agricultural practices in a field follow crop management sequences and depend on cropping systems (pluriannual crop rotation and management). Cropping systems are furthermore spatially distributed on the farm territory according to environmental conditions in fields (e.g. soil type), spatial structure of field patterns on farms and logistic constraints (Thenail and Baudry 2004). As a result, the cropping systems mosaic is highly heterogeneous in space and time. This additional heterogeneity may be of great importance for species (insects, weeds, mammals) using crops for at least part of their life (Vasseur et al. in press).

Studies on the influence of landscape patterns on biodiversity have focused on spatial heterogeneity. On the contrary, temporal heterogeneity has been less studied in landscape ecology (Metzger 2008). Not considering this dimension is a limitation. Past landscape structure can affect present ecological processes, and there is often a time lag between landscape change and responses by organisms (Ernoult et al. 2006; Krauss et al. 2010; Auffret and Cousins 2011). Moreover, in dynamic landscapes the rate of habitat turnover and associated change in landscape structure can sometimes be more important for species survival than the spatial organization of resource patches (Fahrig 1992).

Along this gradient of heterogeneities, the respective role of landscape design (field size, shape, presence of hedgerows etc.) and of cropping practices in the management of biodiversity is a central question. This question is of importance to model population dynamics and to design management plans. A landscape mosaic is built from both design and practices; therefore it is more and more important to foster our capacity to disentangle their effects.

In this chapter, we present how different approaches of landscape heterogeneity, first oriented toward semi natural habitats, and then recognizing the role of crops and their dynamics, may give insights on the fate of biodiversity in agricultural landscapes. We will then discuss their relative efficiency according to landscape and species types. This will give clues for designing agri-environmental schemes for biodiversity.

14.2 The Role of Semi-Natural Elements in Agricultural Landscapes

When looking at the effect of landscape structure on biodiversity in agricultural landscapes the main emphasis has been put on considering the effect of semi natural elements (Billeter et al. 2008; Tscharntke et al. 2005a). They are considered as habitats, refuges, sources, corridors for many species that use crop fields for part of their life cycle and by species which are restricted to them (Deckers et al. 2005; Forman and Baudry 1984). This approach has been the first to be used by landscape ecologists who considered agricultural landscapes as sets of semi-natural elements embedded in a neutral agricultural matrix. We studied the effects of landscape structure, which is defined in this study by the proportion of semi-natural elements, on biodiversity measured with several taxa differing by their way of dispersal and their spatial scale of perception.

It soon appeared that biodiversity responses to this heterogeneity could be linked to farming systems, as it was proved that farming practices also played a major role in the decline of biodiversity. To assess this at the landscape level we compared landscape units, first with similar agricultural systems and contrasted landscape structures, and second, of similar landscape structure and contrasted farming systems.

14.2.1 Comparing Landscapes with Contrasted Landscape Structures

In Brittany, France, as in most places of north western Europe, agricultural landscapes changed dramatically in the 1960s and up to 1980s due to the rapid intensification of agriculture (Robinson and Sutherland 2002). This led to an important decrease in semi natural habitats (Meeus 1993). To assess the effects of these changes on biodiversity we compared the gamma diversity of landscapes that differed by their amount of semi natural elements and where agricultural systems were similar.

The study area, the Long Term Ecological Research (LTER) site "zone atelier Armorique", is located in northern Brittany, south of the Mont Saint Michel Bay, France (48° 36′ N, 1° 32′ W). A hedgerow network (bocage) characterizes the landscape and agriculture is oriented toward milk production. Three units differing by field size, the density of the hedgerow network, and the relative abundance of grassland versus cropland have been delineated (Fig. 14.1). We used global indices such as percentage cover of woodlots, grasslands, crops, hedgerows, heterogeneity (Baudry and Burel 1982) to verify that units are different and have some kind of internal homogeneity (Table 14.1).

We surveyed several groups of organisms which perceive the landscape at different spatial scales, and have different ways of dispersal and different life

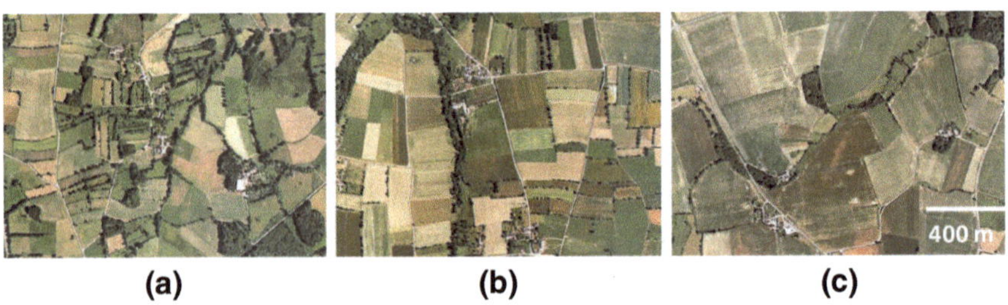

Fig. 14.1 Parts of the three landscape units: the quantity of semi natural areas: woodlots, hedgerows and permanent grassland, decreases from (**a**) to (**b**) and (**c**) as stated in Table 14.1

Table 14.1 Quantity of semi natural areas in each landscape unit

Landscape unit	(a)	(b)	(c)
Woodland and fallow land (%)	15	11	7
Permanent grassland (%)	30	20	18
Hedgerow density (m/ha)	79	63	45

spans. These were two families of *diptera*: *Chironomidae*, and *Empididae*, *carabidae*, herbs, breeding passerines, woody plants and small mammals. All these groups have been sampled using specific sampling methods. Birds were counted according to "IPA" method (Blondel et al. 1970), small mammals were studied by using the pellets of the Barn owl (*Tyto alba*), carabids were caught with interception traps, diptera with yellow attractive traps and plants were identified in hedgerows, with 3014 relevés for woody plants and 455 relevés for herbs.

The results are presented in Table 14.2. They show that the different groups react differently to the changes in landscape structure. Diptera only decrease as semi natural areas decrease, there are few changes but not consistent with the landscape structure gradient for carabids breeding passerines and herbs, and no or almost no differences for woody plants and small mammals. When looking at community similarity between units (a) and (c) Burel et al. (1998) showed that three classes of taxonomic groups could be identified. First, communities of diptera Empididae and Chironomidae lose species from unit (a) to unit (c). For Empididae, species with small wings and a low power of dispersal are not present in landscapes where the distance between water courses and the closest hedgerows, two elements needed to accomplish their life cycle, is high as in unit (c) (Morvan, N. unpublished data). Second, communities of carabidae and herbs do not vary that much in species richness but there is a shift in species composition, some are only present in the site with a high proportion of semi natural areas, while others are only present where this proportion is low. Large apterous carabid forest species characterize unit (a) with a high quantity of semi natural areas, while smaller winged species, adapted to disturbances characterize unit (c) with a high proportion of crops (Aviron et al. 2005). Third, communities of breeding passerines and small mammals have almost the same species in all the units. For small mammals Millan-Pena et al. (2003) showed that if the species were the same their relative abundance varied. Forest species such as the bank vole (*Clethrionomys glareolus*)

Table 14.2 Species richness

	Unit (a)	Unit (b)	Unit (c)
Dipera chironomidae	28	29	15
Diptera empididae	84	82	56
Carabid beetles	55	51	50
Breeding birds	40	35	38
Small mammals	11	11	11
Herbs	189	132	171
Woody plants	40	41	39

are relatively more abundant in dense hedgerow network landscapes, while crop species such as the Field vole (*Microtus agrestis*) are more abundant in sparse ones.

In these bocage landscapes, differing by the proportion of semi natural areas the response of biodiversity varies according to the different groups. For some of them, the high proportion of semi natural areas leads to an increase in species richness which has been shown in other studies for several taxa (Weibull et al. 2000; Schweiger et al. 2005). But according to dispersal ability and longevity of species not all of them react that way. This may be due to the fact that the gradient of landscape structure is not very long, the total proportion of semi natural areas varying only from 25 to 45 % of the total area, contrary to other studies where it may vary from 10 up to 80 % (Gabriel et al. 2005). Our results emphasize that species richness per se is not always a good indicator for measuring a community's response to landscape changes. Similarity indices or relative abundance permit to identify changes in biodiversity even when species richness remains the same. Those indices underline that different groups react differently to the same changes in landscape structure. Nevertheless, for all of them the proportion of semi natural elements had an effect on the structure of the communities.

14.2.2 Comparing Landscapes with Similar Compositions and Contrasted Farming Systems

To assess the role of farming systems at the landscape level we compared biodiversity among landscape units of similar landscape structure but contrasted farming systems.

We studied landscapes located in the Côtes d'Armor, an administrative unit located in the northern part of the Brittany region. Its area is 700,000 ha, 440,000 of which are devoted to agriculture. It is a very dynamic agricultural area specialized in both milk production and hogs and poultry indoor production. Crops sustain mainly husbandry with maize and grassland for cows, and cereals for hogs and poultry (http://draaf.bretagne.agriculture.gouv.fr/Les-Cotes-d-Armor,203). We first selected 11 landscape units which represented the whole diversity of the landscape structures present in the area. They differ by their composition, total length of hedgerows, connectivity of the hedgerow network and heterogeneity of the mosaic. Carabid beetles have been surveyed in order to measure their gamma diversity in hedgerows and we tested for the influence of landscape structure and farming systems on it (Millan-Pena et al. 2003). We then compared sites dominated by cropland, which we split into two groups. They were both characterized by a low proportion of semi natural areas, were similar in landscape composition but differed in landscape configuration. The first one was characterized by large fields, with maize as the dominant crop and a low connectivity of the hedgerow network, while the second was characterized by smaller fields, with wheat and oat as dominant crops and a more connected hedgerow network. Ten hedgerows were sampled per site during the summer of 2001. We compared carabid species composition between the two "cropland" groups. They were characterized by two distinct clusters of species, maize-dominated landscapes hosted species occurring in rather moist and shaded habitats such as *Brachinus scolopeta*, while cereal-dominated landscapes hosted typical crop field species such as *Pterostichus melanarius*.

The abundance of carabids significantly differed between the two types of landscapes ($t = 5{,}82$, $p = 0.01$). The average abundance was 693 individuals per site in the cereal-dominated landscapes, and 333 for the maize-dominated ones. Species assemblages of the most different sites were compared using the ten most abundant species found at each site. These top ten species accounted for 81.6–87.8 % of the total catch, depending on the site. The top ten species present in the maize-dominated landscapes only accounted for 13.7 % in the cereal-dominated ones, reflecting the strong effect of this shift in farming system and change in landscape configuration (Table 14.3).

These results show that farming systems have an effect on biodiversity at the landscape scale. The effects of farming systems on biodiversity have mainly been studied for comparisons between conventional and organic systems and looking at the effect of the surrounding landscape on alpha diversity (Weibull et al. 2000; Purtauf et al. 2005). The main results are that diversity is higher in complex landscapes with a high proportion of semi natural areas whatever the system, and that organic systems enhance diversity in simple landscapes. Our results deal with gamma diversity and systems that are both conventional but with different crop and husbandry productions. They are intensive agricultural systems, but the one dominated by maize and milk production is less favorable for carabid species than the one dominated by cereal crops. This may be due to the differences in farming practices, with higher inputs in maize fields, or to the configuration of the landscape with larger fields and a lower connectivity of the hedgerow network when maize is dominant.

Table 14.3 Abundance of the top ten carabid species in the most different landscape units of the two types: maize and cereal dominated

	Cereal dominated landscape	Maize dominated landscape
Pterostichus madidus (Fabr.)	19	6
Nebria brevicollis (Fabr.)	89	6
Abax parallelepipedus (Piller and Mitterpacher)	0	19
Calathus piceus (Marsham)	81	1
Amaras pp.	28	55
Harpalus rufipes (De Geer)	122	0
Pterostichus melanarius (Illiger)	72	1
Poecilus versicolor (Sturm)	2	3
Agonum dorsale (Pontoppidan)	15	79
Poecilus cupreus (L.)	3	12

14.3 The Role of the Cropping System Mosaic

We have shown that semi-natural elements contribute to produce landscape structures of ecological importance. But, because many species use crops (including grasslands) during their life cycle, the heterogeneity of the cropping systems mosaic is potentially important from an ecological point of view (Kennedy and Storer 2000; Benton 2003). At a given time, this mosaic can be viewed as a spatially heterogeneous mosaic of cropped habitats with varying resources for species (food resources, host plants, shelter), and of disturbances with direct effects on species survival (e.g. insecticide spraying). The cropping system mosaic also generates a spatiotemporal heterogeneity at different time scales (Burel and Baudry 2005). Crop phenology and farming practices lead to fast asynchronous variations of resources from field to field within a year; over several years, crop rotation and management succession result in spatio-temporal variations of resources availability, localization and accessibility, i.e. landscape connectivity for species. In this shifting mosaic, habitat patches are ephemeral regarding the life span of many species. Species survival will therefore depend on their ability to find and colonize new suitable resource patches to supplement or complement habitats and complete their life cycle (Dunning et al. 1992; Wissinger 1997) as well as availability of ephemeral, but suitable habitats over years.

In the following sections, we illustrate how the cropping systems mosaic can influence the movement of organisms, their population dynamics and species diversity at the community level, at infra- and/or- pluriannual time scales. We utilize results from empirical and modeling studies conducted on the LTER "Zone Atelier Armorique" for several insect taxa: two species with contrasted habitat requirements and dispersal abilities, i.e. a grassland butterfly and a carabid beetle of cropped habitats, and the community of wild bees. We will emphasize the temporal dimension by stating the effects of the changing crop mosaic within a year studying insect movements between crops, and looking at the spatial and temporal distribution of organisms during one rotation cycle.

14.3.1 Effects of Crop Phenology and Farming Practices at the Infra-Annual Time Scale

Species movements between habitat patches depend not only on functional landscape connectivity (Kindlmann and Burel 2008), but also on the dynamics of resource quality in patches (Schooley and Branch 2011). For insects using annual crops or grasslands, whether they are phytophageous, nectariferous or predatory, crop cover states control the availability of biotic and abiotic resources (e.g. Alston et al. 1991). These cover states vary throughout the season from crop sowing and growth to harvest. These changes induce quick and frequent changes in insect movements and distributions in the cultivated mosaic (Kennedy and Storer 2000).

We observed these processes for the Meadow brown (*Maniola jurtina* L.), a grassland butterfly species, in response to grassland mowing (Aviron et al. 2007). This common species has no strong host-plants requirements (Vane Wright and Ackery 1981) but has a limited mobility like many endangered butterfly species (Brakefield 1982). We conducted a mark-release-recapture experiment at various herbaceous patches (grasslands, lane banks and road verges) to study butterfly movements and distribution before and after the mowing of two studied grasslands. This survey showed that the mowing of grasslands can lead to changes in butterfly movements between herbaceous patches. This is illustrated by the decreased exchange rates of butterflies between one of the mown grasslands (G4) and surrounding patches (Fig. 14.2a). Mowing also resulted in localized drops of butterfly abundances in mown grasslands (G2 and G4), and a concentration of butterflies in certain unmown, accessible grasslands (G3 and G8; Fig. 14.2b).

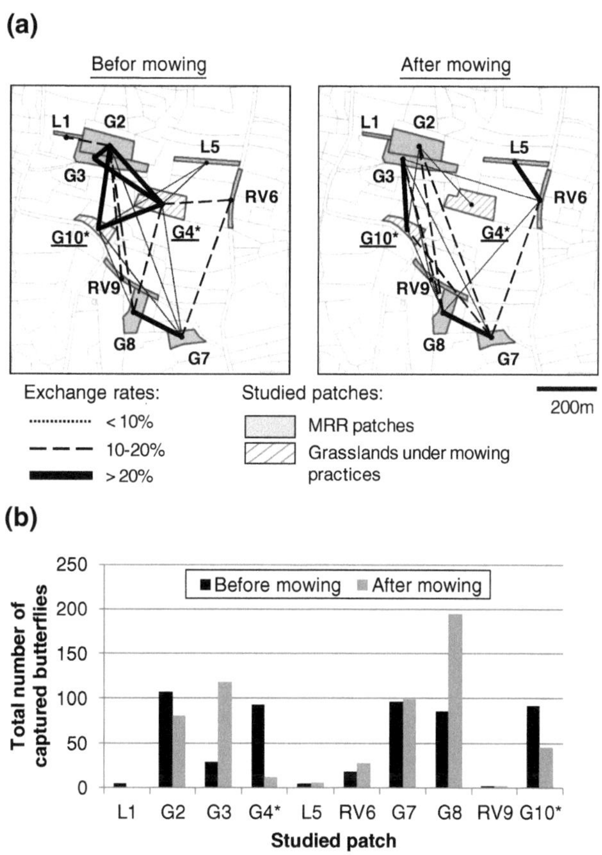

Fig. 14.2 Exchange rates of butterflies between studied herbaceous patches (**a**) and total number of butterflies captured in patches (**b**) before and after grassland mowing (adapted from Aviron et al. 2007)

These localized changes in butterfly abundances in mown and unmown grasslands can probably be explained by a redistribution of butterflies from disturbed grasslands into remnant suitable herbaceous patches.

To compensate for these local and abrupt changes in resource availability on a given patch, the presence of alternative suitable and accessible resource patches will be crucial for species to realize their life cycle (Kennedy and Storer 2000; Men et al. 2004; Carrière et al. 2006; Bressan et al. 2010). Some asynchrony between farming practices and crop cover states might allow to compensate for the ephemeral suitability of crops, by ensuring a temporal continuity of resources for species. A study of a generalist predatory carabid species with limited mobility (*Pterostichus melanarius* Illiger) in annual crops illustrates these processes (Vasseur 2012). Carabid movements were surveyed at the edges between different types of annuals crops (winter cereals, spring maize and pea) with contrasted cover states during the activity-period of carabids. Bidirectional interception traps, adapted from Hawthorne et al. (1998), were used to sample carabid movements between six adjacent crops. The interception traps were open continuously and collected weekly from early May to early September. As an example, Fig. 14.3 displays the orientation of carabid movements at field edges between a pea crop and two adjacent maize fields. It shows that, in the early summer, carabid beetles move more frequently from maize fields (with bare soil at this period) to pea fields (with dense vegetation cover). In late July, an inversion of the orientation of carabid movements at edges is observed, i.e. more movements from pea to maize fields, in relationship with pea harvest and vegetation growth in maize. Thus, *P. melanarius* seems to move throughout its activity-period from crops with unsuitable, sparse vegetation to crops with dense cover. This suggests that adjacent

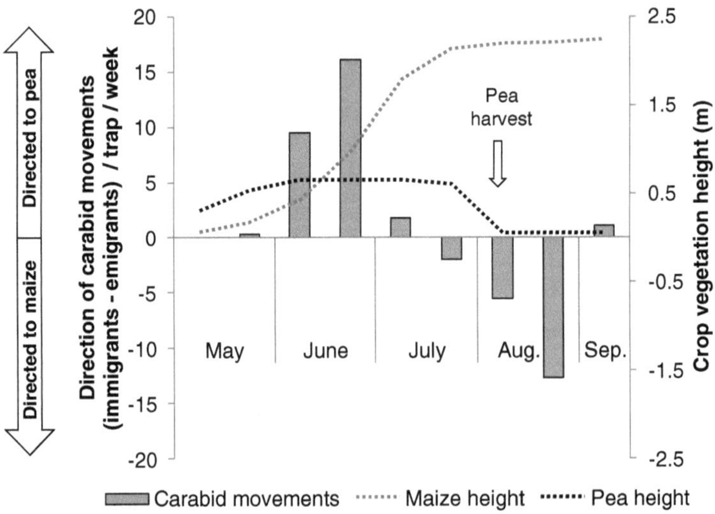

Fig. 14.3 Direction of carabid movements at edges between pea and adjacent maize fields from May to September

annual crops with asynchronous vegetation growth and timing of harvest could provide a temporal continuity of resources for this generalist predatory species. The persistence of carabid populations in a cultivated mosaic within a year might not necessarily require permanent habitats, but complementary cropped habitats that are spatially and temporally connected.

14.3.2 Effect of Crop Rotation and Management Succession Over Years

The pluri-annual effects of the spatio-temporal heterogeneity of the cropping systems mosaic on biodiversity are still mainly unknown. However, crop rotation and management succession over years will determine the temporal availability of suitable cultivated resources patches for species. Moreover, the ability of species to spatially and temporally complement or supplement their resources between cropped habitats during their life cycle will partly drive their survival from one year to another, and therefore over the long term (Rusch et al. 2011; Thorbek and Topping 2005).

14.3.2.1 The Influence of Crop Rotations on Solitary Bees

To persist in a landscape, wild bees require nectar and pollen as food for brood and adults as well as suitable nesting sites (Westrich 1996). Intensive agriculture negatively affects the quality of bee habitat in several ways: (1) increasing crop field area results in the loss of suitable habitats including grasslands that are known to be highly beneficial habitats for bees (Klemm 1996; Steffan-Dewenter et al. 2002); (2) fertilizers, herbicides and intensive grazing reduce floral resources (De Snoo and Van der Poll 1999); (3) harvesting and tillage impede the nesting of most ground-nesting species (Shuler et al. 2005; Morandin et al. 2007); (4) some pesticides induce direct mortality or sublethal effects (Desneux et al. 2007). However, the crop mosaic can offer a great amount of easily available food resources when mass flowering crops such as oilseed rape or sunflower are cultivated (Westphal et al. 2003).

In order to better understand how landscape patterns influence solitary bee[1] communities, we took into account both spatial and temporal heterogeneities of the crop mosaic in addition to the commonly studied semi-natural elements (wooded elements and long-term grasslands) (Le Féon et al. 2011). Thus we considered the proportion of semi-natural elements, of oilseed rape and non-flowering crops at the

---

[1] Wild bees comprise of social species (*Bombus* sp.) and solitary bees (even if different forms of primitive or advanced social behavior exist in some species). Our study only focuses on solitary bees, which represent more than 80 % of wild bee species in Europe.

moment of bee sampling and the proportion of two types of crop rotations. For the last variable, a crop rotation map summarising land-use history over a period of five years was realized. As cereals and grassland are the dominant land uses in the LTER site we distinguished two classes of crop fields: the fields that were sown only with cereals (wheat and maize) during the last five years and fields where the crop rotation included from one to four years of grassland (referred to as "mixed fields"). Solitary bees were trapped on 50 field margins, 15 of which were along oilseed rape fields and the 35 others were randomly located along other fields. Landscape composition was quantified in square windows centered on sampling points. Three window sizes were chosen, covering the range of relevant scales for flight and foraging distances of solitary bees (400, 800 and 1,200 m in width).

We found contrasted effects of non-flowering crops according to the type of margin and the spatial scale (Fig. 14.4). Solitary bee abundance in margins of oilseed rape fields deeply increased with the proportion of non-flowering crops at the moment of bee sampling at the finest spatial scale while it remained unchanged in margins of non-oilseed rape fields (Fig. 14.4a). This result shows that the attractiveness of mass-flowering crops depends on the quality of the surrounding landscape: the use of oilseed rape by solitary bees is higher when the surrounding area provides few floral resources.

Long-term grasslands and crop rotation influence local richness and abundance of bees at large spatial scales. Probably due to a masking effect of mass-flowering crops, these influences are only detected in margins of non-oilseed rape fields. Solitary bee abundance increased with the increasing proportion of long-term grasslands (Fig. 14.4b). Moreover solitary bee abundance and species richness increased with the increasing proportion of "mixed fields" (at least one year of grassland in the past five years) (Fig. 14.4c), whereas the proportion of fields only sown with cereals during the last five years had the opposite effect (Fig. 14.4d). The positive effect of long-term grassland is already known in landscape scale studies on wild bees (e.g. Steffan-Dewenter et al. 2002; Morandin et al. 2007). This type of fields is typically likely to provide wild flowers and suitable nesting sites. The originality in our results is to show that introducing temporary grasslands in cereal rotations is beneficial to bees. As they are generally sown with Poaceae species only and fertilized, the suitability of temporary grasslands for bees remains to be supported by further data. Nevertheless, the introduction of this cover type in cereal rotations could imply a less intensive farming system, potentially beneficial to solitary bees, thanks to (1) reduced pesticides and fertilizer inputs over the whole rotation cycle (2) greater floral resources in properly managed temporary grasslands (3) less disturbed soils better suited for ground-nesting bees. Our result is consistent with Steffan-Dewenter (2001) and Kuussaari et al. (2011) who showed the positive effect of the introduction of set-asides in cereal rotations on pollinator insects.

To sum up our findings, the composition of the landscape at the time of sampling had a direct impact on the spatial distribution of solitary bees only at the finest scale (400 m). On the contrary, when considering the landscape structure over several years (crop rotations and semi-natural elements like long-term grasslands), the effects occurred at the larger scales (800 and 1200 m). Therefore,

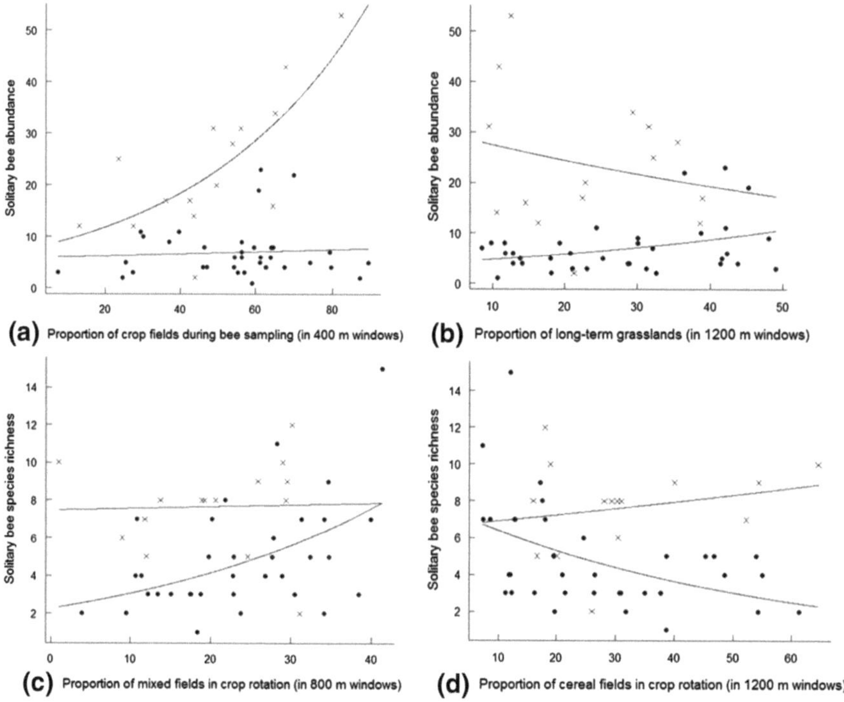

Fig. 14.4 Relationship between **a** solitary bee abundance and proportion of non-flowering crop fields during bee sampling (spring 2007) in 400 × 400 m windows; **b** solitary bee abundance and proportion of long-term grasslands in 1,200 × 1,200 m windows; **c** solitary bee species richness and proportion of "mixed fields" ("Mixed fields" are fields where crop rotation included from one to four years of temporary grassland during the period 2003–2007) in crop rotation in 800 × 800 m windows; **d** solitary bee species richness and proportion of "cereal fields" ("Cereal fields" are fields only sown with cereals (maize and wheat) during the period 2003–2007) in crop rotation in 1,200 × 1,200 m windows. Predictions returned by the Poisson-family models are shown by solid lines for significant relationships only ($P \leq 0.05$, test F). × = oilseed rape field margins. • = non-oilseed rape field margins

our results are in agreement with the hierarchy theory that predicts that spatial and temporal scales are correlated. Phenomena occurring at coarse spatial scales are related to slower processes than phenomena occurring at smaller spatial scales (Allen et al. 1987). The maintenance of populations, a slow process, may be due to "large" spatio-temporal patterns, while feeding behavior, a fast process, is related to fine scale patterns (presence of a mass flowering crop in a given field). Our study showed that examining the heterogeneity of the agricultural mosaic over a whole crop rotation cycle was relevant to better understand the effects of agriculture on solitary bee communities. This approach allowed considering the cumulative effects of field cover and it demonstrated that introducing less intensive covers such as temporary grasslands in cereal rotations positively influences solitary bee communities.

14.3.2.2 Using Models to Predict the Influence of Management Successions and Crop Rotations on Biodiversity

In landscapes where too many habitat patches are simultaneously disturbed each year, the limited habitat complementation/supplementation processes might result in a population decrease or even extinction in the long term (Vasseur et al. in press). We used an existing spatially-explicit model to simulate the yearly and pluri-annual dynamics of populations of the Meadow brown (*Maniola jurtina*) under different scenarios of habitat disturbance extent (i.e. percent cover of grasslands mown in a 1 km^2 landscape) (Aviron et al. 2007). Simulations were run on the landscape unit where empirical data on the effect of mowing on butterfly movements were available, in order to validate the model's predictions. Our results show that when habitat suppression through mowing occurs during the activity period of butterflies (in summer), butterfly populations get rapidly extinct if a large amount of grassland habitats (80 %) is simultaneously disturbed each year (Fig. 14.5). On the contrary, the synchronous disturbance of a lower amount of grasslands (20 %) each year allows population persistence and increase over the years, probably due to higher possibilities of habitat complementation/supplementation for butterflies (Fig. 14.5). Thus, the long-term persistence of butterfly

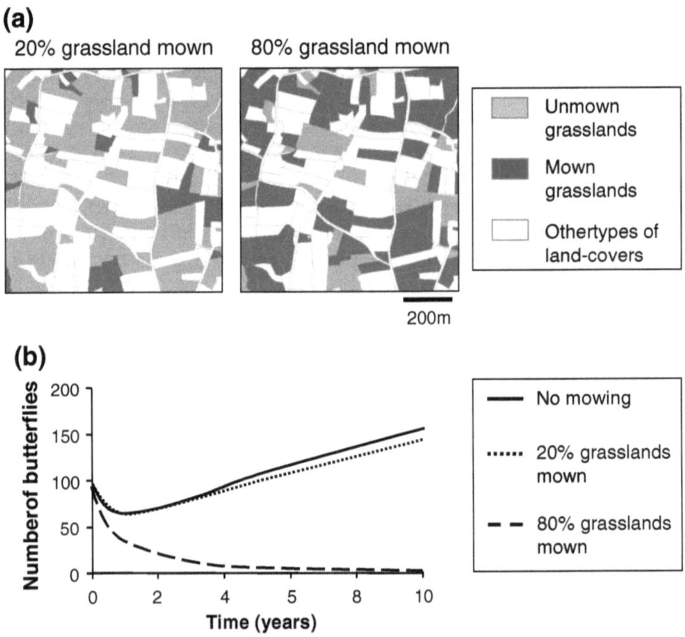

Fig. 14.5 Scenarios of butterfly habitat disturbance (20 and 80 % of grasslands mown each year) (**a**) and predicted evolution of total butterfly abundances over 10 years for the two scenarios of habitat disturbance and in absence of mowing (**b**) (derived from Aviron et al. 2007)

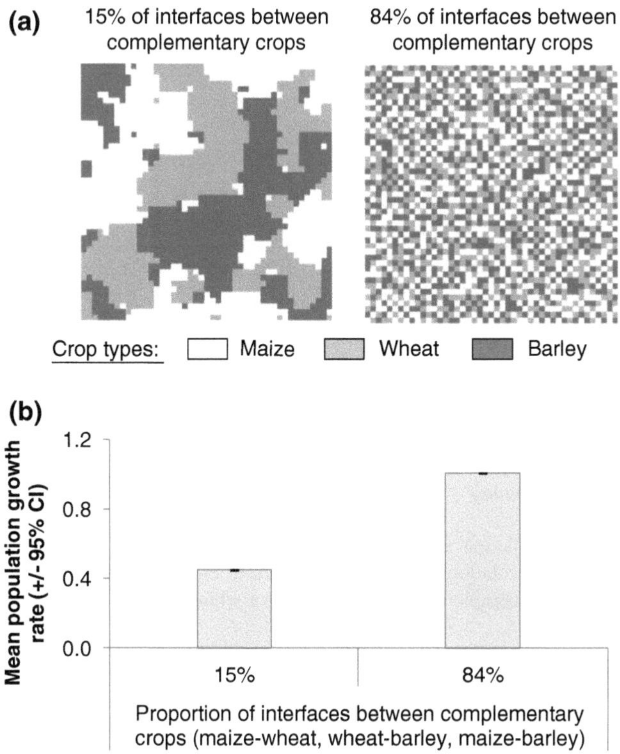

Fig. 14.6 Scenarios of spatial organization of the crop mosaic (15 and 84 % of interfaces between complementary crops: maize-wheat, wheat-barley and maize-barley) (**a**) and predicted growth rate of carabid populations (mean ± 95 % CI) after 30 years of simulation (10 cycles of rotation) for the two scenarios (**b**). (Details of the ANGORA model given in Vasseur 2012)

populations could be strongly affected by the extent of habitat destruction and of direct disturbances caused by mowing.

Complementation of resources by species is not only dependent on the availability of alternative suitable resource patches each year but also on their accessibility (Dunning et al. 1992). Thus, the spatial organization of asynchronous, complementary cropped habitats each year might be crucial for long-term population persistence as well. We used a spatio-temporally explicit model to simulate the dynamics of carabid populations (*P. melanarius*) in a cropping system mosaic characterized by a rotation of three annual crops (maize-wheat-barley) differing mainly by their period of sowing and harvest ("ANGORA" model, Vasseur 2012). Simulations were run on virtual landscapes (grids of 45 × 45 fields) with similar compositions each year (33 % of each crop type) but contrasted spatial organizations of the crop mosaic (i.e. 15 vs. 84 % of total interfaces between complementary crops) (Fig. 14.6a). The results show that, in crops mosaics with similar compositions, the population growth rate over 30 years is higher in mosaics where adjacency between complementary crops, i.e. with asynchronous cover states

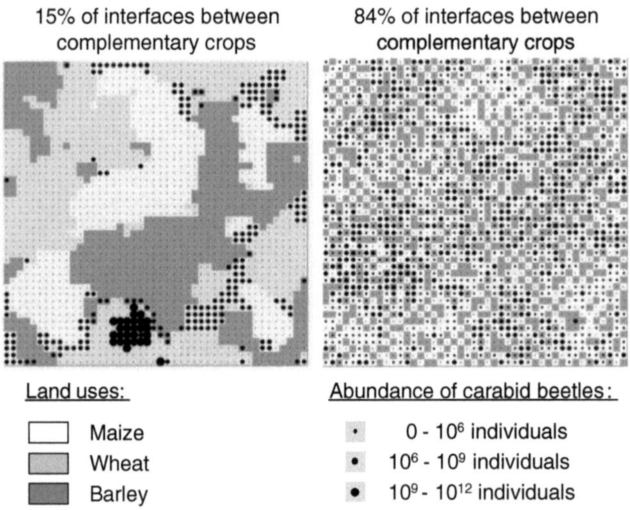

Fig. 14.7 Spatial distribution of carabid populations in the crop mosaic after 30 years of simulation (10 cycles of rotation) for the two scenarios of crop spatial organization (15 and 84 % of interfaces between complementary crops: maize-wheat, wheat-barley and maize-barley) (Vasseur 2012)

(maize-wheat, maize-barley, barley-wheat), is promoted (Fig. 14.6b). Adjacency between complementary crops each year increases survival of carabid beetles until the end of the reproductive period, ensuring a more important renewal of carabid populations. It also allows, in the latter part of carabid activity period, a rapid colonization of new, suitable patches (maize crops). Thus, the spatio-temporal heterogeneity of the cropping systems mosaic is likely to increase landscape spatio-temporal connectivity for this carabid species. Over the long term, this process could result in a homogenization of populations' spatial distribution, and enhance their resilience to frequent local extinctions (Fig. 14.7).

14.4 Discussion

14.4.1 Semi Natural Elements

Our results confirm that semi-natural or more natural elements (Fahrig et al. 2011) have an impact on biodiversity for all considered taxa. This has been shown by previous results on the effect of the amount of these elements, considered as a measure of landscape complexity, on several taxonomic groups. Most of the studies show a positive effect of semi natural elements, and conclude that complex agricultural landscapes favor biodiversity as measured by the number of species (Tscharntke et al. 2005a, b). The use of several taxa permits to discuss this assumption; if the number of some small, short-lived species increases with changes in landscape composition and configuration, in the same landscapes there is no response by longer-lived ones. The first category may perceive the landscape at a finer spatio-temporal scale and thus react to changes that are not perceived by the other categories. Nevertheless, even if the total number of species is not related to the quantity of semi-natural areas, community structure changes. There is a shift in species or in the relative abundance of species. This points out the necessity to adapt the measure of biodiversity to the target species and to the intensity of landscape change. Species richness is a good indicator for strong gradients of landscape changes and for fine spatio-temporal grain species. Otherwise, some more sensitive measures such as the composition of communities or the relative abundance of species are needed to highlight the response of biodiversity to landscape structure changes.

In rural landscapes, all parts of the mosaic are influenced by agricultural activities. Even semi natural areas such as woodlots, hedgerows and even more permanent grassland depend on the farming system. They may be sprayed by pesticides from the crops, enriched by fertilizers from the upper fields; woodlot boundaries and hedgerows are pruned not to shadow crops (Lotfi et al. 2010), etc. It is thus an illusion to draw a strong boundary between semi-natural and productive areas. This is shown by our work on landscapes with similar compositions and different farming systems, with a strong response of species abundance and composition to changes in production type and farming practices. It is of overall importance when comparing landscapes with different structures to explicitly characterize farming activities. This has been done within a European project, green veins, with 24 landscape units, distributed along a double gradient of farming intensity and amount of semi-natural areas. The more important factor to explain biodiversity was then the intensity of the farming system (Billeter et al. 2008).

14.4.2 Cropping System Mosaic

Beyond the effects of farming system intensity, our results show that the spatio-temporal organization of crop covers, farming practices and crop rotations affect biodiversity. Until now, this issue has mainly been addressed for crop pest species (Carrière et al. 2006; Bresson et al. 2010; Kennedy and Storer 2010), but our results on pollinators, predatory arthropods and butterflies show that the heterogeneity of the cropping systems mosaic permits the persistence of beneficial organisms, and of species of conservation interest. For them, the diversity of farming practices and rotations, together with semi-natural elements, ensures habitat complementation and/or supplementation in space and time, and determines landscape connectivity. Over the years, crop rotations will not only control the degree of stability of resources for species, as shown for bee communities, but also their temporal accessibility, as illustrated by our modeling studies. Benefits of the cropping systems mosaic are expected for species in semi-natural elements, especially in landscapes where uncultivated elements are sparse. This has been underlined in previous studies, which showed that landscape connectivity for forest species increases when crops are grown high and dense (Fitzgibbon 1997; Ouin et al. 2000). The relative contribution of semi-natural and cultivated elements for biodiversity is however, likely to vary in time due to crop turnover and associated changing suitability of the cropping systems mosaic (Holzschuh et al. 2011).

14.4.3 The Role of Farming Systems

We have shown that farming systems interact with biodiversity in several ways and at different scales from fields to farm to groups of farms. This is summarized in Table 14.4. At the field scale, only species spending part of their life in crops are concerned. Their populations are driven by the food and microclimatic resources within the field. Processes in the mosaic of a few adjacent fields drive the same type of species, the mosaics control movements from field to field, therefore the possibilities to find food and shelter.

At a wider scale (about 100–1,000 ha), the controlling structure is the crop mosaic and associated semi-natural elements. The different landscape patterns offer different habitats and resources and, therefore, select the species that can thrive. Both species living in semi-natural elements and cropland species are concerned. When a landscape pattern changes by addition or removal of elements or by a new spatial distribution, the species that are not adapted vanish while new ones can come.

Table 14.4 Drivers of biodiversity at different spatial and time scales

Spatial unit	Time unit	Farming/crop processes	Ecological processes
Field	Week/month	Crop growth/crop management	Dynamics of populations of short live field species (1 month/ 1–2 years)
Mosaic of some adjacent fields	Month/year	Heterogeneity of crop management	Field species movement from field to field
Crop mosaic of $\approx$100–1,000 ha	1–10 years	Crop sequences in the different fields Removal/ implementation of semi-natural elements	Differentiation of species assemblages of both fields and semi-natural elements Differentiation of species assemblages and population dynamics at species level
Region	20–50 years	Differentiation of farming systems	Differentiation of species assemblages according to farming systems

At the regional scale, the differentiation of farming systems implies a diversity of production and management practices in terms of inputs and disturbances such as soil tillage and harvesting. This is another major cause of species distribution.

14.4.4 Guidelines for policies

In terms of policies, it has been demonstrated that agri-environmental policies implemented at the field scale only are inefficient (Kleijn et al. 2006). The ignorance of the landscape context explains a large part of this failure (Concepción et al. 2008). An important point that is not integrated in the design of those policies is that the overall (gamma) diversity of a region depends on the diversity of landscapes and farming systems at all scales.

By deciphering the drivers of biodiversity in terms of landscape patterns and farming systems, our research shows that both are important and that field scale processes are controlled by external factors. Therefore, biodiversity objectives must be set at those different scales, taking into account the regional diversity.

In the European Union, policies related to agricultural practices already exist, as in the nitrate directive that makes compulsory the presence of a catch crop in winter in areas where nitrate leaching is a problem. Within the cross-compliance of the Common Agricultural Policy, farmers must record their use of fertilizers and pesticides that must be kept below a certain level. Crop diversification is an objective of the Common Agricultural Policy reform. In France, the implementation of grassy strips along streams is a first step toward a landscape scale management of water and biodiversity. To further enhance biodiversity, policies

Table 14.5 Visible versus Hidden heterogeneity: strength and weaknesses

Visible heterogeneity linked to semi-natural elements	Hidden heterogeneity linked to farming practices and crop phenology
Strength	*Strength*
It is easy to collect data from remote sensing images or field observation, anytime	Take into account all the landscape elements
Stable pattern, generally, for some years	All practices are considered
For public policies, it is relatively easy to add semi-natural elements	Provide a range of variables that can drive biodiversity related processes
Harbor most of threaten and flagship species in rural landscapes	Permit to establish a link between biodiversity and ecosystem services
Weaknesses	*Weaknesses*
Omit the major part of landscapes	Data collection requires a lot of work and need to be redone often
Few consideration for activities of production	Difficult to gather all data on a large area
Provide few evidences on the role of practices in fields	Many variables are correlated
Overemphasize the role of semi-natural elements as a mean to protect and manage biodiversity	Difficult to study the interactions and to decipher the hierarchy of effects

should include a limitation of field size and soil disturbance, such as long-term grassland and minimum tillage. The ban of herbicides in field margin management should be part of the package.

Heterogeneity is an important variable to enhance biodiversity (Benton et al. 2003). In the course of this chapter we show how "visible" heterogeneity linked to land cover, especially semi-natural elements and "hidden" heterogeneity (Vasseur et al. in press) resulting from farming practices play a major role to maintain high levels of biodiversity. In Table 14.5, we present the strengths and weaknesses of these two approaches of biodiversity

14.5 Conclusion

To conclude we may state that until now most regulations to enhance or conserve biodiversity have been aiming at increasing (Aviron et al. 2009) or managing extensively (Kleijn et al. 2011) semi-natural elements. But at an era when food production must increase to feed a growing world population it is important to identify practices at field and landscape levels that will favor biodiversity without retrieving land from production. One may expect a threshold of amount of semi natural areas, 0–5 %, below which biodiversity remains low whatever the practices, as the regional species pool will be low. Above this threshold, crop spatial heterogeneity and environmentally friendly practices will increase biodiversity. For high amounts of semi natural elements, 20 % onwards (Tscharntke et al. 2005b), biodiversity will be high, even simply by keeping the current farming activities (Le-roux et al. 2008). In many parts of the world, agricultural landscapes fall within the second category. It is time to define policies that will encourage agricultural practices and systems that maximize biodiversity for its own sake and for the services it provides to our societies. It must also be acknowledged that all these policies will not increase all species, but may be targeted toward certain groups.

Acknowledgments LTER research is supported by the zone atelier program of INEE-CNRS, additional support is provided by INRA, the ministry of environment, the European Union and the National Research Agency. We thank three anonymous reviewers for their constructive comments.

References

Allen TFH, O'Neill RV, Hoekstra TW. Inter level relations in ecological research and management: some working principles from hierarchy theory. J Appl Syst Anal. 1987;14:63–79.

Alston DG, Bradley JR, Schmitt DP, Coble HD. Relationship of Heliothis zea predators, parasitoids and entomopathogens to canopy development in soybean as affected by Heterodera glycines and weeds. Entomol Exp Appl. 1991;58:279–88.

Auffret AG, Cousins SAO. Past and present management influences the seed bank and seed rain in a rural landscape mosaic. J Appl Ecol. 2011;48:1278–85.

Aviron S, Burel F, Baudry J, Schermann N. Carabid assemblages in agricultural landscapes: impacts of habitat features, landscape context at different spatial scales and farming intensity. Agric Ecosyst Environ. 2005;108:205–17.

Aviron S, Kindlmann P, Burel F. Conservation of butterfly populations in dynamic landscapes: the role of farming practices and landscape mosaic. Ecol Model. 2007;205:135–45.

Aviron S, Nitsch H, Jeanneret P, Buholzer S, Luka H, Pfiffner L, Pozzi S, Schüpbach B, Walter T, Herzog F. Ecological cross compliance promotes farmland biodiversity in Switzerland. Front Ecol Environ. 2009;7:247–52.

Baudry J, Burel F. La mesure de la biodiversité spatiale: utilisation dans les évaluations d'impact. Acta Oecologica Oecologica Applicata. 1982;177–90.

Benton TG, Vickery JA, Wilson JD. Farmland biodiversity: is habitat heterogeneity the key? Trends Ecol Evol. 2003;18:182–8.

Billeter R, Liira J, Bailey D, Bugter R, Arens P, Augenstein I, Aviron S, Baudry J, Bukacek R, Burel F, Cerny M, De Blust G, De Cock R, Diekotter T, Dietz H, Dirksen J, Dormann C, Durka W, Frenzel M, Hamersky R, Hendrickx F, Herzog F, Klotz S, Koolstra B, Lausch A, Le Coeur D, Maelfait JP, Opdam P, Roubalova M, Schermann A, Schermann N, Schmidt T, Schweiger O, Smulders MJM, Speelmans M, Simova P, Verboom J, Van Wingerden WKRE, Zobel M. Indicators for biodiversity in agricultural landscapes: a pan-European study. J Appl Ecol. 2008;45:141–50.

Blondel J, Ferry C, Frochot B. la méthode des indices d'abondance (IPA) ou des relevés d'avifaune par "station d'écoute". Alauda. 1970;38:55–71.

Brakefield PM. Ecological studies on the butterfly Maniola jurtina in Britain. I—Adult behaviour, micro distribution and dispersal. J Anim Ecol. 1982;51:713–26.

Bressan A, Moral García FJ, Sémétey O, Boudon-Padieu E. Spatio-temporal pattern of Pentastiridius leporinus migration in an ephemeral cropping system. Agric For Entomol. 2010;12:59–68.

Burel F, Baudry J, Butet A, Clergeau P, Delettre Y, Le Coeur D, Dubs F, Morvan N, Paillat G, Petit S, Thenail C, Brunel E, Lefeuvre JC. Comparative biodiversity along a gradient of agricultural landscapes. Acta Oecologica. 1998;19:47–60.

Burel F, Baudry J. Landscape ecology: concepts, methods, and applications. Enfield: N.H. Science Publishers; 2003.

Burel F, Baudry J. Habitat quality and connectivity in agricultural landscapes: the role of land use systems at various scales in time. Ecol Ind. 2005;5:305–13.

Carrière Y, Ellsworth PC, Dutilleul P, Ellers-Kirk C, Barkley V, Antilla L. A GIS-based approach for area wide pest management: the scales of Lygus hesperus movements to cotton from alfalfa, weeds, and cotton. Entomol Exp Appl. 2006;118:203–10.

Concepción E, Díaz M, Baquero R. Effects of landscape complexity on the ecological effectiveness of agri-environment schemes. Landscape Ecol. 2008;23:135–48.

Deckers B, De Becker P, Honnay O, Hermy M, Muys B. Sunken roads as habitats for forest plant species in a dynamic agricultural landscape: effects of age and isolation. J Biogeogr. 2005;32:99–109.

De Snoo GR, Van der Poll RJ. Effect of herbicide drift on adjacent boundary vegetation. Agric Ecosyst Environ. 1999;73:1–6.

Desneux N, Decourtye A, Delpuech JM. The sublethal effects of pesticides on beneficial arthropods. Annu Rev Entomol. 2007;52:81–106.

Dunning JB, Danielson BJ, Pulliam HR. Ecological processes that affect populations in complex landscapes. Oikos. 1992;65:169–75.

Ernoult E, Tremauville Y, Cellier D, Margerie P, Langlois E, Alard D. Potential landscape drivers of biodiversity components in a flood plain: Past or present patterns? Biol Conserv. 2006;127:1–17.

Fahrig L. Relative importance of spatial and temporal scales in a patchy environment. Theor Popul Biol. 1992;41:300–14.

Fahrig L, Baudry J, Brotons L, Burel FG, Crist TO, Fuller RJ, Sirami C, Siriwardena GM, Martin J-L. Functional landscape heterogeneity and animal biodiversity in agricultural landscapes. Ecol Lett. 2011;14:101–12.

Fischer J, Brosi B, Daily G, Ehrlich P, Goldman R, Goldstein J, Lindenmayer D, Manning A, Mooney H, Pejchar L, Ranganathan J, Tallis H. Should agricultural policies encourage land sparing or wildlife-friendly farming? Front Ecol Environ. 2008;6(7):380–5.

Forman RTT, Baudry J. Hedgerows and hedgerow networks in landscape ecology. Environ Manage. 1984;8:499–510.

Fitzgibbon CD. Small mammals in farm wood lands: the effects of habitat, isolation and surrounding land-use patterns. J Appl Ecol. 1997;34:530–9.

Gabriel D, Thies C, Tscharntke T. Local diversity of arable weeds increases with landscape complexity. Perspect Plant Ecol, Evol Syst. 2005;7:85–93.

Hawthorne AJ, Hassall M, Sotherton NW. Effects of cereal headland treatments on the abundance and movements of three species of carabid beetles. Appl Soil Ecol. 1998;9:417–22.

Holzschuh A, Dormann CF, Tscharntke T, Steffan-Dewenter I. Expansion of mass-flowering crops leads to transient pollinator dilution and reduced wild plant pollination. Proceedings of the Royal Society B-Biological Sciences 2011;278:3444–3451.

Joannon A, Bro E, Thenail C, Baudry J. Crop patterns and habitat preferences of the grey partridge farmland bird. Agron Sustainable Dev. 2008;28:379–87.

Kennedy GG, Storer NP. Life systems of polyphagous arthropod pests in temporally unstable cropping systems. Ann Rev Entomol. 2000;45:467–493.

Kennedy GG, Storer NP. Life systems of polyphagous arthropod pests in temporally unstable cropping systems. Annu Rev Entomol. 2000;45:467–93.

Kindlmann P, Burel F. Connectivity measures: a review. Landscape Ecol. 2008;23:879–90.

Kleijn D, Baquero RA, Clough Y, Díaz M, Esteban J, Fernández F, Gabriel D, Herzog F, Holzschuh A, Jöhl R, Knop E, Kruess A, Marshall EJP, Steffan-Dewenter I, Tscharntke T, Verhulst J, West TM, Yela JL. Mixed biodiversity benefits of agri-environment schemes in five European countries. Ecol Lett. 2006;9:243–54.

Kleijn D, Rundlã FM, Scheper J, Smith HG, Tscharntke T. Does conservation on farmland contribute to halting the biodiversity decline? Trends Ecol Evol. 2011;26:474–81.

Klemm M. Man-made bee habitats in the anthropogenous landscape of central Europe-substitutes for threatened or destroyed riverine habitats? The conservation of bees. In: Matheson A, Buchmann SL, O'Toole C, Westrich P, Williams IH, editors. 1996. p. 17–34. Academic Press, London.

Krauss J, Bommarco R, Guardiola M, Heikkinen RK, Helm A, Kuussaari M, Lindborg R, Öckinger E, Pärtel M, Pino J, Pöyry J, Raatikainen KM, Sang A, Stefanescu C, Teder T, Zobel M, Steffan-Dewenter I. Habitat fragmentation causes immediate and time delayed biodiversity loss at different trophic levels. Ecol Lett. 2010;13:597–605.

Kuussaari M, Hyvönen T, Härmä O. Pollinator insects benefit from rotational fallows. Agric Ecosyst Environ. 2011;143:28–36.

Féon Le V, Burel F, Chifflet R, Henry M, Ricroch A, Vaissière BE, Baudry J. Solitary bee abundance and species richness in dynamic agricultural landscapes. Agric Ecosyst Environ. 2011. doi:10.1016/j.agee.2011.06.020.

Leroux X, Barbault R, Baudry J, Burel F, Doussan I, Garnier E, Herzog F, Lavorel S, Lifran R, Roger-Estrade J, Sarthou JP, Trommetter M. Agriculture et biodiversité: valoriser les synergies, Paris, éds. Quae. 2008.

Lotfi A, Javelle A, Baudry J, BUREL F. Interdisciplinary analysis of hedgerow network landscapes sustainability. Landscape Res. 2010;35:391–402.

Meeus JHA. The transformation of agricultural landscapes in Western Europe. Sci Total Environ. 1993;129:171–90.

Men XY, Ge F, Yardim EN, Parajulee MN. Evaluation of winter wheat as a potential relay crop for enhancing biological control of cotton aphids in seedling cotton. Biocontrol. 2004;49:701–14.

Metzger JP. Landscape ecology: perspectives based on the 2007 IALE world congress. Landscape Ecol. 2008;23:501–4.
Millan-Pena N, Butet A, Delettre Y, Paillat G, Morand P, Le Du L, Burel F. Response of small mammal community to changes in western French agricultural landscapes. Landscape Ecol. 2003;18:265–78.
Millan De La Pena N, Butet A, Delettre Y, Morant P, Burel F. Landscape context and carabid beetles (Coleoptera: Carabidae) communities of hedgerows in western France. Agric Ecosyst Environ. 2003;94:59–72.
Morandin LA, Winston ML, Abbott VA, Franklin MT. Can pastureland increase wild bee abundance in agriculturally intense areas? Basic Appl Ecol. 2007;8:117–24.
Ouin A, Paillat G, Butet A, Burel F. Spatial dynamics of wood mouse (*Apodemus sylvaticus*) in an agricultural landscape under intensive use in the Mont Saint Michel Bay (France) agriculture. Ecosyst Environ. 2000;78:159–65.
Purtauf T, Roschewitz I, Dauber J, Thies C, Tscharntke T, Wolters V. Landscape context of organic and conventional farms: Influences on carabid beetle diversity. Agric Ecosyst Environ. 2005;108:165–74.
Robinson RA, Sutherland WJ. Post war changes in arable farming and biodiversity in Great Britain. J Appl Ecol. 2002;39:157–76.
Rusch A, Valantin-Morison M, Sarthou J-P, Roger-Estrade J. Multi-scale effects of landscape complexity and crop management on pollen beetle parasitism rate. Landscape Ecol. 2011;26:473–86.
Schooley RL, Branch LC. Habitat quality of source patches and connectivity in fragmented landscapes. Biodivers Conserv. 2011;20:1611–23.
Shuler RE, Roulston TH, Farris GE. Farming practices influence wild pollinator populations on squash and pumpkin. J Econ Entomol. 2005;98:790–5.
Schweiger O, Maelfait JP, Van Wingerden W, Hendricks F, Billeter R, Speelmans M, Augenstein I, Aukema B, Aviron S, Bailey DW, Bukacek R, Burel F, Diekotter T, Dirksen J, Frenzel M, Herzog F, Liira J, Roubalova M, Bugter R. Quantifying the impact of environmental factors on arthropod communities in agricultural landscapes across organizational levels and spatial scales. J Appl Ecol. 2005;42:1129–39.
Steffan-Dewenter I, Tscharntke T. Succession of bee communities on fallows. Ecography. 2001;24:83–93.
Steffan-Dewenter I, Münzenberg U, Bürger C, Thies C, Tscharntke T. Scale-dependent effects of landscape context on three pollinator guilds. Ecology. 2002;83:1421–32.
Thenail C, Baudry J. Variation of farm spatial land use pattern according to the structure of the hedgerow network (bocage) landscape: a case study in northeast Brittany. Agric Ecosyst Environ. 2004;101:53–72.
Thorbek P, Topping CJ. The influence of landscape diversity and heterogeneity on spatial dynamics of agrobiont linyphiid spiders: an individual-based model. Biocontrol. 2005;50:1–33.
Tscharntke T, Klein AM, Kruess A, Steffan-Dewenter I, Thies C. Landscape perspectives on agricultural intensification and biodiversity—ecosystem service management. Ecol Lett. 2005a;8:857–74.
Tscharntke T, Rand TA, Bianchi F. The landscape context of trophic interactions: insect spillover across the crop-non crop interface. Ann Zool Fenn. 2005b;42:421–32.
Vane Wright RI, Ackery PR. The biology of butterflies. British Museum: Academic Press; 1981.
Vasseur C, Joannon A, Aviron S, Burel F, Meynard JM, Baudry J. The cropping systems mosaic: how does the hidden heterogeneity of agricultural landscapes drive populations? Agri Ecosyst Environ (in press).
Vasseur C. Contrôles exercés par les mosaïques de systèmes de culture sur les dynamiques du carabe Pterostichus melanarius Illiger (Coleoptera, Carabidae). Thèse de l'Université de Rennes 1. 2012. p. 205.
Weibull A-C, Bengtsson J, Nohlgren E. Diversity of butterflies in the agricultural landscape: the role of farming system and landscape heterogeneity. Ecography. 2000;23:743–50.

Westphal C, Steffan-Dewenter I, Tscharntke T. Mass flowering crops enhance pollinator densities at a landscape scale. Ecol Lett. 2003;6:961–5.

Westrich P. Habitat requirements of central European bees and the problems of partial habitats. Matheson A, Buchmann SL, O'Toole C, Westrich P, Williams IH, editors. The conservation of bees. London: Academic; 1996. p. 1–16.

Wissinger SA. Cyclic colonization in predictably ephemeral habitats: a template for biological control in annual crop systems. Biol Control. 1997;10:4–15.

Chapter 15
Forest Influences on Climate and Water Resources at the Landscape to Regional Scale

Ge Sun and Yongqiang Liu

Abstract Although it is well known that climate controls the distribution, productivity and functioning of vegetation on earth, our knowledge about the role of forests in regulating regional climate and water resources is lacking. The studies on climate-forests feedbacks have received increasing attention from the climate change and ecohydrology research communities. The goal of this study is to provide an in-depth examination of forest-climate-water interactions by synthesizing recent scientific literature on the influences of forests on climate and water resources from watershed to regional scale. The synthesis paper provides a review of the state of art of our understanding of the mechanisms of interactions of forests and climate and water resources at the landscape and regional scale. The paper presents two case studies that examine the influences of forests on microclimate, watershed hydrology, and regional climate and water resources at a small watershed to region scale using literature from the Coweeta Hydrological Laboratory in the southeast U.S. and a simulation study on the North China Shelter Belt Project. Future research gaps were identified in terms of integrated Earth System modeling to guide forest management for global change mitigation and adaptation.

Keywords Climate change · Forest hydrology · Micrometeorology · Water resources

G. Sun (✉)
Research Hydrologist, Eastern Forest Environmental Threat Assessment Center, Southern Research Station, USDA Forest Service, Raleigh, NC, USA
e-mail: GeSun@fs.fed.us

Y. Liu
Research Meteorologist, Center for Forest Disturbance Science, Southern Research Station, USDA Forest Service, Athens, GA, USA
e-mail: yliu@fed.fs.us

B. Fu and K. B. Jones (eds.), *Landscape Ecology for Sustainable Environment and Culture*, DOI: 10.1007/978-94-007-6530-6_15,

15.1 Introduction

Being different from weather, climate refers to mean atmospheric conditions (i.e., temperature, humidity, wind, precipitation, etc.) over multiple years. The land surface, as represented by vegetation, soil, inland water bodies, is the atmospheric lower boundary where heat, water, momentum, and trace gas exchanges occur. Together with ocean and ice/snow, the atmosphere, and land surface complete the entire climate system. It is well known that climate controls the distribution (Chang 2002, p. 115), productivity and functioning of vegetation on earth (Chapin III et al. 2012) but our knowledge about the role of vegetation, forests in particular, in regulating climate is lacking (Field et al. 2007; Waring and Waring 2007, see p. 302). Forests influence climate through exchanges of energy, water, carbon dioxide, and other chemical species with the atmosphere (Bonan 2008; Cook et al. 2012), thus the studies on climate-forests feedbacks have received increasing attention from the climate change and ecohydrology research communities (Jackson et al. 2001, 2009; Vose et al. 2011; Lee et al. 2011).

Forests cover about 4.17×10^9 ha or 31.8 % of the Earth's land surface (Chang 2002). Natural forests in a large area are generally found in regions where annual precipitation exceeds 400 mm and net radiation exceeds 20 kcal (or 27 W/m^2) below which grasslands or shrublands may dominate the landscape (Sun et al. 2011a). In forested areas, precipitation often exceeds evapotranspiration rate, and thus forests are sources of surface water resources and sinks of nutrients and carbon. For example, it is estimated that over 60 % of water supply comes from forest lands in the United States (Brown et al. 2008). Forest soils are regarded as 'sponge', and soil erosion is rare in forests, and thus forests provide the best water quality among all land uses. Forests can affect micro-climate by altering solar radiation and precipitation redistribution through large forest canopies (Lee 1981). However, our understanding of the influences of forests on climate is limited although much progress in this research topic has been made in the past few decades with advances in climate modeling (e.g., Charney et al. 1977; Shukla and Mintz 1982; Sud et al. 1988; Lean and Warrilow 1989; Shukla et al. 1990; Nobre et al. 1991; Dickinson and Kennedy 1992). These modeling studies show that deforestation could influence air temperature (increase or decrease) through altering land surface albedo and energy partitioning in which plants play an active role. Detailed summary on the impacts of land use change on climate are found by Pielke and Avissar (1990) and Pielke et al. (2007, 2011).

In history, humans have long recognized the roles of trees in providing shading, shelters, fibres, and ideal micro-climate, and other amenities. Forest environment is frequently viewed as 'pleasant, peaceful, sublime, and salutary' (Lee 1981). The earliest most influential publications that specifically address forest-climate relationships can be traced back to *Forests and Moisture: or Effect of Forests on Humidity of Climate* by John Brown published in 1877, *The Earth as Modified by Human Action* by G. Marsh, 1864 publish in 1874, Forests and Water in the Light of Scientific Investigation (Zon 1927), and *Forest Influences* (Kittredge 1948). These early publications were mostly propelled by disaster preventions from forest clearing by the American colonists in the nineteenth century. The U.S. Forest Service, formerly Forestry Division of the US Department of Agriculture, started to examine forest influences in the late 1800s amid public concerns of large scale deforestation that was believed to link to large floods, landslides and soil erosion Fernow, 1893. In a transmittal letter dated Nov 1, 1892 from the Division Chief B.E. Fernow to USDA Secretary regarding the state of art findings on forest-climate-water relations, Fernow wrote: '...a review of meteorological observations which have been made mostly in foreign countries, for the purpose of determining whether and to what extent forests influence climate, together with a discussion of manner in which forests affect water conditions of the earth and other matter illustrating the question of forest influences in general'. Subsequently, beginning from the 1930s, the USFS began to establish permanent forest experimental stations across the nation with initial goals to quantify influences of forest deforestation on watershed hydrology using a 'paired watershed' approach. Many of these stations have become the core Long Term Ecological Research (LTER) sites, such Coweeta, H.J. Andrew, and Hubbard Brooks, designated for long term process-based ecosystem research (Adams et al. 2008). In recognizing the close coupling among carbon, water, and energy cycles, since the early 1990s, a series of networks (i.e., FLUXNET) have been established globally to quantify flux exchanges between land surfaces and the atmosphere (Baldocchi et al. 2001). The accumulated data in the past two decades have greatly advanced our understanding global carbon and water balances under a changing climate (Law et al. 2002; Jung et al. 2010; Sun et al. 2011a, b). These long-term worldwide studies provided much of our understanding the relationships between forest covers, micrometeorology, and headwater watershed hydrology.

Historically, there is full of misconception and debate on the true influences of forests on local and regional climate and water resources around the world (Chang 2002; Andreasian 2004; Sun et al. 2006). Traditional wisdoms suggest that forests bring rains, and thus forests provide abundant water and removing forests result in droughts, loss of springs, and desertification. This perception even resulted in law suits toward the forest industry in the North Pacific of the United States where how to manage old growth Douglass fir forests has been controversial in terms of the hydrologic and ecological consequences of deforestation and forest management. To some extent, the debates are still going among scientists due to our limited understanding of the complex interactions of physical and biological process within the atmosphere-land interface and the earth systems as a whole. A good example is by Ellison et al. (2012) who argue that forest influence on climate and water resources must be evaluated at large context and the 'negative' effects of water use by trees on water yield are local and should not be exaggerated to minimize the overall 'positive' influences of forests on regional distribution of precipitation and air temperature.

During the past few decades, forests have been confirmed about their large capacity of carbon sequestration (Ryan et al. 2010; Pan et al. 2011), thus their role in slowing down the current trend of global warming (Bonan 2008). Now, we begin to know that forests influence redistribution of global solar energy and energy budget on earth, thus play a key role in the global hydrologic cycle (Jung et al. 2010). It has become possible to trace the movements of water vapor and atmospheric gases to develop a clearer idea of the role played by forests in moderating or regulating rainfall in different part of the world. For example, in temperate regions and tropical regions such as Southeast Asia, the main source of water vapor in the atmosphere is from evaporation at the surface of the oceans. In the Amazon Basin, however, nearly 50 % of water vapor in the atmosphere in the region of Manaus and Belém appears to be 'recycled' from the forest. Oyebande (1988), Eltahir and Bras (1993), and Dickinson et al. (1993) provide good summaries of the effects of forests on rainfall and water yield in the tropics. Garcia-Carreras and Parker (2011) recently reports deforestation may intensify rainfall in cut area and decrease rainfall of the surrounding areas and threats remaining rain forests in western Africa.

In general, majority of our knowledge on forest-climate-water relations is derived from small watershed studies, thus the influences of forests on local and global climate and water supply are still open for debate (Bonan 2008; Ellison et al. 2012; van der Ent et al. 2012). Our current knowledge about forests' role in moderating climate and water resources at a large watershed (Wei et al. 2008; Lin and Wei 2008; Wei and Zhang 2010) or regional scale is limited, and a broad understanding of forest-water-climate interactions is needed for determining forest management strategies in climate change mitigation and adaptation.

The overall goal of this study is to provide an in depth understanding of forest-climate-water interactions at regional to global scale by synthesizing recent scientific literature on the influences of forests on climate and water resources. Specific objectives are to: (1) present state of art of our understanding of the mechanisms of interactions of forests and climate and water resources at the landscape and regional scale, (2) present two case studies that examine the influence of forests on climate and water resources, and (3) identify research gaps that help guide future studies that can help forest management for global change mitigation and adaptation.

15.2 Principles of Forest Influences on Climate and Water Resources

Forests influence climate and water resources through their physical and biological functions that affect the energy, water, and biogeochemical balances (Zhao and Pitman, 2010). Key mechanisms and processes within a forest landscape are illustrated to demonstrate the close interactions between atmosphere and land surfaces and the tight forest-climate-water relationship (Fig. 15.1).

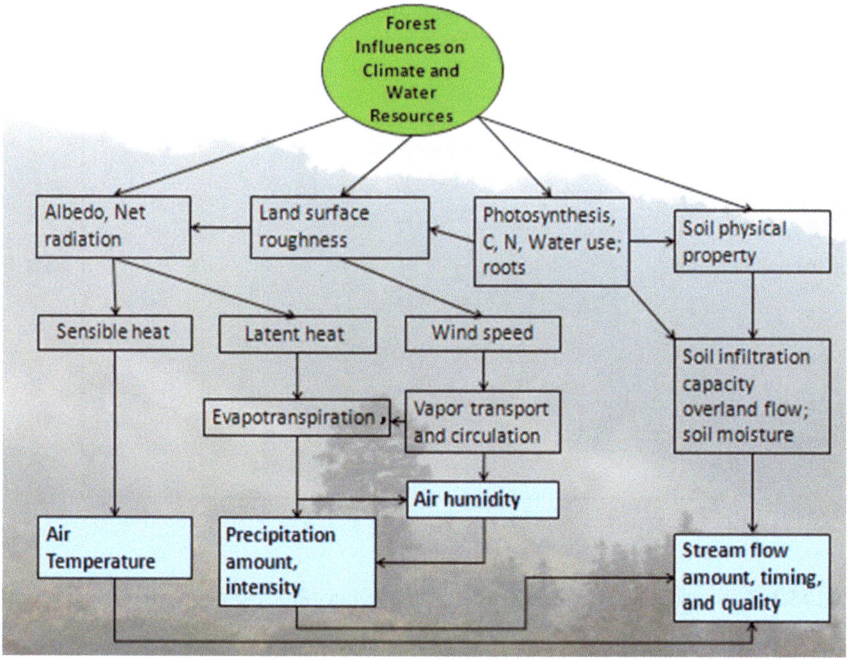

Fig. 15.1 A conceptual model describing the interactions of forests, climate, and streamflow at multiple scales (modified from Liu et al. 2008, 2010)

15.2.1 Unique Physical Characteristics

Forests are distinctly different from other land surfaces in physical properties both above ground and below ground in terms of the ability of light absorption and reflectance (Albedo), leaf and root biomass, surface roughness, and soil characteristics. These properties have profound influences on the energy and water balances from the ecosystem to global scale.

15.2.2 Albedo

Surface Albedo or light reflectance, is an important parameter that affects energy balance of ecosystems, and can be as important as greenhouse gases in affecting climate change (Betts 2000). Because forests have higher leaf area and biomass than grass or other short crops, forests generally have lower surface albedo. A lower albedo value means more solar energy available (higher net radiation) for sensible heat and latent heat, i.e., evapotranspiration. A comparison of albedo and net radiation measured for a mid-rotation (15-year-old) and young loblolly pine

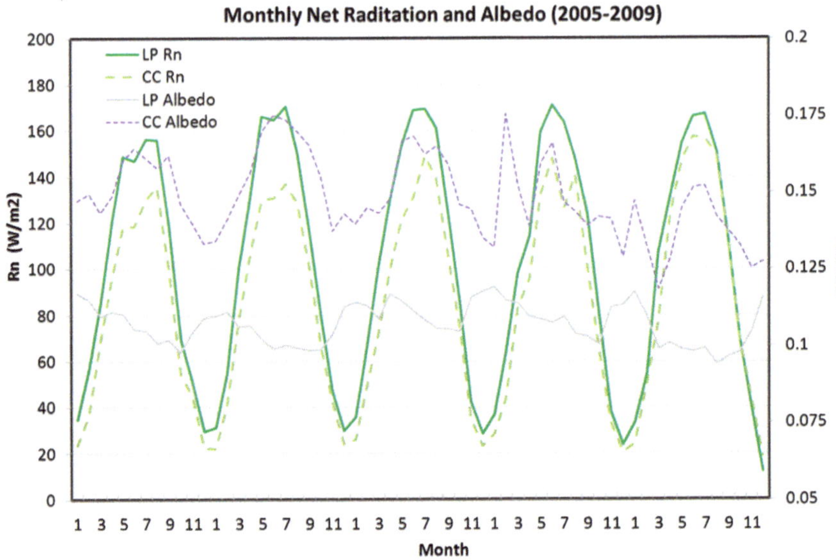

Fig. 15.2 A comparison of monthly mean net radiation (*Rn*) and albedo for a mid-rotation loblolly pine plantation (*LP*) and a clear-cut (young plantation) (*CC*) sites on the lower coastal plain in North Carolina, USA (see more data in Sun et al. 2010)

forest shows that albedo and net ration fluctuate seasonal and change over time due to the climate variation and plant development resulting a decrease in albedo and an increase in net radiation (Fig. 15.2) (Sun et al. 2010).

15.2.3 Surface Roughness

Due to uneven canopies, lands covered by tress or other vegetation have larger surface roughness than bare ground, leading to stronger turbulence and therefore smaller aerodynamic resistance for air and water vapor mixing. The measured roughness by Liu et al. (2007), for example, was 0.0058 and 0.0259 m for bare and maize soil, respectively, with the corresponding aerodynamic resistance ranges of 30–130 and 10–90 s m^{-1} during the day time. Lower aerodynamic resistance for vegetated soil suggests that water loss from lands would be higher if other meteorological conditions are the same. Higher surface roughness also means lower wind speed. Recent observed global trend of decreasing wind speed is believed to do with increase in surface roughness due to increase in plant biomass and reforestation in some cases (Liu et al. 2008; Vautard et al. 2010; McVicar et al. 2012).

15.2.4 Leaf Area Index and Rooting Depth

Forests have larger leaf area, deeper roots, and biomass and therefore forests can generally intercept more precipitation and transfer more water from soils to the atmosphere through evaporation and transpiration when compared to bare land or vegetated surfaces with short crops. Indeed, Leaf Area Index (LAI) (total leaf area per unit of ground surface area) is a very important land surface characteristic that controls seasonal evapotranspiration dynamics (Fig. 15.3) (Sun et al. 2011a). Indeed, leaf area index dynamics reflect not only the amount and heath of biota but also the environmental conditions such as light, water, nutrient associated with the biota. Larger LAI means higher canopy conductance and higher capacity to transfer more water from the soils to the atmosphere.

Similarly, forests generally have deep and massive rooting systems, 'the underground forests', that are advantageous over vegetated covers to extract water from soil moisture reservoirs even groundwater to meet water demand. Deep roots allow trees to adjust to droughts and stabilize water use under water stress conditions. The active functions are import machinists of climate change adaptation and ecological feedbacks to climate change (Jones et al. 2012).

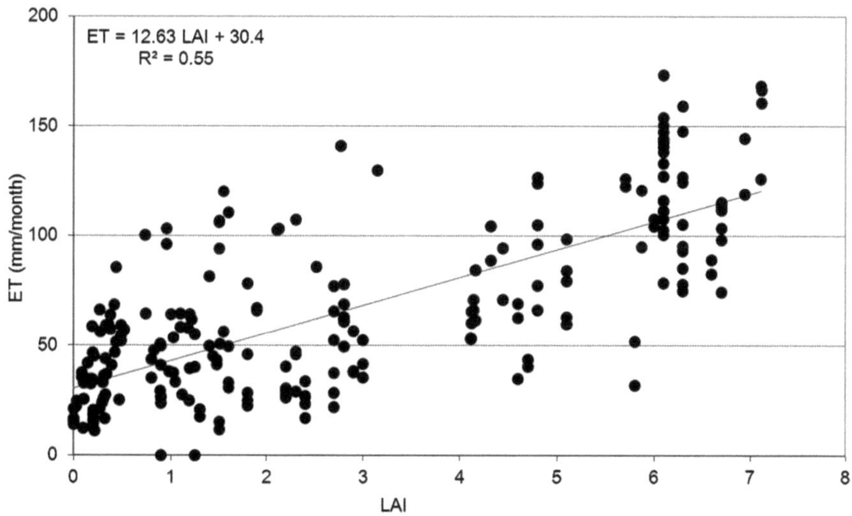

Fig. 15.3 Leaf area index (*LAI*) is a major control for seasonal evapotranspiration (*ET*). Data are derived from 13 eddy flux and sapflow measurement sites across a large climatic and ecosystem gradient in China, US, and Australia (Sun et al. 2011a)

15.2.5 Soil Physical Properties

Compared to soils of croplands or of other land uses, forest soils have much higher infiltration capacity Bruijnzeel, 2004. Forest soils are known for their high organic matter content derived from plant litter fall above ground and dead roots below ground that are conducive to high activities of soil organisms, high soil porosity, and higher hydraulic conductivity. For example, saturated hydraulic conductivity of the top soil layer in a mature loblolly pine forest on the North Carolina coastal plain can be as high as 700 cm/h (Diggs 2004). Compaction caused by forest harvest operation increased bulk density from 0.22 to 0.27 g/cm^3, decreased saturated hydraulic conductivity from 397 to 82 cm/h (Grace et al. 2006). Soil disturbances in intensive agriculture and forestry such as plowing, bedded, harvesting activities can dramatically degrade soil hydraulic properties. For example, deforestation and subsequently tillage practices in Iran resulted in almost a 20 % increase in bulk density, 50 % decrease in organic matter and total nitrogen, a 10–15 % decrease in soluble ions comparing to the undisturbed forest soil (Hajabbasi et al. 1997). The unique soil physical properties of forests explain the high soil water retention, high soil infiltration rates, minor overland flow rate, low streamflow, and high groundwater recharge commonly found in forests Zhou et al., 2010.

15.2.6 Interactions Between Forests, Climate, and Streamflow

As illustrated in Fig. 15.1 and the following energy balance equation, forests affect the redistribution of solar radiation into sensible and latent heat fluxes through passive (i.e., light reflection) and active physiological processes (photosynthesis, transpiration etc.). Latent heat is the energy source for evapotranspiration, a key component of the hydrologic cycle. The changes in sensible heat flux and evapotranspiration, which is accompanied with latent heat change, will modify air temperature and humidity. The change in air temperature, together with the changes in turbulence and wind, will modify atmospheric circulation. Precipitation will be affected due to the changes in circulation, temperature, and humidity.

$$\mathrm{Rn} = (1 - \alpha) * \mathrm{S} + \mathrm{L}, \text{ and } \mathrm{Rn} = \mathrm{LE} + \mathrm{H} + \mathrm{G}$$

where Rn is net radiation, α is albedo, S is incoming shortwave radiation, L is net long wave radiation. LE and H are latent and sensible heat flux, respectively.

According the principle of water balance below, water yield or streamflow (Q) is largely controlled by ET, or LE, at a long term scale (e.g., a few years) when the change in soil water storage is negligible. However, at short temporal scale (e.g., 1 day, 1 month), soil water storage can be significant water source for Q and ET. In this case, both soil water storage and ET are important.

$$Q = P - ET \pm S$$

In general, when compared to un-vegetated or less vegetated land surfaces, ET rates of trees or forests are higher, and thus streamflow is lower in forest dominated watersheds Whitehead and Robinson (1993). Worldwide 'paired watershed' experiments have confirmed this general conclusion: that is deforestation will decrease ET and increase streamflow, but afforestation or reforestation will increase ET and decrease streamflow (Zhang et al. 2001; Andreassian 2004; Brown et al. 2005; Jackson et al. 2005). However, these experiments do not track the lateral water vapor exchange in the atmosphere above the topographical watershed boundaries. It is also worthy of noting that there is a large variability for the general forest-water relationships. For example, clearing a fully forested upland watershed may increase flow by as high as 700 mm/year in the rainforest region, but the same forest management activity may not have much effect on a wetland-dominated forested watershed (Sun et al. 2001). This large variability is presumably due to the variability of the type, extent and magnitude of forest disturbances (Sun et al. 2001; Sun et al. 2008), climatic regime including precipitation form (snow vs. rain) and distribution (Jones et al. 2012), watershed aspects and altitude (Ford et al. 2011), geology, soil depth (Scott et al. 2005), and forest types (conifer vs. deciduous) treated (Swank and Douglass 1974).

15.3 Case Study 1: Effects of Forest Management on Water Yield at a Small Watershed Scale—The Coweeta Experiments

The Coweeta Hydrological Laboratory is located near the town of Otto in western North Carolina in the southeastern U.S. (Fig. 15.4). Coweeta presents one of the oldest forest hydrology research sites in the world. This outdoor Lab is managed by the US Department of Agriculture Forest Service Southern Research Station and serves as one of the core Long Term Ecological Research (LTER) sites. Numerous 'paired watershed' studies have been conducted over the past 78 years at Coweeta for examining the hydrologic and ecological impacts of natural and human disturbances and design best watershed management practices. In the fall of 2009, Coweeta celebrated its 75th anniversary of establishment in 1934. Theoretical and applied research continues at Coweeta to this day. The typical experimental design followed at Coweeta is based upon the paired watershed concept in which a control watershed and a treatment watershed represent the experimental domain. The watershed pair is selected because the watersheds are known to have similar hydrological characteristics. The undisrupted continuous watershed research contributes much of our understanding of forest-climate-water relationship in the humid southern Appalachian Mountains and is an important source of our global knowledge in forest hydrology. The results from these

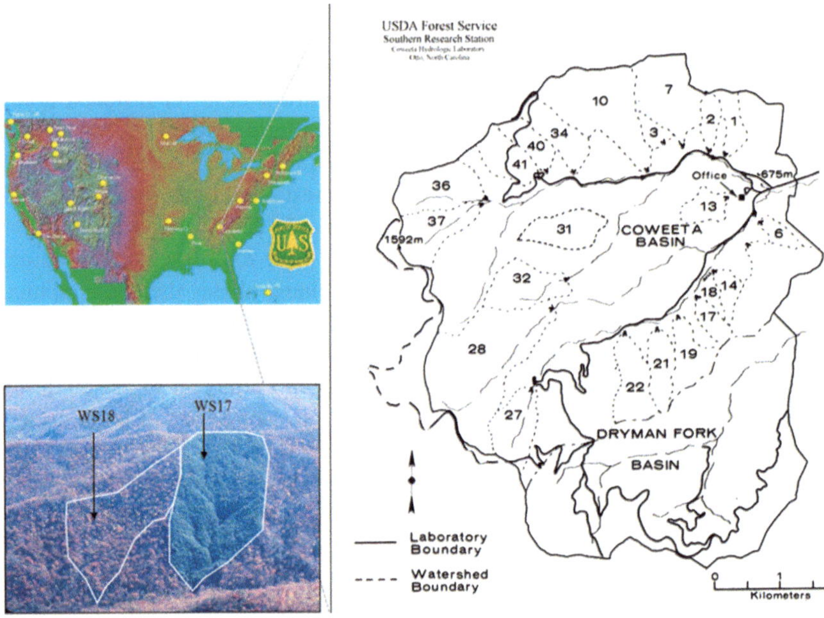

Fig. 15.4 Location of the Coweeta Hydrologic Lab that is intensively instrumented for paired watershed studies. Insert photos are to show typical watershed manipulation experiments. In this case, a tree species conversion experiment used *Watershed 18* as the control (deciduous forests) and *Watershed 17* as treatment (converted native deciduous to evergreen white pine forests)

experiments have been previously reported (Swank and Vose 1994; Swank et al. 1988, 2001). Below is a synopsis of key findings in terms of forest influences on water and micrometeorology at a small watersheds scale.

The climate of Coweeta is characterized as subtropical, marine climate with moderate temperatures (13 °C) and high abundant precipitation (1,800 mm/year at low elevation but greater than 3,000 mm at high elevation). The parent rocks are gneiss and schistorigins and weathered residual soil mantles are deep, up to 2 m at foothills and 1 m on hillslopes. The region was heavily logged before 1923 and hardwood trees (Oak-Hickory) dominate the second growth forests.

15.3.1 Micrometeorology

It is well known forests affect micro-meteorology such as humidity and radiation redistribution (Swift 1972). Clearing riparian forests increases radiation reaching the forest floor thus elevate stream water temperature (Swift and Messer 1971; Swift 1973, 1982). Conifer forests (i.e., white pine plantations) have lower albedo than deciduous hardwood forests (Swift et al. 1975), thus more energy is available for evapotranspiration, partially explaining the 20 % lower water yield observed for one watershed that was converted from native southern hardwoods to pine forests. Actual forest ET at Coweeta is generally higher than potential ET (PET) as estimated with references to water or grass surfaces (Rao et al. 2011).

15.3.2 Seasonal and Annual Water Yield

At Coweeta, various watershed manipulations experiments have been conducted to demonstrate and quantify the effects of forest management practices on water quantity and water quality. Although these studies were conducted in watershed less than 200 ha, they provide the basis for understanding the forest-water relations at a landscape scale and beyond, such as a regional scale (i.e., southern Appalachian Mountains).

15.3.2.1 Mountain Farming

Mountain farming experiments (Watershed 3) were conducted in 1940s to demonstrate the impacts of common farming practices in the steep southern Appalachian Mountain regions on water resources. Watershed monitoring data show that mountain farming that involved tree felling, brush burning, cattle grazing, plowing and cultivation for corns, severely reduced surface soil infiltration capacity, thus increased overland flow, peak flow rate (over 8 times higher compared to before treatment), and sediment loading rate (increase 2–80 times). Crop yield without fertilizer use was low due to intense storm and wildlife damages in the studied watershed (USFS 1948).

15.3.2.2 Mountain Grazing

About 20 % of the land area in the Coweeta area was intensively grazed with fenced cattle for local economic support. Woodlands grazing experiments (Watershed 7) show that soil compactions are significant. Within the first year, soil macro-porosity of top 10 cm soil decreased 10 % (USFS 1948). The loss of understory (palatable seedlings) lowed wind to blow litter out of the forests and reduced organic matter, thus eliminating the hydrological functions of forests (Munns 1947).

15.3.2.3 Clear Cutting Forests

To demonstrate that the significant effects of forest clear-cutting only (i.e., no trees removed from the sites) on evapotranspiration and water yield, Coweeta conducted a long-term repeated cutting experiment starting from 1939 (Watershed 13 on north facing slope; Watershed 17 on north facing slope). The first year following treatment in Watershed 13, water yield increased by 36 cm (60 %) (Meginnis 1959) (Fig. 15.5). In 1964, the 24-year-old stand on Watershed 13 was re-cut. The first year water yield increase for the second cutting was 38 cm, a 40 % increase in water yield. On watershed 17, water yield was 41 cm (65 %) higher than the effect of the south facing watershed. The differential hydrologic response to the same forest cutting activity was explained by the energy availability in the two watersheds. For the south facing watersheds, the changes in received solar energy for evapotranspiration were small before and after tree removal. In contrast, for the north facing watershed, the solar energy was only effective for evapotranspiration prior to removing the fore canopies when taller trees at the bottom of the slope transferred energy received to the soil reservoirs (Black, 1996, p. 126).

Both Watershed 13 and Watershed 17 were low elevation watersheds (outlets at 725 and 760 msl) where temperature was significantly higher and precipitation was significantly lower than the high elevation watersheds. Watershed 37, a steep, high elevation (watershed outlet at 1,033 msl) watershed was clear-cut in 1963

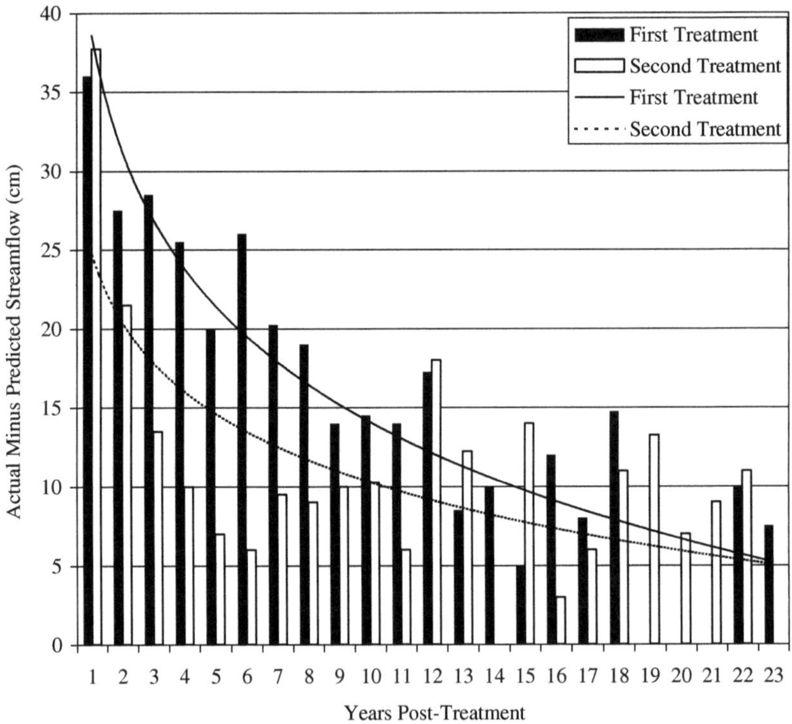

Fig. 15.5 Effects of clear-cutting on annual water yield (*Watershed 13*). All woody vegetation cut in 1939 and allowed to regrow until 1962 when the watershed was again clear-cut; no products removed in either treatment (Data from Coweeta Hydrologic Lab, USDA forest service)

(Swift and Swank 1981; Swank and Helvey 1970). This produced a water yield increase of 26 cm the first year after the treatment. Clearly, climate regimes had influences on the hydrologic effects of forest treatments.

15.3.2.4 Tree Species Conversion

Coweeta is one the few sites that have examined the hydrologic responses to tree species change at the watershed scale (Swank and Douglass 1974; Komatsu et al. 2007). Paired watershed experiment studies (Watershed 17 and Watershed 18) concluded that converting native deciduous forests to white pine plantations has reduced flow by 20 % in the 17th year after treatment (i.e., planting white pine) in 1956 (Swank and Douglass 1974) and ET increased by 40 % in the mid-2000s (Ford et al. 2007) (Fig. 15.6). The major reason was that conifers had a higher leaf areas index (up to 7.1) than the control watershed (Peak LAI for the deciduous forest less than 6.5), thus higher interception water loss and transpire water year round (Swank and Douglass 1974). Seasonally, the largest streamflow differences between the evergreen forest watershed and the deciduous forest watershed were found in the dormant winter season.

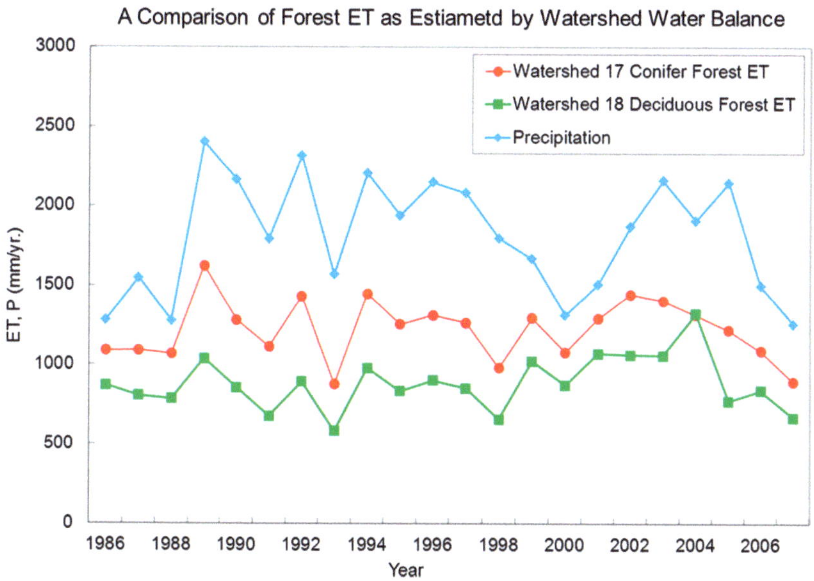

Fig. 15.6 Paired watershed experiments (*Watershed 18* is a control, no management imposed) show that the watershed dominated by white pine plantations have higher evapotranspiration (*ET*) and lower water yield (*Q*) than the adjacent watershed dominated by native deciduous second growth forests in the Coweeta Hydrologic Laboratory in North Carolina, USA. All woody vegetation was cut in *Watershed 17* in 1940 and regrowth cut annually thereafter in most years until 1955; no products removed. White pine trees planted in 1956 and released from hardwood competition as required with cutting or chemicals

15.4 Case Study 2: Potential Climate Influences of Large Scale Reforestation at Regional Scale—The Green Great Wall Project in Northern China

The Green Great Wall (GGW) forest shelterbelt project in northern China was initiated in 1978 and still continues this day. The mass reforestation project aimed at curbing the southward expansion of the desert, improving climate conditions, and protecting the natural environments in the arid region. The forest shelterbelt is about 7,000 km long zonally and 400–1,700 km wide (Fig. 15.7). It stands along the southern edge of the sandy lands, closely paralleling to the Great Wall, thereby gaining the name of the Green Great Wall (GGW) (SFA 2006). The project target was to cover 60 % of the project areas by 2000, 85 % by 2020, and 100 % by 2050. When the GGW project is completed, forest coverage in the region will increase from 5 % to 15 %. Until now, 25.07 million hectares of forests have been planted. However, few studies have comprehensively evaluated the regional environmental and ecological influences of this large effort (Liu et al. 2008a, b).

The regional effects of GGW on climate and water were examined using a modeling approach (Liu et al. 2008a, b). Two simulations were conducted using the National Center for Atmospheric Research (NCAR) regional climate model (Version 3) (RegCM3) (Giorgi et al. 1993a, b) for the period from January 1987 to February 1988. One control simulation used present IGBP land cover data and the

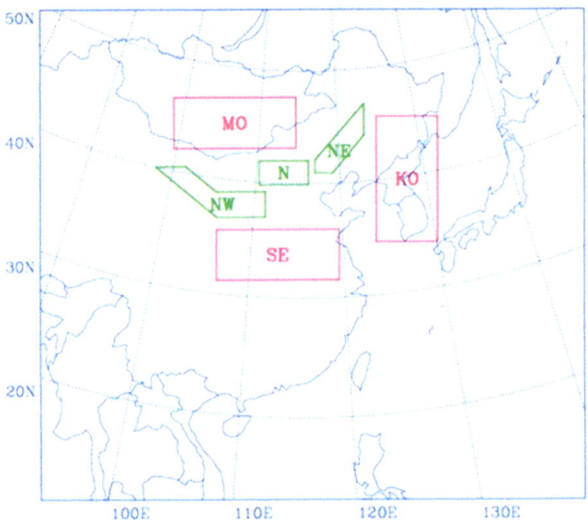

Fig. 15.7 Location of Green Great Wall project and simulation domains. The three *green boxes* in the *middle* represent the afforested areas in northwest (*NW*), north (*N*) and northeast (*NE*) of China. The *pink boxes* are the surrounding areas in southeast (*SE*) China, Mongolia (*MO*) and Korea (*KO*)

other experimental simulation used hypothetical land covers assuming all croplands, grass lands, and sandy lands are replaced by evergreen needle pine forests—the major forest type for reforestation in the region. The model operates at a 4 min time step and 50 km spatial resolution for 10 years.

Simulation results show that afforestation leads to overall increases in precipitation, soil moisture and air relative humidity, and decreases in wind speed and air temperature in the afforested areas. In addition, the results also show significant influences outside the afforested areas, suggesting a role of afforestation in changing the climate conditions in surrounding regions.

Simulated precipitation changes as a result of GGW were lumped into six major areas as outlined (Fig. 15.8). Precipitation increases from spring to summer, then decreases in fall, and decreases further in winter. In each season, precipitation change is generally the highest in SE, lowest in NW and MO, and in-between in N, NE, and KO. Precipitation disturbance is positive in all afforestation areas with the

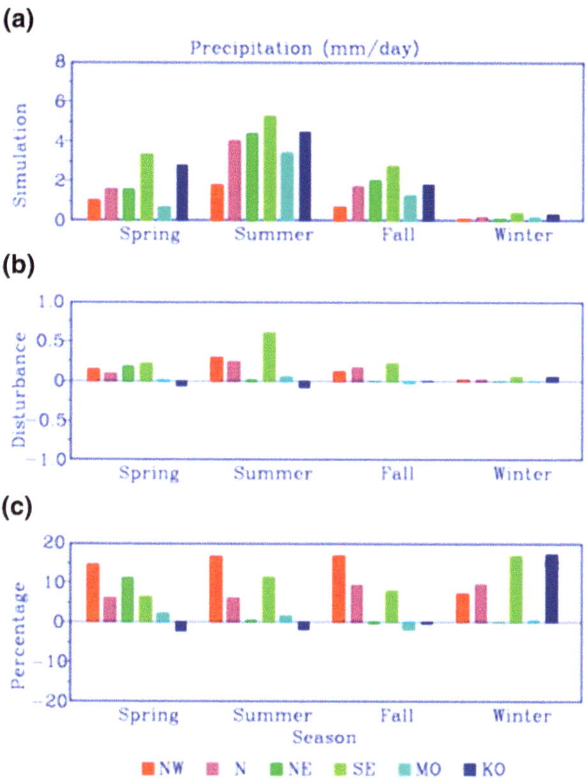

Fig. 15.8 Regional averages of precipitation (mm/day). **a–c** are control simulation, disturbance (the difference between experiment and control simulations), and ratio of the difference to simulation (%). Four seasons are shown from *left* to *right*, each with the afforested areas in Northwest (*NW*), North (*N*), and Northeast (*NE*) China, and the surrounding areas in Southeast China (*SE*), Mongolia (*NO*), and Korea (*KO*)

largest relative change in NW in all seasons except winter, and the lowest relative change in NE. Precipitation disturbance is also consistently positive in SE. The disturbance in KO is positive with a large value in winter but very small in other seasons. The disturbance is small in all seasons in MO.

The spatial patterns of precipitation changes in spring and summer are characterized by overall positive in the afforestation areas and in south of the afforestation areas over the southern North China region, which is surrounded by a negative disturbance, mainly east of the afforestation areas along the China border. The disturbance turns positive again with the most significant increase over the oceanic region south to the Korea peninsula. The spatial pattern in the fall is closer to that in the spring. In winter, a positive disturbance is dominant and occurs mostly southeast of the afforestation areas.

In comparison with the precipitation disturbance, evapotranspiration disturbance is larger in NW and N, but smaller in NE, SE and MO in spring. The same magnitude is found in NW and N, but it turns negative in NE. The disturbance is slightly positive in autumn and small in winter in the three afforested areas. Disturbance in air relative humidity is positive in all areas and all seasons except summer in NE. Air temperature is increased in winter for all areas, but varies among areas in other seasons. It is reduced by nearly 0.5 °C in NW, and slightly reduced in the other afforested areas with the exception of a large positive disturbance that occurs during summer in NE.

Effects of reforestation differ from precipitation and evapotranspiration in the afforested areas in that it is negative in all seasons except autumn in N. Disturbance outside of the afforestation areas can be either positive or negative, but mostly the former for the disturbance with large magnitude. These results indicate that runoff is mostly decreased in and outside the afforested areas. Disturbance in soil moisture of the surface layer varies across the regions. It is positive during all seasons for NW and SE, and positive in spring but negative during the three other seasons for N and NE. The depth of the rooting layers increases after afforestation. As a result, soil moisture increases in the afforested areas.

15.5 Knowledge Gaps and Future Research Needs

Fossil fuel burning and land use change (e.g., deforestation) are two of the top factors that have directly contributed the ongoing climate change (IPCC 2007). Climate change is the most serious environmental problem that humans will face for a long time. Mitigating and adapting to climate change requires a comprehensive integrated approach that must consider the interactions and tradeoffs of the options.

15.5.1 Understanding Feedbacks Within the Forest-Climate-Human Systems

Our knowledge about how forests will respond to climate change is limited and we know little about the consequences of management options designed to combat climate change. For example, schemes to increase carbon sequestrations in man-made forests through REDD+ (reducing greenhouse gas emissions from deforestation and forest degradation) could negatively impacts biodiversity (Parrotta 2012) and water resources (Jackson et al. 2005) and the environment (Cao 2008) if not properly implemented. We need to better understand the feedbacks between climate change and human response and actions using a system approach (Avila et al. 2012). Current climate change models are not reliable for local predictions and the Earth system models that incorporate land surface processes and atmospheric processes and human influences are still in their infancy (Bonan 2008; Angelini et al. 2011). Forests can affect regional climate processes and variability at long time scales (Notaro and Liu, 2006; 2008). Similar to oceans, the land has the capacity to retain anomalous signals over a much longer period than the atmosphere. Land surface processes could contribute to long-term atmospheric variability by passing their relatively slow anomalous signals to the atmosphere (e.g., Yeh et al. 1984; Dickinson and Handerson-Sellers 1988; Delworth and Manabe 1988; Vinnikov et al. 1996; Liu and Avissar 1999; Koster and Suarez 1995).

Forest-climate interactions contribute to local and regional climate variability at interannual and decadal scales (Zeng et al. 2000). Landscapes in Amazon, the Sahel, western Africa tropical rainfall forests, northwest China have changed dramatically since the 1970s as a result of deforestation and over-cultivation (Sampson 2004). These changes have been linked to some regional climate disasters such as the prolonged drought in northern Africa during the 1970s (Charney 1975; Xue and Shukla 1993) and flooding and dust storms in China. The declined vegetation converge in the southern United States in the 19th century and early 20th century due to agriculture and industrial activities might be a contributor to the drying climate and severe dust bowls during the 1930s. Future global climate change suggests ted that a large portion of the temperate deciduous forests in the Southeast would be replaced with temperate deciduous savanna in response to the projected climate change (Neilson et al. 1998, 2005). Notaro et al. (2007) indicated that, for the projected future climate change due to the greenhouse effect, tree coverage is expected to increase in many regions, including southeastern U.S. Studying land–atmosphere interaction has emerged as one of the most active research areas in atmospheric and hydrological sciences in the past decades, partly due to the increasing attention to human activity related to regional environmental changes (Bonna 2008).

15.5.2 Improving Earth System Modeling Capacity: Bridging Landscape Processes and Regional Climate and Hydrology

Climate is the ultimate driving force for landscape-scale hydrologic processes which are naturally linked to local climate. Traditional watershed or landscape hydrological studies largely assume climate as a stationary external force, and hydrologic processes have no influences on local climate. For example, we know that reforestation will increase ET at the watershed scale, but we rarely tract how far and where the water vapor will travel across the physical watershed boundaries. Scaling empirical observations at the landscape scale to regional scale is still a difficult task and remains to be an active research area in landscape ecology and regional and global hydrology.

Simulating the true interactions and feedbacks between land surface processes such as forest vegetation functions and climate systems requires the tight coupling of regional climate models and landscape vegetation dynamics, and global circulation climate models or regional climate models (Phipps et al. 2011). Existing integrated dynamic vegetation models (DGVMs) have the capacity to simulate natural forest vegetation dynamics and the influences of external disturbances such as climate variability (e.g., drought and flood) and physical and chemical climate effects (e.g., greenhouse gases), species invasion, wildfire, insect outbreak on ecological processes (i.e., water and carbon cycles). DGVMs simulate daily or monthly carbon, water and nitrogen cycles driven by the changes in atmospheric chemistry including ozone, nitrogen deposition, CO_2 concentration, climate, land-use and land-cover types and disturbances. DGVMs usually include four core components of biophysics, plant physiology, soil biogeochemistry, and dynamic vegetation and land-use. Examples of DGVMs include HYBRIDS (Friend et al. 1997), MC1 (Bachelet et al. 2001), the Lund-Potsdam-Jena (LPJ) (Sitch et al. 2003), CLM (Levis et al. 2004), IBIS (Foley et al. 2005), and the DELM (Tian et al. 2009).

Efforts have been made to couple DGVMs into Global Circulation Models (GCMs) and Regional Climate Model (RCMs). For example, CLM is fully coupled with the National Center for Atmosphere Research's Community Earth System Model (CESM) and WRF (Jin et al. 2010), respectively. The coupled models are able to simulate the impacts on and feedbacks to climate from dynamic changes in forests. They will be especially useful for understanding the roles of afforestation in mitigating the impacts of climate change discussed above. For further assessing the mitigation roles and making management plans, comprehensive modeling systems such as the integrated Regional Earth System Model (iRESM) (http://www.pnl.gov/atmospheric/iresm/) are needed. iRESM is a modeling framework developed in the Pacific Northwest National Laboratory (PNNL) to address regional human-environmental system interactions in response to climate change and the uncertainties therein. The framework consists of a suite of integrated models representing regional climate change, regional climate policy, and the regional economy.

15.5.3 Understanding the Roles of Afforestation in Mitigating Negative Effects of Climate Change

Forest ecosystems are large carbon sinks (Pan et al. 2011) and thus could play an important role in mitigating climate change. Sustainable forest management strategy that aims at maintaining or increasing forest carbon stocks will not only produce sustained yield of timber or energy but also will generate the largest sustained mitigation benefit. For example, a large afforestation effort that plans to plant about 18 million acres of new trees to replace pasture and farming lands by 2020 are being implemented in the southeastern US, as well as in Great Lake states and the Corn Belt states (Watson 2009) in the U.S. The project would be even larger than the one carried out by the Civilian Conservation Corps during the Great Depression, which planted 3 billion trees from 1933 to 1942.

Forests also can modulate climate by controlling energy and water transfers. If warmer conditions increase vegetation coverage, for example, evapotranspiration and solar radiation absorbed on the surface will increase. The change in evapotranspiration, which often plays a more important role, will lead to cooling. The feedback from evapotranspiration would partially offset any greenhouse warming. Through such feedback mechanisms, ecosystems influence their local environment and combined with their ability to sequester atmospheric CO_2, can act to mitigate climate change impacts.

Small watershed studies worldwide clearly show that forests are 'biological water pumps', (Makarieva et al. 2009) and they consume large amount of water to realize other ecosystem services (e.g., carbon sequestration and moderating climate). This has been confirmed worldwide (Scott and Lesch 1997; Robinson et al. 2003; Ice and Stednick 2004). Thus, when other conditions are equal, compared to other land uses, such as grasslands and urban lands, forested watersheds have lower total water yield (Bosh and Hewlett 1982; Zhang et al. 2008a, b; Wang et al. 2009; 2011) and peakflow rates/floodings (Eisenbies et al. 2007; Alila et al. 2009), and thus reforestation can help mitigating the negative impacts of extreme storm events (Ford et al. 2011; Vose et al. 2011) in addition to achieving carbon sequestration benefits. Forests protect water quality (e.g., preventing soil erosion and sediment loading in streams) under a changing climate that increases rainfall intensity in some regions. However, these basic understanding of forest-water relationships are based on watershed studies and data at large basin and regional scales are still lacking. While the important role of forests in mitigating global change through modifying the carbon cycle has been widely recognized, their importance to mitigate extreme climate and hydrology (floods and droughts) through the land-atmosphere interaction has yet to be fully explored and quantified (van der Ent et al. 2010; Vose et al. 2012). This is exemplified by the recent debate on forests influences on regional water supply (Ellison et al. 2012; van der Ent et al. 2012) and forests' role in flood controls (Calder et al. 2007; Bradshaw et al. 2007; Laurence 2007; Van Dijk et al. 2009). Coupled climate-vegetation-hydrology models should be useful tools for understanding the role of the vegetation in regional and global climate and water cycles and design management strategies and options to adapt to a new environment.

Acknowledgments Funding for this study was supported by the USDA Forest Service Southern Research Station and also by the Chinese Academy of Science CAS/SAFEA International Partnership Program for Creative Research Teams of 'Ecosystem Processes and Services'.

References

Alila Y, Kuras PK, Schnorbus M, Hudson R. Forests and floods: a new paradigm sheds light on age-old controversies. Water Resour Res. 2009;45:W08416. doi:10.1029/2008WR007207.

Adams MB, Loughry LH, Plaugher LL. Experimental forests and ranges of the USDA forest service, revised. US For Serv Gen Tech Rep NE-GTR-321; 2008. p. 183.

Andreassian V. Waters and forests: from historical controversy to scientific debate. J Hydrol. 2004;291:1–27.

Angelini IM, Garstang M, Davis RE, Hayden B, Fitzjarrald DR, Legates DR, Greco S, Macko S, Connors V. On the coupling between vegetation and the atmosphere. Theor Appl Climatol. 2011. doi:10.1007/s00704-010-0377-5.

Avila FB, Pitman AJ, Donat MG, Alexander LV, Abramowitz G. Climate model simulated changes in temperature extremes due to land cover change. J Geophys Res. 2012;117:D04108. doi:10.1029/2011JD016382.

Bachelet D, Lenihan JM, Daly C, Neilson RP, Ojima DS, Parton WJ. MC1: a dynamic vegetation model for estimating the distribution of vegetation and associated carbon, nutrients, and water. Technical documentation, version 1.0. Gen Tech Rep PNW-GTR-508. Portland: US Department of Agriculture, Forest Service, Pacific Northwest Research Station; 2001.

Betts RA. Offset of the potential carbon sink from boreal forestation by decreases in surface albedo. Nature. 2000;408:187–90.

Black P. Watershed hydrology. 2nd ed. CRC Press: Boca Raton; 1996. p. 449.

Baldocchi D, Falge E, Gu L, Olson R, Hollinger D, Running S, Anthoni P, Bernhofer C, Davis KJ, Evans R, Fuentes J, Goldstein A, Katul G, Law BE, Lee Z, Malhi Y, Meyers T, Munger JW, Oechel W, Paw U KT, Pilegaard K, Schmid HP, Valentini R, Verma S, Vesala T, Wilson KB, Wofsy S. FLUXNET: a new tool to study the temporal and spatial variability of ecosystem-scale carbon dioxide, water vapor, and energy flux densities. Bull Am Meteorol Soc. 2001;82(11):2415–34.

Bonan GB. Forests and climate change: forcings, feedbacks, and the climate benefits of forests. Science. 2008;320:1444–9.

Bradshaw CJA, Sodhi NS, Peh KS-H, Brook BW. Global evidence that deforestation amplifies flood risk and severity in the developing world. Global Change Biol. 2007;13:2379–95. doi:10.1111/j.1365-2486.2007.01446.x.

Bosch JM, Hewlett JD. A review of catchment experiments to determine the effect of vegetation changes on water yield and evapotranspiration. J Hydrol. 1982;55:3–23.

Brown AE, Zhang L, McMahon TA. A review of paired catchment studies for determining changes in water yield resulting from alterations in vegetation. J Hydrol. 2005;310:1–34.

Brown TC, Hobbins MT, Ramirez JA. Spatial distribution of water supply in the coterminous United States. J Am Water Resour Assoc. 2008;6(44):1474–87.

Bruijnzeel LA. Hydrological functions of tropical forests: not seeing the soils for the trees. Agric Ecosyst Environ. 2004;104:185–228.

Charney JG. Dynamics of deserts and drought in the Sahel. Quart J Roy Meteor Soc. 1975;101:193–202.

Calder IR, Smyle J, Aylward B. Debate over flood-proofing effects of planting forests. Nature. 2007;450:945. doi:10.1038/450945b.

Cao S. Why large-scale afforestation efforts in China have failed to solve the desertification problem. Environ Sci Technol. 2008;42:1826–31.

Chang M. Forest hydrology: an introduction to water and forests. BocaRaton: CRC Press; 2002. p. 373.

Chapin III FS, Matso PA, Vitousek PM, Chapin MC. Principles of terrestrial ecosystem ecology. 2nd ed. New York: Springer; 2012. p. 544.

Charney JG, Quirk WJ, Chow SH, Kornfield J. A comparative study of the effects of albedo change on drought in semi-arid regions. J Atmos Sci. 1977;34:1366–85.

Cook BI, Anchukaitis KJ, Kaplan JO, Puma MJ, Kelley M, Gueyffier D. Pre-Columbian deforestation as an amplifier of drought in Mesoamerica. Geophys Res Lett. 2012;39:L16706. doi:10.1029/2012GL052565.

Delworth T, Manabe S. The influence of potential evaporation on the variabilities of simulated soil wetness and climate. J. Climate. 1988;1:523–47.

Dickinson RE, Henderson-Sellers A. Modeling tropical deforestation: a study of GCM land–surface parameterizations. Quart J Roy Meteor Soc. 1988;114:439–62.

Dickinson RE, Kennedy P. Impacts on regional climate of Amazon deforestation. Geophy Res Lett. 1992;19(19):1947–50.

Dickinson RE, Durbudge TB, Kennedy PJ, McGuffie K, Pitman AJ 1993 Tropical deforestation: modeling local- to regional-scale climate change. J Geophy Res. 1993;98(D4):7289–315.

Diggs JA. 2004. Simulation of nitrogen and hydrology loading of forested fields in eastern North Carolina using DRAINMOD-N 11. M.S. Thesis, North Carolina State University, Raleigh, NC; 2004. p. 155.

Eisenbies MH, Aust WM, Burger JA, Adams MB. Forest operations, extreme events, and considerations for hydrologic modeling in the Appalachians—A review. For Ecol Manage. 2007;242:77–98. doi:10.1016/j.foreco.2007.01.051.

Ellison D, Futter MN, Bishop K. On the forest cover–water yield debate: from demand- to supply-side thinking. Global Change Biol. 2012;18:806–820. doi:10.1111/j.1365-2486.2011.02589.x.

Eltahir EAB, Bras RL. On the response of the tropical atmosphere to large-scale deforestation. Quart J Roy Meteor Soc. 1993;119:779–93.

Field CB, Lobell DB, Peters HA, Chiariello NR. Feedbacks of terrestrial ecosystems to climate change. Annu Rev Environ Resour. 2007;32:1–29.

Foley JA, DeFries R, Asner GP, Barford C, Bonan G, Carpenter SR, Chapin FS, Coe MT, Daily GC, Gibbs HK, Helkowski JH, Holloway T, Howard EA, Kucharik CJ, Monfreda C, Patz JA, Prentice IC, Ramankutty N, Snyder PK. Global consequences of land use. Science. 2005;309:570–4.

Ford CR, Hubbard RM, Kloeppel BD, Vose JM. A comparison of sap flux-based evapotranspiration estimates with catchment-scale water balance. Agric For Meteorol. 2007;145:176–85. doi:10.1016/j.agrformet.2007.04.010.

Ford CR, Laseter SH, Swank WT, Vose JM. Can forest management be used to sustain water-based ecosystem services in the face of climate change? Ecol Appl. 2011;21:2049–67.

Fernow BE. Forests influences (1972 reprint version). No. 7 Bulletin of U.S. Forest service; 1893.

Friend A, Stevens A, Knox R, Cannell M. A process-based, terrestrial biosphere model of ecosystem dynamics (Hybrid v3.0). Ecol Model. 1997;95:249–87.

Garcia-Carreras L, Parker DJ. How does local tropical deforestation affect rainfall? Geophys Res Lett. 2011;38:L19802. doi:10.1029/2011GL049099.

Grace MJ III, Skaggs RW, Cassel DK. Soil physical changes associated with forest harvesting operations on an organic soil. Soil Sci Soc Am J. 2006;70:503–9.

Giorgi F, Marinucci MR, De Canio G, Bates GT. Development of a second generation regional climate model (RegCM2), I, boundary layer and radiative transfer processes. Mon Weather Rev. 1993a;121:2794–813.

Giorgi F, Marinucci MR, De Canio G, Bates GT. Development of a second generation regional climate model (RegCM2), II, Convective processes and assimilation of lateral boundary conditions. Mon Weather Rev. 1993b;121:2814–32.

Hajabbasi MA, Jalalian A, Karimzadeh HR. Deforestation effects on soil physical and chemical properties, Lordegan, Iran. Plant Soil. 1997;1997(190):301–8. doi:10.1023/A:1004243702208 .

Ice GG, Stednick JD. A century of forest and wildland watershed lessons. Soc Am Foresters. 2004;287.

Intergovernmental Panel on Climate Change (IPCC), Climate Change 2007.

Jackson RB, Carpenter SR, Dahm CN, McKnight DM, Naiman RJ, Postel SL, Running SW. Water in a changing world. Ecol Appl. 2001;11:1027–45. doi:10.1890/1051-0761(2001)011[1027:WIACW]2.0.CO;2.

Jackson RB, Jobbagy EG, Avissar R, Roy SB, Barrett DJ, Cook CW, Farley KA, Le Maitre DC, McCarl BA, Murray BC. Trading water for carbon with biological carbon sequestration. Science. 2005;310:1944–7. doi:10.1126/science.1119282.

Jackson RB, Jobbágy EG, Nosetto MD. Ecohydrology in a human-dominated landscape. Ecohydrology. 2009;2:383–9. doi:10.1002/eco.81.

Jones JA, Creed IF, Hatcher KL, Warren RJ, Adams MB, Benson MH, Boose E, Brown WA, Campbell JL, Covich A, Clow DW, Dahm CN, Elder K, Ford CR, Grimm NB, Henshaw DL, Larson KL, Miles ES, Miles KM, Sebestyen SD, Spargo AT, Stone AB, Vose JM, Williams MW. Ecosystem processes and human influences regulate streamflow response to climate change at long-term ecological research sites. BioScience. 2012;62(4):390–404. doi:10.1525/bio.2012.62.4.10.

Jin J, Miller NL, Schlegel N. Sensitivity Study of Four Land Surface Schemes in the WRF Model. Adv Meteorol. 2010 doi: 10.1155/2010/167436.

Jung M, Reichstein M, Ciais P, Seneviratne SI et al. Recent decline in the global land evapotranspiration trend due to limited moisture supply. Nature. 2010;467:951–54.

Kittredge J. Forest influences. New York: McGraw-Hill Book Co.; 1948. p. 99–114.

Komatsu H, Tanak N, Kume T. Do coniferous forests evaporate more water than broad-leaved forests in Japan? J Hydrol. 2007;336:361–75.

Koster RD, Suarez MJ. The relative contributions of land and ocean processes to precipitation variability. J Geophys Res. 1995;100(D7):13775–790.

Laurence WF. Environmental science: forests and floods. Nature. 2007;449:409–10.

Levis S, Bonan GB, Vertenstein M, Oleson KW. The Community Land Model's dynamic global vegetation model (CLM-DGVM): Technical description and user's guide. NCAR Tech. Note TN-459+IA, 2004;50 pp.

Law BE, Falge E, Gu L, Baldocchi DD, Bakwin P, Berbigier P, Davis K, Dolman AJ, Falk M, Fuentes JD, Goldstein A, Granier A, Grelle A, Hollinger D, Janssens IA, Jarvis P, Jensen NO, Katul G, Mahli Y, Matteucci G, Meyers T, Monson R, Munger W, Oechel W, Olson R, Pilegaard K, Paw UKT, Thorgeirsson H, Valentini R, Verma S, Vesala T, Wilson K, Wofsy S. Environmental controls over carbon dioxide and water vapor exchange of terrestrial vegetation. Agric For Meteorol. 2002;113:97–120. doi:10.1016/S0168-1923(02)00104-1.

Lean J, Warrilow DA. Simulation of the regional climatic impact of Amazon deforestation. Nature. 1989;342:411–13.

Lee R. Forest hydrology. New York: Columbia University Press; 1981. pp. 498–509.

Lee X, Goulden ML, Hollinger DY, Barr A, Black TA, Bohrer G, Bracho R, Drake B, Goldstein A, Gu L, Katul G, Kolb T, Law B, Margolis H, Meyers T, Monson R, Munger W, Oren R, Paw U KT, Richardson AD, Schmid HP, Staebler R, Wofsy S, Zhao L. Observed increase in local cooling effect of deforestation at higher latitudes. Nature. 2011;479:384–87.

Lin Y, Wei X. The impact of large-scale forest harvesting on hydrology in the Willow Watershed of Central British Columbia. J Hydrol. 2008;359:141–9.

Liu S, Lu L, Mao D, Jia L. Evaluating parameterizations of aerodynamic resistance to heat transfer using field measurements. Hydrol Earth Syst Sci. 2007;11:769–83.

Liu Y-Q, Avissar R. A study of persistence in the land–atmosphere system with a fourth–order analytical model. J. Climate. 1999;12:2154–68.
Liu Y-Q, Stanturf J, Lu H. Modeling the potential of the NorthernChina forest shelterbelt in improving hydroclimate conditions. J Am Water Resour Assoc. 2008;44(5):1176–92.
Liu Y-Q. A numerical study on hydrological impacts of forestrestoration in the southern United States. Ecohydrology. 2010;4:299–314.
Liu JG, Li SX, Ouyang ZY, Tam C, Chen XD. Ecological and socioeconomic effects of China's policies for ecosystem services. Proc Natl Acad 894 Sci USA. 2008b ;105(28):9477–482.
Makarieva AM, Gorshkov VG, Li B-L. Precipitation on land versus distance from the ocean: evidence for a forest pump of atmospheric moisture. Ecol Complex. 2009;6:302–7.
Marsh GP. Man and nature; or, physical geography as modified by human action. New York: C. Scribner; 1864.
Meginnis HG. Increasing water yields by cutting forest vegetation. In: Symposium of Hannoversch-Munden. Publ. 48. Louvain, Belgium: International Association of Scientific Hydrology: 1959;59–68.
McVicar TR, Roderick ML, Donohue RJ, Li LT, Van Niel TG, Thomas A, Grieser J, Jhajharia D, Himri Y, Mahowald NM, Mescherskaya AV, Kruger AC,Rehman S, Dinpashoh Y. Global review and synthesis of trends in observed terrestrial near-surface wind speeds: implications for evaporation. J Hydrol. 2012;416–7:182–205.
Munns EN. Forest hydrology in the Appalachians. J Soil Water Conserv. 1947;2:71–6.
Neilson RP, Prentice IC, Smith B, Kittel TGF, Viner D. Simulated changes in vegetation distribution under global warming. In: Watson RT, Zinyowera MC, Moss RH, Dokken DJ, editors. The regional impacts of climate change: an assessment of vulnerability. Cambridge: Cambridge University Press; 1998. p. 439–56.
Neilson RP, Pitelka LF, Solomon A, Nathan R, Midgley GF, Fragoso J, Lischke H, Thompson K. Forecasting regional to global plant migration in response to climate change: challenges and directions. Bioscience. 2005;55:749–59.
Nobre CA, Sellers PJ, Shukla J. Amazonian deforestation and regional climate change. J Clim. 1991;4:957–88.
Notaro M, Vavrus S, Liu Z. Global vegetation and climate change due to future increases in CO2 as projected by a fully coupled model with dynamic vegetation. J Clim. 2007;20:70–90.
Notaro M, Liu Z. Observed vegetation–climate feedbacks in the United States. J Clim. 2006;19:763–86.
Notaro M, Liu Z. Statistical and dynamical assessment of vegetation feedbacks on climate over the boreal forests. Clim Dyn. 2008;31:691–712.
Oyebande L. Effects of tropical forests on water yield. In: Reynolds ERC, Reynolds FB, editors. Forest, climate, and hydrology: regional impacts. Japan: The United Nations University; 1988. p. 16–50.
Pan Y, Birdsey RA, Fang J, Houghton R, Kauppi PE, Kurz WA, Phillips OL, Shvidenko A, Lewis S, Canadell JG, Ciais P, Jackson RB, Pacala SW, McGuire AD, Piao S, Rautiainen A, Sitch S, Hayes D. A large and persistent carbon sink in the world's forests. Science. 2011;333:988–93.
Parrotta J. Understanding Relationships between biodiversity, Carbon, forests and people: The key to achieving REDD+Objectives. A global assessment report. IUFRO on behalf of the Collaborative Partnership on Forests (CPF), Global Forest Expert Panel on Biodiversity, Forest Management and REDD+; 2012.
Phipps SJ, Rotstayn LD, Gordon HB, Roberts JL, Hirst AC, Budd WF. The CSIRO Mk3L climate system modelversion 1.0—Part 1: description and evaluation. Geosci Model Dev Discuss. 2011;4(1):219–287. doi:10.5194/gmdd-4-219-2011.
Pielke RA, Avissar R. Influence of landscape structure on local andregional climate. Landscape Ecol. 1990;4:133–55.
Pielke RA, Adegoke J, Beltr A, An-Przekurat CA, Hiemstra J, Lin J, Nair US, Niyogi D, Nobis TE. An overview of regional land-use and land-cover impacts on rainfall. Tellus. 2007;59B:587–601.

Pielke RA Sr, et al. Land use/land cover changes and climate: modeling analysis and observational evidence, Wiley Interdiscip. Rev Clim Change. 2011;2:828–50. doi:10.1002/wcc.144.

Rao L, Sun G, Ford CR, Vose J. Modeling potential evapotranspiration of two forested watersheds in the Southern Appalachians, USA. Trans ASABE. 2011;54(6):2067–78.

Ryan MG, Harmon ME, Birdsey RA, Giardina CP, Heath LS, Houghton RB, Jackson RB, McKinley DC, Morrison J, Murray BC, Pataki DE, Skog KE. A synthesis of the science on forests and carbon for U.S. Forests. Issue 13 of Issues in Ecology; 2010.

Robinson M, Cognard-Plancq AL, Cosandey C et al. Studies of the impact of forests on peak flows and baseflows: a European perspective. For Ecol Manage. 2003;186:85–97.

Sampson RN. Southern forests: Yesterday, Today, and Tomorrow, In: Michael Rauscher H, Johnsen K, editors. Southern forest science: Past, Present, and Future. Department of Agriculture Forest Service, General Technical Report SRS-75; 2004. p. 408.

Scott DF, Lesch W. Streamflow responses to afforestation with Eucalyptus grandis and pinuspetula and to felling in the Mokobulaan experimental catchments, South African. J Hydrol. 1997;199:360–77.

Scott DF, Bruijnzeel LA, Mackensen J. The hydrologic and soil impacts of reforestation. In: Bonell M, Bruijnzeel LA, editors. Forests, water and people in the humid tropics. Cambridge University Press; 2005. pp. 622–51.

SFA (China State Forest Administration). China's Forest Shelter Project Dubbed ''Green Great Wall.'' 2006. http://www.chinaview.cn, accessed on July 9, 2006.

Sitch S, Smith B, Prentice IC, Arneth A, Bondeau A, Cramer W, Kaplan JO, Levis S, Lucht W, Sykes MT, Thonicke K, Venevsky S. Evaluation of ecosystem dynamics, plant geography and terrestrial carbon cycling in the LPJ dynamic global vegetation model. Glob Change Biol. 2003;9:161–85.

Shukla J, Mintz Y. Influence of land–surface evapotranspiration on the earth's climate. Science. 1982;215:1498–501.

Shukla J, Nobre CA, Sellers PJ. Amazonia deforestation andclimate change. Science. 1990;247:1322–5.

Sud YC, Shukla J, Mintz Y. Influence of land–surface roughness on atmospheric circulation and precipitation: a sensitivity study with a general circulation model. J Appl Meteor. 1988;27:1036–54.

Sun G, McNulty SG, Shepard JP, Amatya DM, Riekerk H, Comerford NB, Skaggs RW, Swift L Jr. Effects of timber management on wetland hydrology in the southern United States. For Ecol Manage. 2001;143:227–36.

Sun G, Zhou G, Zhang Z, Wei X, McNulty SG, Vose JM. Potential water yield reduction due to reforestation across China. J Hydrol. 2006;328:548–58.

Sun G, Zuo C, Liu S, Liu M, McNulty SG, Vose JM. Watershed evapotranspiration increased due to changes in vegetation composition and structure under a subtropical climate. J Am Water Resour Assoc. 2008;44(5):1164–75.

Sun G, Noormets A, Gavazzi1 MJ, McNulty SG, Chen J, Domec J-C, King JS, Amatya DM, Skaggs RW. Energy and water Balance of two contrasting Loblolly pine plantations on the lower coastal plain of North Carolina, USA. For Ecol Manage. 2010;259:1299–310.

Sun G, Alstad K, Chen J, Chen S, Ford CR, Lin G, Liu C, Lu N, McNulty SG, Miao H, Noormets A, Vose JM, Wilske B, Zeppel M, Zhang Y, Zhang Z. A general predictive model for estimating monthly ecosystem evapotranspiration. Ecohydrology. 2011a;4:245–55. doi:10.1002/eco.194.

Sun G, Caldwell P, Noormets A, Cohen E, McNulty SG, Treasure E, Domec JC, Mu Q, Xiao J, John R, Chen J. Upscaling key ecosystem functions across the conterminous United States by a water-centric ecosystem model. J Geophys Res. 2011b;116:G00J05. doi:10.1029/2010JG001573.

Swank WT, Douglass JE. Streamflow greatly reduced by converting hardwoods to pine. Science. 1974;185:857–9.

Swank WT, Swift LW, Douglass JE. Streamflow changes associated with forest cutting, species conversion, and natural disturbances. In: Swank WT, Crossley DA, editors. Forest hydrology and ecology at Coweeta. Ecological studies, vol. 66. New York: Springer; 1988. p. 297–312.

Swank WT, Vose JM. Long-term hydrologic and stream chemistry responses of southern Appalachian catchments following conversion from mixed hardwoods to white pine. In: Landolt R, editor. HydrologiekleinerEinzugsgebiete: Gedenkschrift Hans M. Keller. Bietrag-ezurHydrologie der Schweiz 35. Bern, Schweizerische: SchweizerischeGesellshacft fur Hydrologie und Limnologie; 1994. p. 164–72.

Swank WT, Vose JM, Elliot KJ. Long-term hydrologic and water quality responses following commercial clearcutting of mixed hardwoods on a Southern Appalachian catchment. For Ecol Manage. 2001;143:163–78.

Swift LW. Effect of forest cover and mountain physiography on the radiant energy balance, D.F. Thesis. Durham: Duke University; 1972.

Swift LW Jr, Messer JB. Forest cuttings raise temperatures of small streams in the Southern Appalachians. J Soil Water Conserv. 1971;26(3):111–6.

Swift LW Jr. Lower water temperatures within a streamside buffer strip. Research Note SE–193. Asheville, NC: U.S. Department of Agriculture, Forest Service, Southeastern Forest Experiment Station; 1973. p. 7.

Swift LW Jr. Duration of stream temperature increases following forest cutting in the southern Appalachian mountains. In: International symposium on hydrometeorology. American Water Resources Association; 1982.

Swift LW, Swank WT, Mankin JB, Luxmoore RJ, Goldstein RA. Simulation of evapotranspiration and drainage from mature and clear-cut deciduous forests and young pine plantation. Water Resour Res. 1975;11:667–73.

Swift LW Jr, Swank WT. Long term responses of streamflow following clearcutting and regrowth. Hydrol Sci Bull. 1981;26(3):245–56.

Swank WT, Helvey JD. Reduction of streamflow increases following regrowth of clearcut hardwood forest. In: Proceedings of the symposium on the results of research on representative and experimental basins, December 1970, Wellington, New Zealand. Publication 96. United Nations Educational, Scientific and Cultural Organization—International Association of Scientific Hydrology, Leuven, Belgium; 1970. p. 346–60.

Tian H, Chen G, Liu M, Zhang C, Sun G, Lu C, Xu X, Ren W, Pan S, Chappelka A. Ecosystem net primary productivity, evapotranspiration, and water use efficiency in the southern United States during 1895–2007. For Ecol Manage. 2009;259:1311–27.

United States Department of Agriculture Forest Service (USFS). Watershed management research Coweeta experiment forest, Macro County, North Carolina, Southeastern Forest Experiment Station, Asheville, NC; 1948.

van der Ent RJ, Savenije HHG, Schaefli B, Steele-Dunne SC. Origin and fate of atmospheric moisture over continents. Water Resour Res. 2010;46:1–12.

van der Ent RJ, Coenders-gerrits AMJ, Nikoli R, Savenije HG. The importance of proper hydrology in the forest cover-water yield debate: commentary on Ellison et al. (2012). Glob Change Biol. 2012;18:806–20.

Van Dijk AIJM, Van Noordwijk M, Calder IR, BruijnzeeL SLA, Schellekens J, Chappell NA. Forest–flood relation still tenuous—comment on 'Global evidence that deforestation amplifies flood risk and severity in the developing world' by CBradshaw CJA, Sodi NS, Peh KS-H, Brook BW. Glob Change Biol. 2009;15:110–5. doi:10.1111/j.1365-2486.2008.01708.

Vautard R, Cattiaux J, Yiou P, Thépaut J-N, Ciais P. Northern Hemisphere 1124 atmospheric stilling partly attributed to increased surface roughness. Nat Geosci. 2010;1125(3):756–61.

Vinnikov K, Robock A, Speranskaya NA, Schlosser CA. Scales of temporal and spatial variability of midlatitude soil moisture. J Geophys Res.1996;101(D3):7163–74.

Vose J, Sun G, Ford CR, Bredemeier M, Ostsuki K, Wei X, Zhang Z, Zhang L. Forest ecohydrological research in the 21st century: what are the critical needs? Ecohydrology. 2011;4(2):146–58.

Vose JM, Ford CR, Laseter S, Dymond S, Sun G, Adams MB, Sebestyen S, Campbell J, Luce C, Amatya D, Elder K, Heartsill-Scalley T. Can forest watershed management mitigate climate change impacts on water resources? Book chapter In: IAHS Revisiting experimental catchment studies in forest hydrology (proceedings of a workshop held during the 25 IUGG general assembly in Melbourne, June–July 2011), IAHS Publ.;2012. p. 353.

Wang JF, Chagnon FJF, Williams ER, Betts AK, Renno NO, Machado LAT, Bisht G, Knox R, Brase RL. Impact of deforestation in the Amazon basin on cloud climatology. Proc Natl Acad Sci USA. 2009;106(10):3670–4.

Wang YP, Kowalczyk E, Leuning R, Abramowitz G, Raupach MR, Pak B, van Gorsel E, Luhar A. Diagnosing errors in a landsurface model (CABLE) in the time and frequency domains. J Geophys Res. 2011;116:G01034. doi:10.1029/2010JG001385.

Waring SW, Waring RH. Forest ecosystems: analysis at multiple scales. 3rd ed. Academic Press; 2007. p. 440.

Watson T. Climate plan calls for forest expansion, USA TODAY. www.usatoday.com/news/nation/environment/2009-08-19-forest_N.htm (2009).

Wei X, Sun G, Liu S, Hong J, Zhou G, Dai L. The forest-streamflow relationship in China: a 40-year retrospect. J Am Water Resour Assoc. 2008;44(5):1076–85.

Wei X, Zhang M. Quantifying+stream flow change caused by forest disturbance at a large spatial scale: a single watershed study. Water Resour Res. 2010;46:W12525. doi:10.1029/2010WR009250.

Whitehead PG, Robinson M. Experimental basin studies—an international and historical perspective of forest impacts. J Hydrol. 1993;145:217–30.

Xue Y, Shukla J. The influence of land surface properties on Sahel climate, part I: desertification. J Clim. 1993;6:2232–45.

Yeh TC, Wetherald RT, Manabe S. The effect of soil moisture on the short-term climate and hydrology change—a numerical experiment. Mon Weather Rev. 1984;112:474–90.

Zeng X, Dickinson RE, Walker A, Shaikh M, DeFries RS, Qi J. Derivation and evaluation of global 1 km fractional vegetation cover data for land modeling. J Appl Meteor. 2000;39:826–39.

Zhang L, Dawes WR, Walker GR. Response of mean annual evapotranspiration to vegetation changes at catchment scale. Water Resour Res. 2001;37:701–8.

Zhang XP, Zhang L, McVicar TR, Van Niel TG, Li LT, Li R, Yang QK, Liang W. Modeling the impact of afforestation on mean annual streamflow in the Loess Plateau, China. Hydrol Process. 2008a;22:1996–2004.

Zhang X. P., L. Zhang, J. Zhao, P. Rustomjiand P. Hairsine. 2008b. Responses of streamflow to changes in climate and land use/cover in the Loess Plateau, China. Water Resour Res. 2008b;44:W00A07. doi:10.1029/2007WR006711.

Zhou, Wei GX, et al. Forest recovery and river discharge at the regional scale of Guangdong province, China. Water Resour Res. 2010;46:w09503. doi:10.1029/2009WR008829.

Zon R. Forests and water in the light of scientific investigation, USDA forest service, Washington, DC. Senate document 469, 62nd congress, 2nd session; 1927.

Chapter 16
Are We Destroying Our Insurance Policy? The Effects of Alien Invasion and Subsequent Restoration

A Case Study of the Kromme River System, South Africa

Alanna J. Rebelo, David le Maitre, Karen J. Esler and Richard M. Cowling

Abstract Securing sufficient and reliable water supply is a priority for many countries worldwide, but their efforts are hindered by widespread landscape degradation and uncertainty around future climate change. We used historical aerial photographs and mapping techniques to investigate how a South African landscape has changed over the past century. The Kromme River Catchment, a valuable water-providing catchment for the Nelson Mandela Bay metropolitan hub, has become heavily degraded. The floodplain wetlands, which historically occupied the entire valley floor, have been almost completely replaced by agriculture or invaded by the alien tree *Acacia mearnsii*. Some efforts have been made to restore the wetlands and control the invasive plants, but our results show that at the current rate of clearing it would take 30 years before *A. mearnsii* would be brought under control. We recommend that investment should be made, as a type of insurance for natural capital, in restoring resilience in important water-providing catchments to hedge against future climatic uncertainties.

Keywords Floodplain wetlands · Ecosystem goods and services · Degradation · Restoration · *Prionium serratum* · Land-cover change · Climate change

A. J. Rebelo (✉) · K. J. Esler
Department of Conservation Ecology and Entomology, Stellenbosch University, Stellenbosch, South Africa
e-mail: ARebelo@sun.ac.za

D. le Maitre
Natural Resources and the Environment, Council for Scientific and Industrial Research (CSIR), Stellenbosch, South Africa

K. J. Esler
Centre for Invasion Biology (C.I.B), Stellenbosch, South Africa

R. M. Cowling
Department of Botany, Nelson Mandela Metropolitan University, Port Elizabeth, South Africa

B. Fu and K. B. Jones (eds.), *Landscape Ecology for Sustainable Environment and Culture*, DOI: 10.1007/978-94-007-6530-6_16,

Diary Extract: C. I. La Trobe 1816
Upper Langkloof
'First impression of Langkloof: "a vale of perhaps 100 miles enclosed by mountains of different heights. On entering it we felt not a little disappointed ... we saw a long ridge of comparatively low hills, divided by narrow parallel kloofs, without wood or water, skirting a dull uncultivated vale...."'. (Skead 2009)
-The Langkloof is the local name for the Kromme River valley

16.1 Introduction

The concept of insurance, using financial capital as a store for future undesirable and yet possible events, is well entrenched within modern society. It is considered wise to invest in and protect manufactured and human capital but the concept of protecting and investing in natural capital is met with great resistance. Natural capital is the ecosystem infrastructure that provides humankind with essential ecosystem goods and services (Aronson et al. 2007; Mander et al. 2010). The Kromme River system has valuable natural capital, including wetlands and aquifers, which provide water as well as regulating and storing services to the downstream Nelson Mandela Bay metropolitan hub (Haigh et al. 2002; Raymer 2008). The Nelson Mandela Bay metropolitan hub has a history of struggling to match water supply with demand, largely due to its rapidly increasing population size as well as economic development (Raymer 2008). There is a strong correlation of 0.91 between population size and water demand, and it is already evident that water supply is set to outstrip demand for the Nelson Mandela Bay metropolitan hub in the near future (Eberhard 2009). This ever increasing demand for resources coupled with the uncertainty surrounding the predicted changes in climate, will be a recipe for disaster if it is not mitigated by investing in and ensuring that natural capital is maintained (Fig. 16.1)

Natural Capital:
"*Natural Capital is an economic metaphor for the stock of physical and biological natural resources that consist of*:
Renewable natural capital (*living species and ecosystems*);
Nonrenewable natural capital (*subsoil assets, e.g. petroleum, coal, diamonds*);
Replenishable natural capital (*e.g., the atmosphere, potable water, fertile soils*); *and*
Cultivated natural capital (*e.g., crops and forestplantations*)."
(Aronson et al. 2007)

The ongoing efforts to maintain economic growth and development are driving the intensification of agriculture and other ways of using landscapes. Land transformation, or habitat loss, is currently the major factor endangering species (Pimm and Raven 2000; Raimondo et al. 2009) and ecosystems (Rouget 2004; Millennium Ecosystem Assessment 2006). This effect is exacerbated in semi-arid

Fig. 16.1 The Kromme River in the Eastern Cape of South Africa as seen from its headwaters towards the coast. The Kromme is a narrow river valley, hence the name 'Langkloof', 100 km in length, bordered on each side by steep mountain ranges. The Kromme River has been heavily transformed by agriculture as shown in the foreground

Mediterranean-type environments with irregular rainfall, as there is often a mismatch between seasons when water is needed and when rainfall occurs (Kondolf 2011). This leads to measures aimed at capturing and using every drop of water available. Ecosystems need resilience to persist and humankind is making itself vulnerable by stripping ecosystems of this resilience by compromising their structural and functional integrity (eroding natural capital). This is particularly apparent in wetland and riparian ecosystems. Despite the uncertainty surrounding climate change, there is general agreement that it is likely to result in water shortages and an increase in floods in southern Africa (Midgley et al. 2005; Schulze 2005; Bates et al. 2008; Le Maitre et al. 2009). Alternatives to traditional infrastructure (such as dams and inter-basin transfers), such as using the natural infrastructure it provides (such as wetlands and aquifers) are likely to prove more effective in mitigating the effects of climate change and water scarcity (Matthews et al. 2011).

Ecosystem Goods and Services:
Ecosystem goods and services are the benefits that society derives either directly or indirectly from ecosystem functions (Daily et al. 2000; de Groot et al. 2002).
These goods and services can be classified into three main groups:
Provisioning services (e.g. water, food, fuel),

Regulatory services (carbon sequestration, water filtration, crop pollination),
Cultural services (fulfillment of human needs: spiritual, cultural, aesthetic, intellectual)
(Aronson et al. 2007)

Wetlands continue to be destroyed worldwide, as well as in South Africa, a trend which sacrifices long term societal benefit for short term private gains (Ashton 2002). Major threats to wetlands and associated river systems are agriculture, forestry, invasive alien species and poor land and fire management (Mooney et al. 1986; Rowntree 1991; Groombridge 1992; Rejmánek and Randall 1994; Grundling et al. 1999; David et al. 2000; Brinson and Malvárez 2002; Collins 2005; Kotze et al. 2009). Invasive woody alien trees, such as *Acacia mearnsii*, commonly known as Black Wattle, are one of the greatest threats to South Africa's water supply because of high water consumption rates (Dye and Jarmain 2004). Alien plants had invaded about 10.1 million ha of South Africa and Lesotho to various degrees by 1996, resulting in the loss of an estimated 3,300 million m^3 of water per annum (Le Maitre et al. 2000) (Figs. 16.2, 16.3).

***Acacia mearnsii*(Black Wattle)**
Black Wattle is arguably one of South Africa's most aggressive alien invasive plants. It is a tall woody tree, a competitive invader with extremely rapid growth rates, high seed yield and drought tolerance (Crous et al. 2011). It transpires large volumes of water and, together with other woody alien invasive plants, has been shown to decrease river flow, base flow and yield of South African River Systems (Bosch and Hewlett 1982; Dye 1996; Le Maitre et al. 2009). It has shallow root systems and thus is not able to withstand flood waters, resulting in trees being ripped out which causing significant channel instability and erosion in river systems (Scott et al. 2004; Grenfell et al. 2005). Black Wattle shades out native plant species, such as the wetland plant, palmiet (*Prionium serratum*) (Boucher and Withers 2004; van Wilgen et al. 2008). Black Wattle originates from Australia and as such has adapted to fire. Consequently it is very difficult to eradicate as burning simply stimulates the germination of its sizeable seed banks and many trees resprout. Black Wattle poses a significant threat to attaining water security in South Africa.

Fig. 16.2 The flowers of a Black Wattle (*Acacia mearnsii*) tree

Fig. 16.3 An aerial photograph taken from a helicopter of Black Wattle (*Acacia mearnsii*) invading the Kromme River System, South Africa. A large expanse of the river has been cleared (foreground and right) but re-growth with the next flood or fire is inevitable due to accumulated seed-banks

Restoration has had a highly successful return on investment worldwide as it has repeatedly been shown to improve the delivery of many ecosystem services (Aronson et al. 2007) and increase biodiversity (Aronson et al. 2007; Blignaut and Aronson 2008). In the 1990s South Africa recognized the threat to ecosystems and the economy by alien plants and have acknowledged the impact of alien invasion and poor management (van Wilgen et al. 1998). In 1996 a restoration programme called Working for Water commenced the clearing of invasive alien plants. It has been found to be economically viable and competitive to restore natural capital and infrastructure rather than using expensive, traditional engineering techniques (van Wilgen et al. 1998, 2008). The Kromme River Catchment, a Mediterranean-type climate catchment in the Eastern Cape of South Africa, was selected as a priority location for Working for Water, because of its importance in water provision for the Nelson Mandela Bay metropolitan hub. The water use of this metropolitan hub is predicted to increase from of 100 million m^3 per annum in 2007 to about 130 million m^3 per annum by 2017 (Murray et al. 2008). A major aim of the Working for Water project is to make more water available by removing invasive alien plants with high water consumption rates (McConnachie et al. 2012). However Working for Water's ability to cope with the scale of the problem and its efficiency over the past 15 years have been called into question (van Wilgen et al. 1998, 2012; Hobbs 2004; McConnachie et al. 2012) (Fig. 16.4).

Fig. 16.4 Workers from South Africa's Working for Water Programme. Working for Water, besides restoring the landscape by clearing invasive alien plants, also empowers local people by creating jobs for unskilled workers

Working for Water

In the 1990s South African scientists recognized the widespread damage to the landscape by alien invasion and acknowledged the urgent need for restoration. In 1996, the government's Working for Water programme began clearing the invasive alien trees in the Kromme River System (McConnachie et al. 2012). Working for Water aims to make more water available by clearing invasive alien plants that use high amounts of water. It is run through the Department of Water Affairs. Since Working for Water started in 1995, more than one million hectares of invasive alien plants have been cleared throughout the country. Working for Water has also provided jobs and training to about 20,000 people a year. These people are drawn from the most marginalized areas, and of the total, 52 % are women. Currently there are 300 projects in all nine South African provinces. (Department of Water Affairs and Forestry 2006)

Here we assess how changes in the South African landscape as a result of increased 'progress and development' have affected the Kromme Catchment. We ask what changes are likely to happen in the future, not only in terms of continued land transformation, but coupled with climate change. Have recent attempts to restore this landscape been successful? Are current restoration programs efficient? We attempt to discover the main driver of these changes to answer the question: how can these complex systems be managed in such a way that they become our insurance against climate change?

16.2 Methods

16.2.1 Study Site

The Kromme River (33°S, 24°E) is located in the Eastern Cape Province of South Africa (Fig. 16.5). It is about 100 km in length from its upper reaches (550 m above sea level) to its estuary. The catchment is narrow and steep, bordered by the

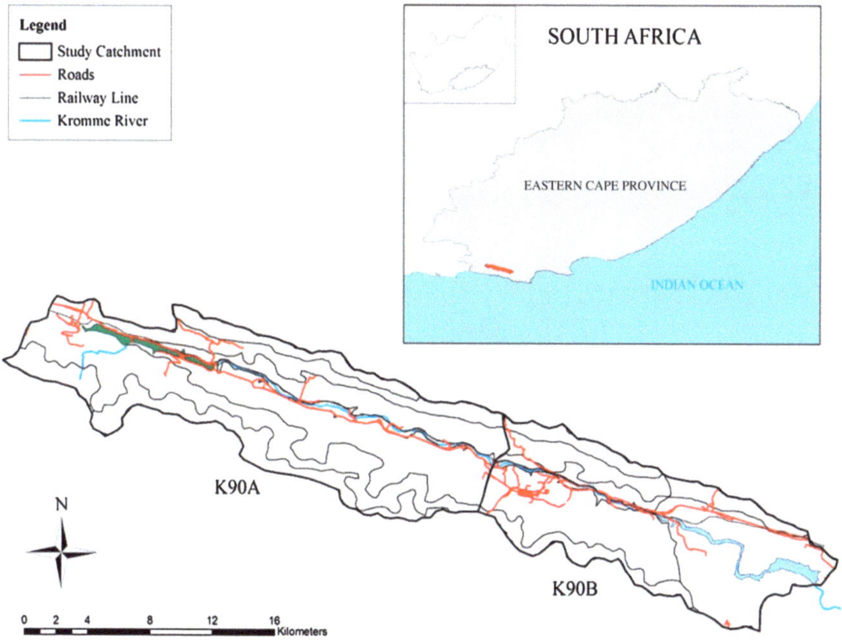

Fig. 16.5 The location of the Kromme River study catchments (K90A and K90B) and the position of the Churchill Dam in the South Eastern corner of the catchment. The Kromme River is located in the Eastern Cape Province of South Africa and the Nelson Mandela Bay metropolitan area receives 24 % of its water from the Churchill Dam. Solid black lines within the catchment delineate 11 subcatchments, ■ the remaining palmiet wetlands, ■ Churchill Dam

Suuranys Mountains (±1,050 m) to the north, and the Tsitsikamma Mountains (±1500 m) to the south, both running from east to west.

Rainfall in the region is unpredictable, but tends to exhibit a bimodal pattern, with maximums in spring and autumn (Midgley et al. 1994). Mean annual precipitation (MAP) for the entire catchment is ±614 mm. Mean annual runoff (MAR) for the entire catchment is ±75 mm which is ±11 % of the rainfall (WR2005).

The catchment has been heavily transformed by agriculture and alien invasion (Fig. 16.6a). Groundwater recharge rates are estimated to be fairly high despite the relatively low rainfall, largely because of the shallow soils in the mountain slopes and the low water-use of fynbos (Fig. 16.6b). Kareedouw (population under 1,000) is the only town in the catchment (Fig. 16.6c). The catchment consists predominantly of shales and sandstones of the Cape Supergroup (Toerien and Hill 1989) (Fig. 16.6d). The Cape Fold Belt is part of an intensely folded range with dipping beds forming a trellis drainage pattern (Lewis 2008). There are six large and five minor tributaries entering from the southern mountain range, and seven large and numerous minor tributaries entering from the drier northern mountain range in the upper catchment (Haigh et al. 2002). River flow from the northern tributaries is mostly seasonal. Several of the tributaries have alluvial fans which limit the extent

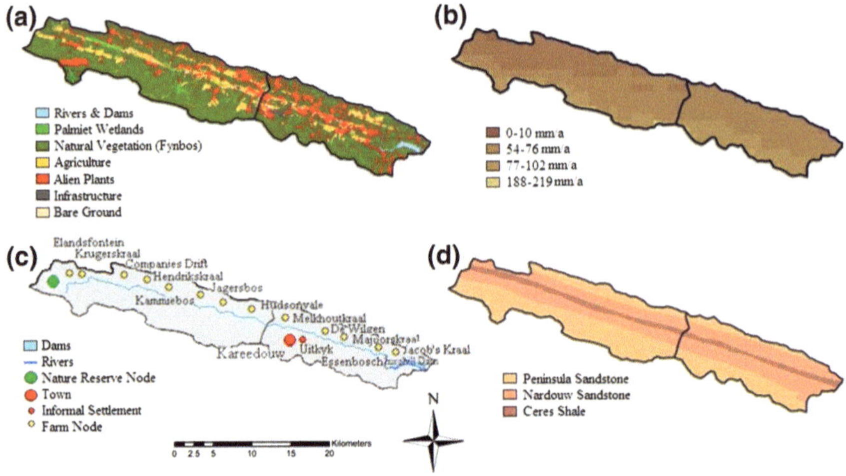

Fig. 16.6 Characteristics of the upper Kromme Catchments: **a** vegetation and land-use, **b** ground water recharge (mm/annum), **c** towns and farms and **d** geology (Middleton and Bailey 2008)

of the palmiet (*Prionium serratum*) wetlands. Historically, the palmiet stabilized the floodplain alluvium, forming peat basins which would have covered a large area of the floodplain (Haigh et al. 2002).

16.2.2 Land-Use Changes

Land-use in the Kromme was mapped at high resolution (~5 m) spanning four decades: 1954, 1969, 1986, 2007. These time-slices were supplemented with a reference state based on a reconstruction of the land-cover of the Kromme system prior to European occupation, and using the Garden Route Initiatives Vegetation Map (Vlok et al. 2008). Land-use mapping was done using 1:20,000 aerial photographs in a GIS system (ArcMap) to divide the area into 15 pre-selected land-cover categories (Table 16.1). The Fynbos was divided into 10 types by Vlok et al. (2008) but these were amalgamated into two categories for the modelling study based on whether the vegetation type classified as productive or unproductive for livestock. Areas of fynbos that was mapped as heavily degraded by over-grazing or altered fire regimes were kept as a separated class.

Maps are currently in the WGS 84 geographic co-ordinate system, and projected using Transverse Mercator Projection. Areas that were invaded by alien plants were only distinguished if they had reached maturity and the density was greater than that of 80 %. Canopy cover values lower than this are difficult to distinguish from indigenous vegetation in aerial photographs. There is a small degree of error, which varies depending on the difficulty of identifying and mapping each land-use category. The most difficult land-cover type to map was the

Table 16.1 The key to the different land-uses mapped in the Kromme Catchment, using aerial photograph based polygons captured using the ArcMAPsoftware

-	-	Land-use	Description
-	1	Dams	Including small farm dams and a large municipal dam
-	2	Mountain Seep Wetlands	High altitude/gradient wetlands on the mountain slopes
-	3	Palmiet Wetlands	Wetlands in the valley, dominated by Prionium serratum
-	4	Riparian Vegetation	Woody vegetation in ravines, either thicket or Afromontane forest
-	5	Unproductive Fynbos	7 Different unproductive fynbos and renosterveld vegetation types
-	6	Productive Fynbos	3 Different productive fynbos and renosterveld vegetation types
-	7	Degraded Fynbos	Degraded by heavy grazing or poor fire management
-	8	Irrigated Fields	Any agriculture that is not irrigated
-	9	Dryland Farming	Agriculture that has an irrigation system (sprinkler or central pivot)
-	10	Orchards	Orchards with irrigation systems
-	11	*Acacia mearnsii*	The dominant woody invasive alien plant in the catchment
-	12	Pinus Sp	The 2nd most common woody invasive alien plant in the catchment
-	13	Alien Plants	All other woody invasive plants, mainly Eucalyptus Sp
-	14	Infrastructure	All unnatural structures: houses, roads, railway lines, quarries
-	15	Open Soil	Open soil, sites of erosion or deposition in the river valley

mountain seep wetlands, as it is difficult to distinguish them from surrounding dryland fynbos. Indeed the sizes of these seep wetlands are likely to fluctuate seasonally and to be different each year. However this error was justified in that mapping was done for hydrological modelling purposes and the hydrological differences between fynbos and seep wetlands would be marginal when compared to other land-use types such as floodplain wetlands.

Mapping done using the most recent photographs was ground-truthed by mapping land-cover adjacent to the road that traverses the catchment. Some additional areas were verified using photographs and observations made during a helicopter trip over the catchment. The 2007 aerial photographs and map were used to cross-check identifications made from historical aerial photographs where the mapped classes could not be verified. Additional verification was done using maps compiled by different organizations and individuals: National Land Cover (NLC) (Van den Berg et al. 2008), maps showing extent, clearing and follow up done by Working for Water, land-use maps for the Baviaansklo of Mega Reserve, and land-use maps for the Garden Route Initiative (GRI) (Vlok et al. 2008).

16.2.3 Geomorphology Changes

The total active channel length was measured along the center line of the Kromme River from the aerial photographs from each of the four time slices.

16.3 Results

16.3.1 Historical Record and Meteorological Setting

Diary Extract: C. I. La Trobe 1816
East of Jagersbos
"… this country, unproductive as it generally is in means of subsistence for man and beast [is clothed] with an astonishing profusion of vegetable beauty. Hardly a spot exists upon which some curious and beautiful plant does not rear its head in its proper season; and in the midst of this brown desert we see the magnificent chandelier or red-star flower, measuring from four to five inches, to a foot and a half in the spread of its rays growing luxuriantly among the stones [Brunsvigia littoralis]". (Skead 2009)

This historical overview of the Kromme was compiled using two sources: Haigh et al. (2002) and Raymer (2008). The earliest record of agriculture in the catchment was in 1775 when a Mr Ferreira applied for grazing rights at Jagersbos. By this stage, settlers had already occupied the eastern part of the catchment. Orchards and grazing were the most common forms of land-use until 1930. In 1931 a

Fig. 16.7 A damaged tributary in the Kromme River. The headcut moved backwards up the hill, eroding away sediment and vegetation

particularly large flood destroyed many orchards along the river banks, causing severe erosion. After this, many farmers turned to pasture, dairy and meat production. This is also when Black Wattle appeared for the first time along a stretch of the Kromme River. From 1931 to 1934, good rainfall years ensured Black Wattle establishment. After the war ended in 1942, agricultural pressure increased. After orchards were swept away again in flood in 1965, the farmers raised the banks of the river in an attempt to contain future floods. This caused significant channel erosion. By 1986, more than half of the valley floor had been converted to agriculture and Black Wattle had formed dense stands on the floodplains. In 1996 Working for Water began clearing the Black Wattle, revealing the extent of the damage to the wetlands. In 2000 Working for Wetlands began building a series of weirs to prevent headcuts from eroding further upstream.

Diary Extract: C. J. F. Bunbury 30 March 1838
Langkloof
"The country was extremely arid except along the course of the little streams, and on the hills near the younger Kamper's residence the bushes had been burnt to a considerable extent, a practise general in this country and advantageous to the cattle but very provoking to the botanist". (Skead 2009)

Fig. 16.8 An aerial photograph taken from a helicopter of the main floodplain of the Kromme River after a large flood event. The floodplain was once covered by large palmiet wetlands, specially adapted withstand the force of the flood waters. The removal of the palmiet wetlands has destabilized the system, causing massive headcuts which lower the water table and reduce the agricultural potential of the land

Fig. 16.9 Cement weirs built along the Kromme River, Eastern Cape, South Africa. These weirs are built to restore the river by stopping the headcut from proceeding backwards up the river. This traps sediments and allows vegetation, such as palmiet in this photograph, and eventually wetlands to recover

Since 1931, the first recorded flood, there has been a major flood approximately every decade, the exceptions being the 1940s and the 1970s. In the 1980s there were two major flood episodes, the first being a series of three consecutive floods in 1981 and the second in 1983. The 1996 floods were described as the largest ever experienced in the catchment. In the past decade three major flood events have been recorded: 2004, 2006, and 2007 (Figs. 16.7, 16.8, 16.9).

Erosion Damage in the Kromme River

In the Kromme River, headcuts formed as a result of activities which disturbed the Kromme River's path. Examples from the Kromme were the building of a provincial road (the R62), the building of the railway line through the wetlands, river or floodplain. The damage done was exacerbated by farmers allowing animals to graze in these disturbed areas, or ploughing these areas up for agriculture. These activities created 'nick-points' or weaknesses which lead to rapid erosion and the loss of the eroded sand and gravel downstream. The nick-points migrate upstream and create progressively wider and deeper head-cuts and dongas over time. This process was rapidly accelerated during the large floods in the Kromme Catchment. The channels formed by the headcuts are detrimental because they drain groundwater from the surrounding alluvium, drying it out and reducing

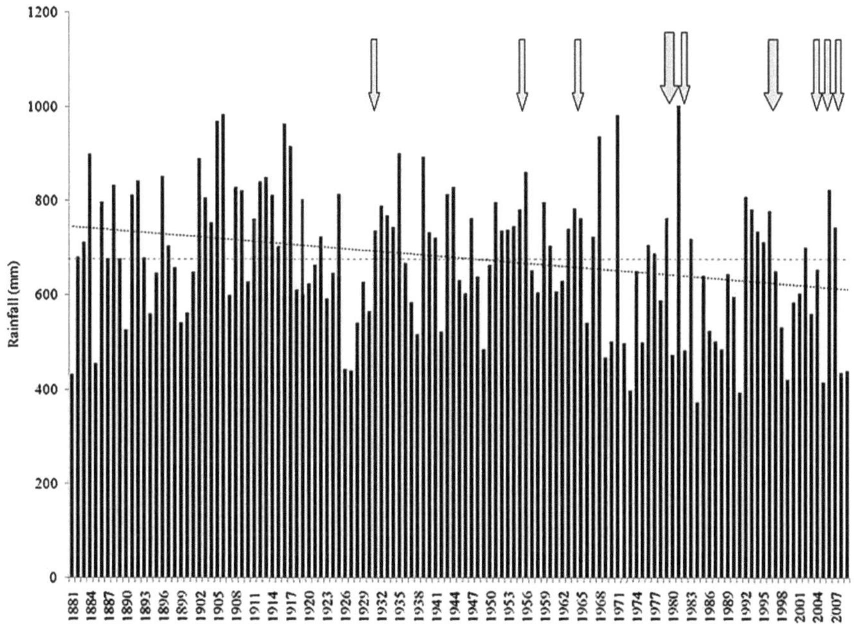

Fig. 16.10 Annual precipitation in the upper Kromme from 1881 to 2007. The *black stippled line* is the trend line for rainfall. The *horizontal gray stippled line* indicates the overall mean rainfall per annum for this period (678 mm). *Gray arrows* indicate the occurrence of flood events according to the historical record (which only begins in the late 1920s), *larger arrows* represent larger floods

the lands productivity for agriculture or grazing. This process also destroys wetlands – which in a healthy state provide many services to society.

Overall it appears as though annual rainfall has decreased, albeit not significantly ($R^2 = 0.071$), over the past century (Fig. 16.10), with the seven lowest rainfall years all occurring during the past 40 years. Furthermore, the annual rainfall has not exceeded 823 mm in the last 30 years, compared with nineteen times in the preceding century. It appears as though extreme rainfall events are increasing in frequency, despite the fact that there is a decrease in annual rainfall in the past 40 years.

16.3.2 Land-Use Changes

Over time, both productive and unproductive fynbos vegetation groups have become degraded as a result of increasing grazing pressures and increases in fire

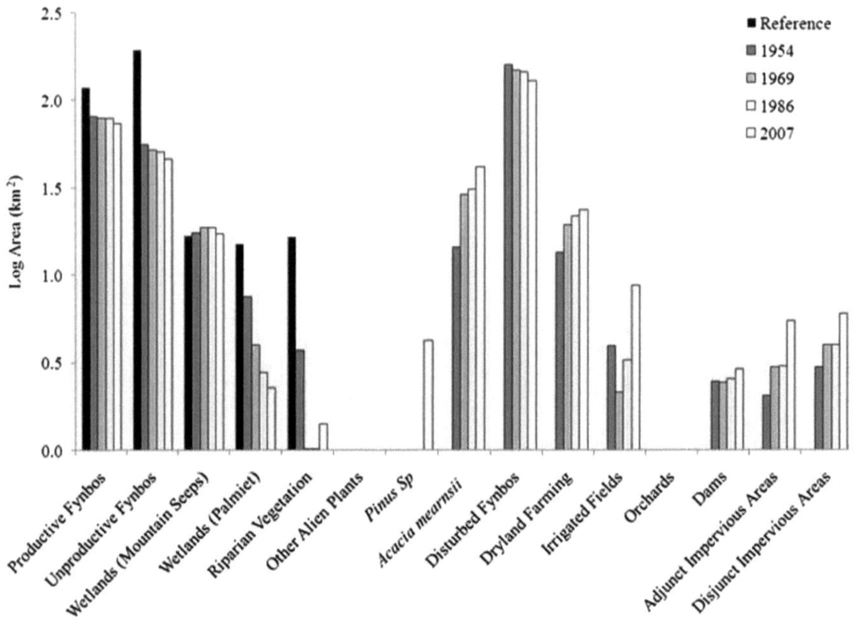

Fig. 16.11 Land-cover change in the Kromme Catchment from before 1954 to 2007

frequency (Fig. 16.11) Mountain seep wetlands have remained relatively unchanged over time, although they have become invaded by alien plants in some places. The most significant changes are in the relatively fertile floodplains which were dominated by palmiet wetlands. These wetlands have largely been replaced by agriculture, both irrigated and dryland, and also to a large extent by Black Wattle invasion. Black Wattlehas also invaded ravines, replacing indigenous Afromontane forest.

16.3.2.1 Urban Sprawl

The infrastructure of the town of Kareedouw has increased steadily from 1954 to the 1980s, however there is a dramatic change between 1986 and 2007 (Fig. 16.12). This is the result of the ending of the Apartheid Regime in South Africa in 1994, which brought in new land tenure and labour laws, leading to a movement of people from farms into nearby towns. During this time, four new townships were established around Kareedouw. Black Wattle invasion increased steadily from 1954 to 2007. The palmiet wetlands in close proximity to the town had completely disappeared by 2007.

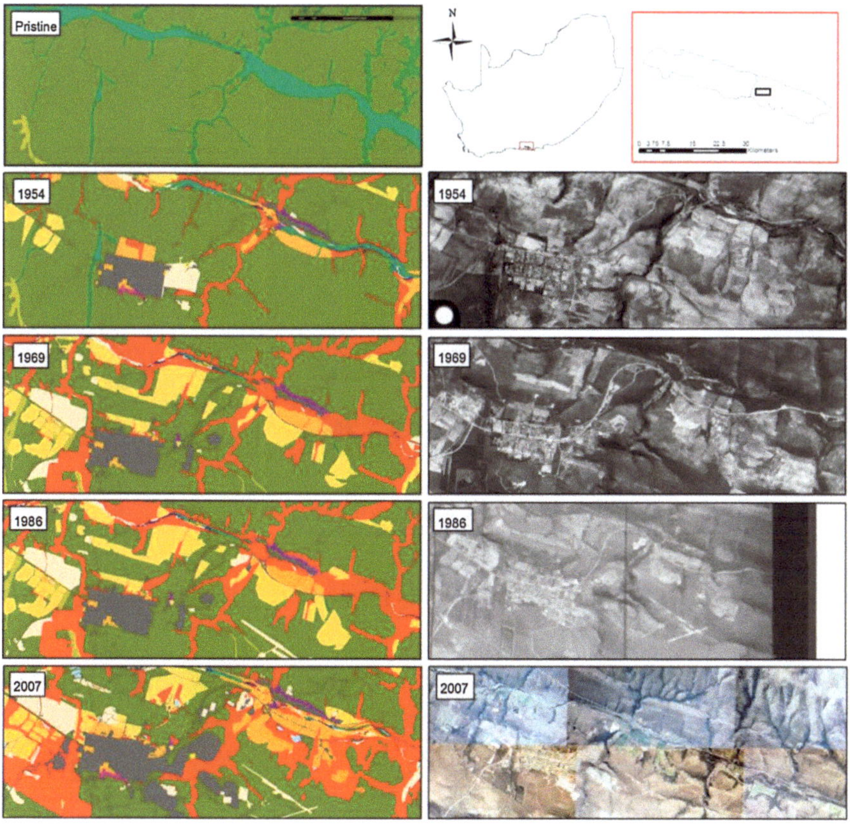

Fig. 16.12 Snapshots capturing the spread of a rural town, Kareedouw from 1954 to 2007. Kareedouw is the only town in the Kromme River Catchment. Important land-use changes include: ■ Black Wattle, ■ infrastructure, ■ dryland agriculture, ■ irrigated agriculture, ■ exposed soil, ■ palmiet wetlands and ■ fynbos

16.3.2.2 Increase in Agriculture

There was an increase in agriculture from 1954 to 2007, almost completely replacing palmiet wetlands along the entire length of the upper Kromme River (Fig. 16.13). Large, functional palmiet wetlands remain at only one location along the Kromme, mostly displaced by agriculture. More recently there has been a shift from pasture crops to orchards by some farmers, especially in the Jagersbos area (Fig. 16.13). Black Wattle has invaded the area not claimed by agriculture. However by 2007 it had been largely removed from the main channel and floodplains themselves. Where Black Wattle is cleared by Working for Water, it is often immediately replaced by agriculture.

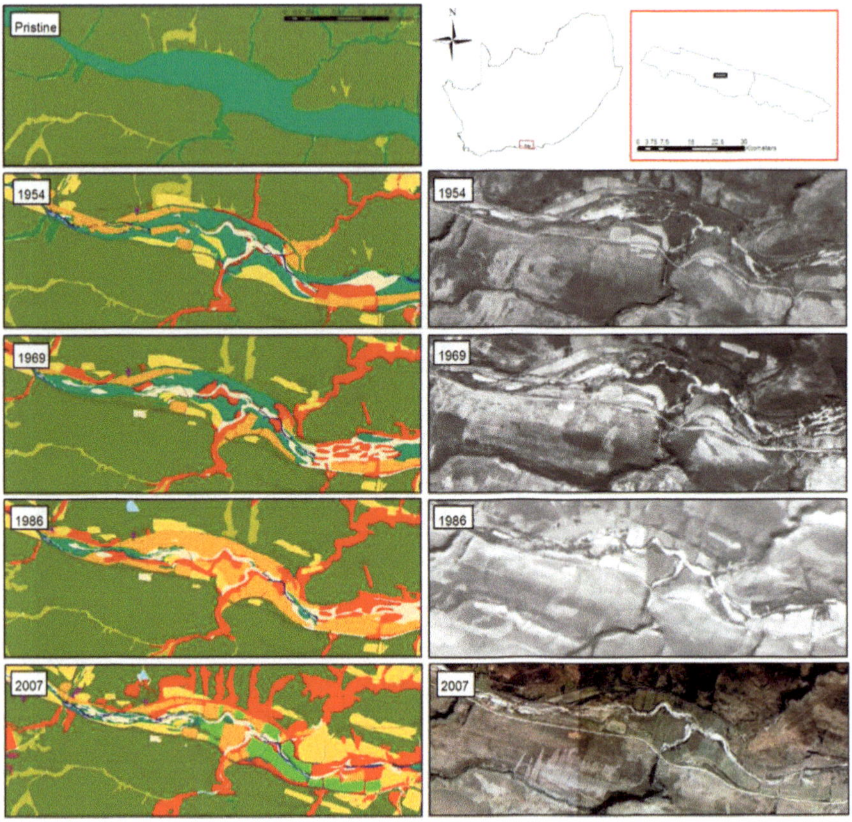

Fig. 16.13 Snapshots capturing the increase in agriculture in the fertile floodplains at a farm called Jagersbos from 1954 to 2007 in the Kromme River Catchment. Important land-use changes include: ■ palmiet wetlands, ■ orchards, ■ dryland agriculture, ■ irrigated agriculture, ■ Black Wattle, ■ exposed soil, ■ dams, and ■ fynbos

16.3.2.3 Impact on Palmiet Wetlands

Palmiet Wetlands (Fig. 16.15)
Wetlands dominated by *Prionium serratum* (palmiet) (Fig. 16.14) have been neglected and under studied. They are widely distributed in the acid waters of the Fynbos Biome, from the Gifberg to Port Elizabeth, and have outliers in the Eastern Cape and southern KwaZulu-Natal (Rogers 1997; Boucher and Withers 2004). They are generally non-channeled or channeled valley bottom wetlands (Collins 2005). Palmiet wetlands are often underlain by a layer of peat, built up over thousands of years (Grundling 2004). *Prionium* is a monotypic genus, recently moved from the family Juncaceae to Prionaceae (Munro and Linder 1997; Boucher and Withers 2004). Palmiet grows in dense stands that may appear to be separate plants, but are often clonal systems. Growth occurs throughout the year, flowering in spring and summer and fruit appears in March. Palmiet is completely salt and shade intolerant.

Fig. 16.14 Palmiet, *Prionium serratum*, is a unique South African wetland plant, the only species in its family Prioniaceae. Palmiet has long, strap-like leaves and plants grow up to two meters tall

Palmiet has adapted to fire, but alien plants invading wetlands cause palmiet stems to lengthen in search of sunlight, which exposes it to increased fire damage. Palmiet is perceived by landowners to block rivers and is often removed in favor of agriculture. This causes destabilization of rivers and wetlands (Boucher and Withers 2004).

The last remaining intact palmiet wetlands are located on a farm named Krugersland (Figs. 16.16, 16.17). These particular wetlands have been placed under protection and they are not permitted to be removed for agriculture. But the change in native land-cover to different land-uses surrounding these wetlands and encroaching upon these wetlands over time is pronounced.

16.3.2.4 Land-Use Change Around the Churchill Dam

The snapshots in Fig. 16.19 show the change of land-use on municipal property surrounding the Churchill Dam. On the far right it is possible to see the plantations of Black Wattle and Eucalyptus trees that were planted by the authorities themselves. This was later recognized to be a conflict of interest, and the plantations were removed. However the municipality has failed to take responsibility for these alien plants and they have spread.

The Churchill Dam (Fig. 16.18)
The building of the Churchill Dam began in 1940 and in 1943, the construction of the multi-arched Churchill Dam (able to hold 2.961 billion litres of water), was completed. The first test of the Churchill Dam took place in 1944, which was a high rainfall year. The dam filled overnight to a depth of 27 m and a few days later overflowed (Raymer 2008). Today it is a very important water supply for the Nelson Mandela Metropolitan hub in the

Fig. 16.15 A typical palmiet wetland in Jonkershoek, in the Western Cape of South Africa. Palmiet wetlands are often underlain by a thick layer of peat, built up over thousands of years which perform many important functions including water storage and filtration

Eastern Cape of South Africa as it provides approximately 40 % of the city's water supply. The Eastern Cape has been in a drought with the Churchill Dam being less than 30 % full for the past few years.

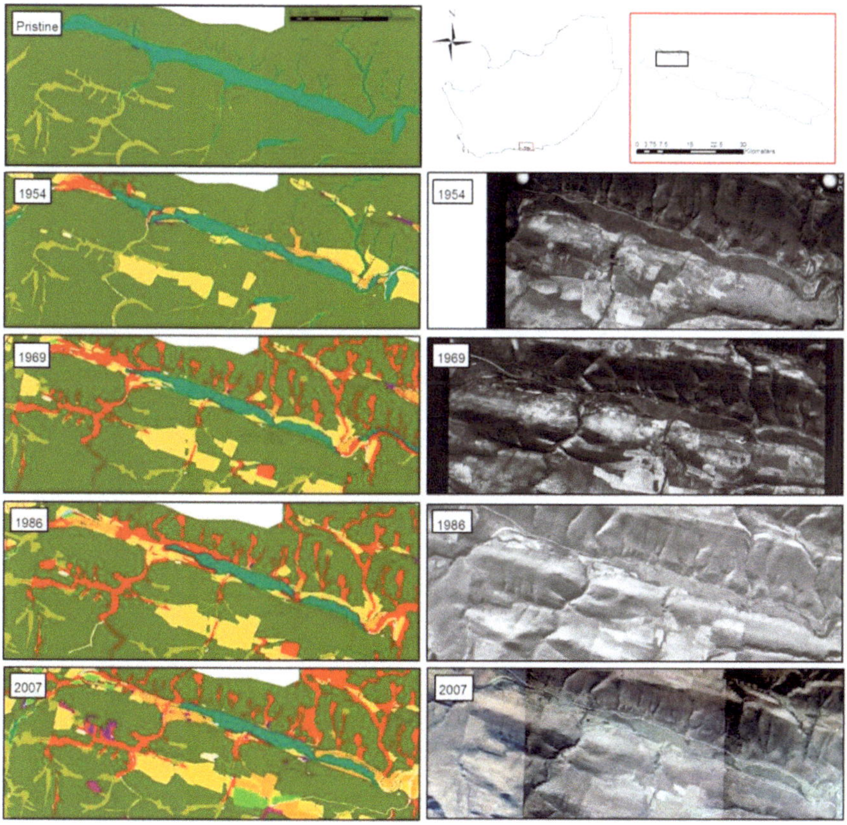

Fig. 16.16 Snapshots capturing the change in extent and surrounding land-use of valuable palmiet peat wetlands due to an increase in agriculture and invasion of Black Wattle from 1954 to 2007. These wetlands occur on Krugersland Farm and are currently the last existing wetlands in the Kromme River. Important land-use changes include: ■ palmiet wetlands, ■ Black Wattle, ■ dryland agriculture, ■ irrigated agriculture, and ■ fynbos

16.3.2.5 Geomorphological Changes: Changes in Channel Length

The total channel length has increased from 1954 to 2007 (Fig. 16.20). Very little change in the total channel length occurred between 1954 and 1986 with the most marked changes happening before and after this period. This implies that the river channel has become more meandering with time, which may be a result in decreased channel stability brought about by the invasion of Black Wattle.

Fig. 16.17 The last remaining functional palmiet wetlands and peat beds in the Kromme River, Eastern Cape, South Africa. The wetlands would have historically covered the floodplains of the Kromme Riverbed, but have largely been removed in favor of agriculture

16.3.2.6 Rate of Spread of Black Wattle

The most dramatic increase in Black Wattle invasion is 10.37 ha during the period 1986–2007 (Fig. 16.21). However this is also the greatest time-step, with 24 years between the respective aerial photographs.

The rate of invasion was greatest in the period between 1954 and 1969 at about 96 ha/a, which was its initial invasion phase. After this time, the invasion ratesslowed to 12 ha/a perhaps indicating some threshold was reached. Between 1986 and 2007 the rate of invasion increases again to 43 ha/a, possibly indicating the stage when began to invade old lands (Table 16.2).

16.3.2.7 Effectiveness of Mitigation Attempts: Rate of Clearing by Working for Water

Working for Water began alien clearing in the Kromme Catchment in 1996. The control programme involves an initial treatments followed by several, sometimes

Fig. 16.18 The multi-arched Churchill Dam on the Kromme River in the Eastern Cape of South Africa

up to eight, further treatments before a given invasion is reduced to acceptable levels (McConnachie et al. 2012) (Table 16.3).

Thus Working for Water is clearing aliens at three times the rate of invasion.

16.4 Discussion

Over the past half century wetlands, floodplains and fertile riverbeds, the areas in the catchment that are the most vital in terms of providing essential services to mankind, have been the most heavily impacted and transformed. Research has shown the transition from intact indigenous vegetation to landscapes heavily transformed by agriculture and invasive alien plants results in significant hydrological changes (Prinsloo and Scott 1999; Jackson et al. 2001; Jewit 2002; Gleick 2003; Shiklomanov and Rodda 2003; Allan 2004; Scanlon et al. 2007; Gleick et al. 2011). The main drivers of land-use change and wetland transformation in the Kromme appear to be unsustainable agricultural practices and alien invasion of the riparian zone. These drivers cause erosion and headcuts, lowering of the water table, decreased river flow due to increased transpiration and irrigation, greater flood damage, decreased base flow and a decrease in water quality (Hibbert 1971;

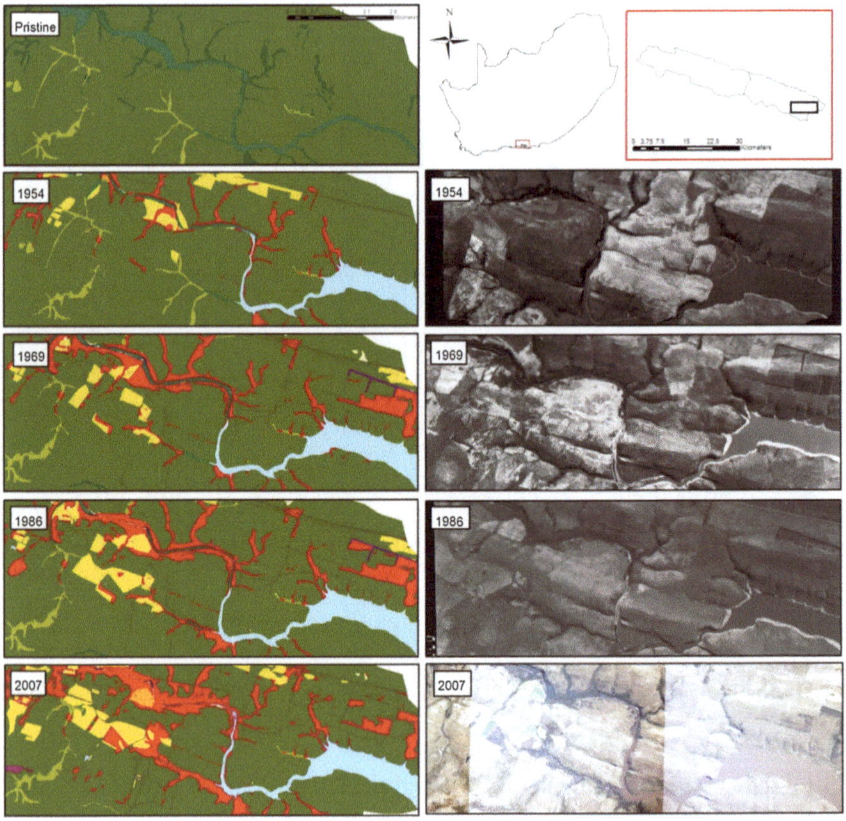

Fig. 16.19 Snapshots capturing the change in agriculture and invasion of Black Wattle in the vicinity of the Churchill Dam from 1954 to 2007. The Kromme River is an important water source for the nearby city of Port Elizabeth. Important land-use changes include: ■ dams, ■ Black Wattle, ■ palmiet wetlands, ■ dryland agriculture, and ■ fynbos

McGuinness and Harrold 1971; Bosch and Hewlett 1982; Rowntree 1991; Smith and Scott 1992; Dye 1996; Scott and Lesch 1997; Scott 1999; Le Maitre and Görgens 2001; Le Maitre et al. 2002; Andreassian 2004; Dye and Jarmain 2004; Scott et al. 2004; Calder 2005; Grenfell et al. 2005).

Land-use change over large areas has been shown to alter erosion intensity, causing channel lengthening and a decrease in active channel width (Michalková et al. 2011). In South Africa and Australia, IAP have been found to impact the geomorphology of rivers systems by modifying the river channel, while subsequent removal can lead to significant channel instability and mobilization of sediment (Beyers 1991; Rowntree 1991; Richardson et al. 1997; Bunn et al. 1998). With the destruction of wetlands in the Kromme, the river appears to have become more braided and sinuous with time. It is difficult to be certain about channel

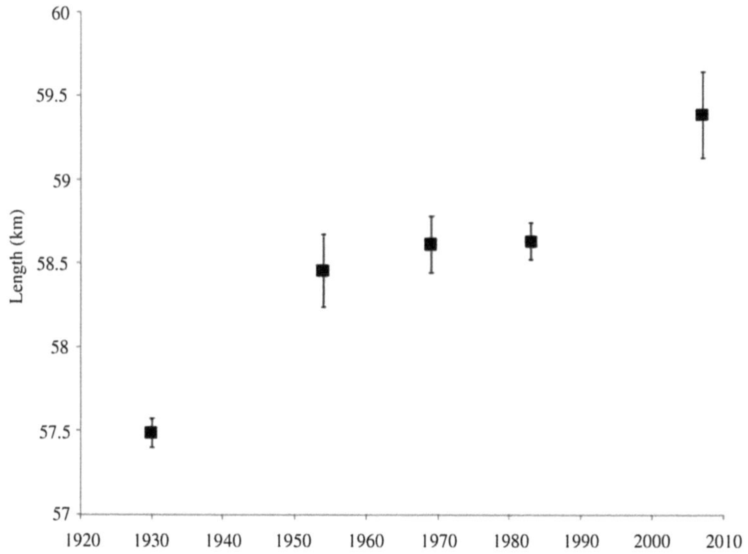

Fig. 16.20 Change in the mean (±standard deviation) total channel length from 1954 to 2007 measured along the centerline for the length of the upper Kromme River

change because the Kromme River was originally a valley-bottom wetland with no channel visible at the surface. That the Kromme River channel length is decreasing over time is likely to be a result of alien invasion and concomitant wetland destruction.

What has happened in the Kromme may be a reasonable reflection of what is happening in other South African catchments (Mander et al. 2010). This damage to natural capital in catchments that are valuable for water provision is counter-intuitive. In the case of the Kromme, farming is marginal and it is not an important agricultural catchment (Haigh et al. 2002). In such cases we would recommend prioritizing land-use at a catchment scale in terms of which ecosystem goods or services it is to provide. It is clear that high quality water-related ecosystem goods and services are not compatible with intensive agriculture in the floodplains of the same catchment. Yet this is the model that South Africa appears to follow.

In the face of climate change, water resources in South Africa are likely to become scarcer and less predictable over time (World Water Assessment Program 2009; Matthews et al. 2011). Specific predictions include an increase in summer rainfall, a decrease in winter rainfall, an increase in rainfall intensity in the east, a monthly rainfall change of 10 mm or more, and an increase in air temperature (mainly minimum temperature) by up to 2–3 °C (Midgley et al. 2005). Furthermore, climate change is likely to result in an increase in floods and drought (Midgley et al. 2005). In the Kromme, rainfall is decreasing over time while major floods appear to be increasing. These predictions indicate a likely increase in extreme events, which a healthy, resilient, functioning river system may be able to

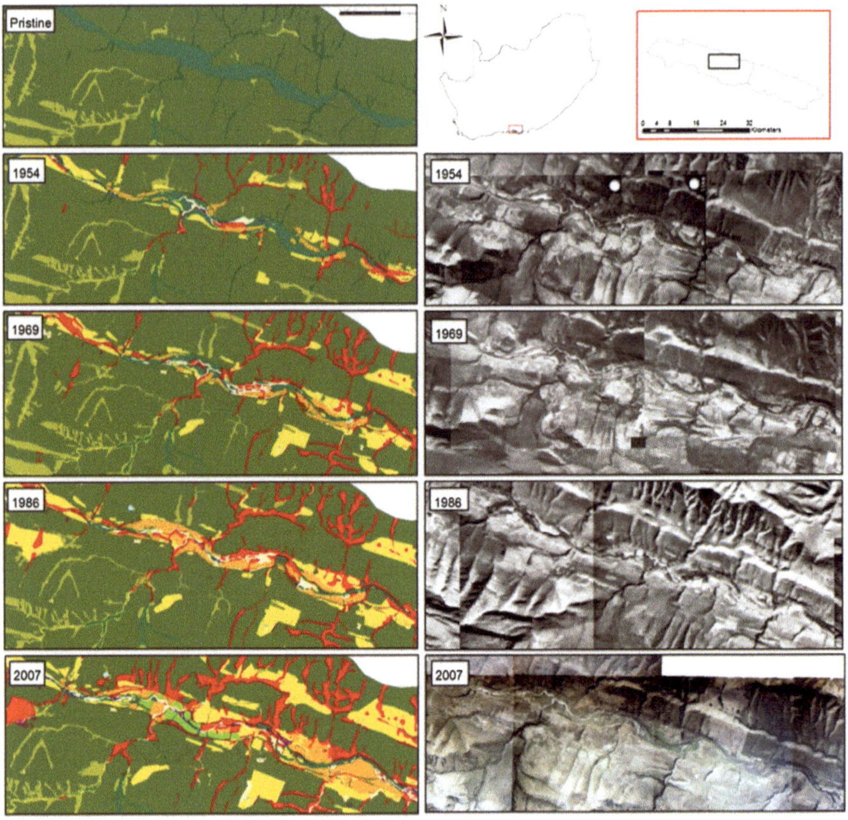

Fig. 16.21 Snapshots capturing the rate of the invasion of Black Wattle from 1954 to 2007 along the Kromme River. Important land-use changes include: ■ palmiet wetlands, ■ Black Wattle, ■ dryland agriculture, ■ irrigated agriculture, ■ orchards, ■ exposed soil, ■ seep wetlands and ■ fynbos

Table 16.2 The expansion and rate of invasion by Black Wattle from 1954 to 2007 along the Kromme River

	Reference	1954	1969	1986	2007	Mean
Acacia mearnsii (ha)	0.00	1,440	2,886	3,097	4,134	–
Rate of change (ha/decade)	–	1,440	1,447	211	1,037	–
Number of years	–	54	15	17	21	–
Rate of change (ha/a)	–	27	96	12	49	46.0

absorb. However, with most of the wetlands in the Kromme transformed, the catchment may have lost its buffering ability and may no longer be able to absorb these extreme events. The wetlands that do remain are located upstream, near the headwaters, and as a result have no ability to filter and purify water downstream

Table 16.3 The extent of the Black Wattle alien invasion clearing by Working for Water along the Kromme River from 2002 to 2010

Years	Area cleared (ha/a)
2002	93.61
2003	57.18
2004	77.19
2005	155.79
2006	143.7
2007	149.24
2008	147.08
2009	151.89
2010	269.39
Mean	138.3411

(Fig. 16.22). However if palmiet was restored further downstream, services that are crucial to downstream stakeholders –including water purification and flood attenuation, may be recovered with time (Aronson et al. 2007; Blignaut and Aronson 2008).

In an attempt to restore the Kromme Catchment, Working for Water is clearing Black Wattle at three times the average rate of invasion. However the available data do not differentiate between initial clearing and follow-up, so it is possible

Fig. 16.22 The headwaters of the Kromme River are in a pristine condition as they fall within the Formosa Nature Reserve, Eastern Cape, South Africa

that their rate of clearing is slower than these data indicate. At their current rate of clearing it would take Working for Water another 30 years to clear the Kromme, and this is just one of many South African catchments. Such a large investment in the Kromme over a long period of time with such a low rate of progress would suggest that the WfWater Programme could do with improvement (Hosking and du Preez 2004; McConnachie et al. 2012). Part of this may be the lack of communication: the failure to bridge the gap between managers, implementing agents and landowners and society at large (Cowling et al. 2008).

16.5 Conclusion

The question remains as to how these complex systems can be managed so that they are insured against future climate change. Managing the functioning of rivers requires holistic, integrated catchment management approaches as well as interdisciplinary co-operation (Dollar et al. 2007; Nel et al. 2007, 2009). Riparian systems have been described as complex adaptive systems and both a social learning process and an adaptive management approach is needed (Pahl-Wostl 2007). We recommend that important water providing catchments, where agriculture is marginal, should be prioritized for provision of water-related ecosystem services alone. Investment into improving the resilience of these systems as insurance against future climate changes is essential. This should be in the form of prohibiting unsustainable land management practices and enforcing the laws that protect rivers and wetlands, eradication of invasive alien plants and rehabilitation of the river and wetlands. This investment in restoration of an important water-providing catchment cannot be done without education and a social learning process (Pahl-Wostl 2007; Cowling et al. 2008).

Acknowledgments Thank-you to the following people for invaluable support and guidance: Katie Gull, Mr Japie Buckle, Mr Pierre Joubert, Prof James Blignaut, Mr Mike Powell, GIS Consultancy Garth Stephenson. Thank-you to my Lord and Saviour, Jesus Christ. All my work is dedicated to You. Thank-you to the following organizations for funding and support: ASSET Research, the Water Research Commission of South Africa, the Fynbos Forum Innovation Scholarship (Table Mountain Fund and WWF), the HB Thom Fund, Dudley D'Ewes Scholarship (Cape Tercentenary Foundation), the Ernst and Ethel Eriksen Trust.

References

Allan JD. Landscapes and riverscapes: the influence of land use on stream ecosystems. Annu Rev Ecol Evol Syst. 2004;2004(35):257–84.

Andreassian V. Waters and forests: from historical controversy to scientific debate. J Hydrol. 2004;291:1–27.

Aronson J, Milton SJ, Blignaut JN, editors. Restoring natural capital: science, business, and practise. Washington, DC: Island Press; 2007.

Ashton PG. Avoiding conflicts over Africa's water resources. Ambio. 2002;31(3):236–42.
Bates BC, Kundzewicz ZW, Wu S, Palutik JP, editors. Climate change and water. Technical Paper of the Intergovernmental Panel on Climate Change. Geneva: IPCC Secretariat; 2008. p. 210.
Beyers G. A preliminary assessment of the geomorphological impact of alien vegetation in the Baviaanskloof region. Grahamstown: Third Year Project, Department of Geography, Rhodes University; 1991.
Blignaut J, Aronson J. Getting serious about maintaining biodiversity. Conserv Lett. 2008;1(1):12–7.
Bosch JM, Hewlett JD. A review of catchment experiments to determine the effect of vegetation changes on water yield and evapotranspiration. J Hydrol. 1982;55:3–23.
Boucher C, Withers MJ. Notes on the ecology of *Prionium serratum* (palmiet), a Cape river specialist. Veld Flora. 2004;90(1):26–8.
Brinson MM, Malvárez AI. Temperate freshwater wetlands: types, status and threats. Environ Conserv. 2002;29:115–33.
Bunn SE, Davies PM, Kellaway DM, Prosser IP. Influence of invasive macrophytes on channel morphology and hydrology in an open tropical lowland stream, and potential control by riparian shading. Freshw Biol. 1998;39:171–8.
Calder IR. Blue revolution II: land use and integrated water resources management. London: Earthscan; 2005.
Collins NB (2005) Wetlands: the basics and some more. Free State Department of Tourism, Environmental and Economic Affairs. http://www.csir.co.za/rhp/.
Cowling RM, Egoh B, Knight AT, O'Farrell PJ, Reyers B, Rouget M, Roux DJ, Welz A, Wilhelm-Rechman A. An operational model for mainstreaming ecosystem services for implementation. Proc Natl Acad Sci. 2008;105:9483–8.
Crous CJ, Jacobs SM, Esler KJ. Drought-tolerance of an invasive alien tree, *Acacia mearnsii* and two native competitors in fynbos riparian ecotones. Biol Invas. 2011;14:619–31.
Daily GC, Soderquist T, Aniyar S, Arrow K, Dasgupta P, Ehrlich PR, Folke C, Jansson AM, Jansson BO, Kautsky N, Levin S, Lubchenco J, Maler KG, David S, Starrett D, Tilman D, Walker B. The value of nature and the nature of value. Science. 2000;289:395–6.
David PB, Leslie BB, Kristine AC, Andrew TW. Water Environment Federation. Vancouver, British Columbia: Proceedings of the Watershed Conference; 2000.
de Groot RS, Wilson MA, Boumans RMJ. A typology for the classification, description and valuation of ecosystem functions, goods and services. Ecol Econ. 2002;41:393–408.
Department of Water Affairs and Forestry, South Africa: Working for water information pamphlet. http://www.dwaf.gov.za/wfw/Docs/WfWadvertOct06.pdf. Accessed Oct 2006.
Dollar ESJ, James CS, Rogers KH, Thoms MC. A framework for interdisciplinary understanding of rivers as ecosystems. Geomorphology. 2007;89:147–62.
Dye PJ. Climate, forest and stream flow relationships in South African afforested catchments. Commonwealth For Rev. 1996;75(1):31–8.
Dye P, Jarmain C. Water use by black wattle (*Acacia mearnsii*): implications for the link between removal of invading trees and catchment streamflow response. S Afr J Sci. 2004;100:40–4.
Eberhard R: Urban water demand scenarios for Nelson Mandela Bay Area. Presentation to study steering committee by Palmer Development Group. 2009.
Gleick P. Global freshwater resources: soft-path solutions for the 21st century. Science. 2003;302:1524–8.
Gleick PH, Christian-Smith J, Cooley H. Water-use efficiency and productivity: rethinking the basin approach. Water Int. 2011;36(7):784–98.
Grenfell MC, Ellery WN, Preston-Whyte RA. Wetlands as early warning (eco)systems for water resource management. Water SA. 2005;31(4):465–72.
Groombridge B. Global biodiversity, status of the earth's living resources. Chapman and Hall, London: World Conservation Monitoring Centre; 1992. p. 297.
Grundling PL: Peatlands on the brink of the final phase in the extinction of a unique wetland type in southern Africa. Working for Wetlands Programme, National Botanical Institute, Pretoria.

IMGG Scientific Symposium, September 2004. International Mire Conservation Group. 2004, p. 5.
Grundling PL, Marnewick GC: Mapping, characterization and monitoring of the Highveld peatlands: compilation of existing data and evaluation of inventory methodology. Compiled for the Department of Agriculture, Directorate Land and Resources, Management on behalf of the Agricultural Research Council: Institute for Soil, Water and Climate. 1999.
Haigh EA, Grundling PL, Illgner PM: Report on the scoping study on the status of the Kromme River peatland complex and recommended mitigatory measures. Department of Water Affairs and Forestry Project X832633. 2002.
Hibbert AR. Increases in streamflow after converting chaparral to grass. Water Resour Res. 1971;7(1):71–80.
Hobbs RJ. The working for water programme in South Africa: the science behind the success. Divers Distrib. 2004;10:501–3.
Hosking SG, du Preez M. A cost-benefit analysis of the *working for water programme* on selected sites in South Africa. Water SA. 2004;30(2):143–52.
Jackson RB, Carpenter SR, Dahm CN, Mcknight DM, Naiman RJ, Postel SL, Running SW. Water in a changing world. Ecol Appl. 2001;11(4):1027–45.
Jewitt G: Can integrated water resources sustain the provision of ecosystem goods and services? Physics and Chemistry of the Earth. 2002; 27:887–896. http://www.sciencedirect.com/science?_ob=ArticleURL&_udi=B6X1W-4HC0JFN-6&_user=613892&_coverDate=12%2F31%2F2005&_rdoc=1&_fmt=high&_orig=search&_sort=d&_docanchor=&view=c&_rerunOrigin=scholar.google&_acct=C000032099&_version=1&_urlVersion=0&_userid=613892&md5=d145d833410ee9948a29880c7842142b-bbib16.
Kondolf GM: Mediterranean-climate episodic channels: long periods of boredom, brief moments of terror. The international mediterranean ecosystems conference MEDECOS XII. 6–9 Sep 2011.
Kotze DC, Breen CM, Nxele IZ, Kareko J: WET-ManagementReview: the impact of natural resource management programmes on wetlands in South Africa. In: Breen C, Dini J, Ellery W, Mitchell S, Uys M , editors. Water Research Commission, Pretoria, South Africa. WRC Report No TT 335/09. 2009.
Le Maitre DC, Görgens A: Potential impacts of invasive alien plants on reservoir yields in South Africa. Tenth South African National Hydrology Symposium: 26–28 Sep 2001.
Le Maitre DC, Van Wilgen BW, Gelderblom CM, Bailey C, Chapman RA, Nel JA. Invasive alien trees and water resources in South Africa: case studies of the costs and benefits of management. For Ecol Manage. 2002;160(3):143–59.
Le Maitre DC, Colvin C, Maherry A. Water resources in the Klein Karoo: the challenge of sustainable development in a water-scarce area. S Afr J Sci. 2009;105:39–48.
Le Maitre DC, Versfeld DB, Chapman RA. The impact of invading alien plants on surface water resources in South Africa: a preliminary assessment. Water SA. 2000;26(3):397–408.
Lewis CA. Geomorphology of the Eastern Cape, South Africa. 2nd ed. Grahamstown: NISC; 2008. p. 21–8.
Mander M, Blignaut J, Van Niekerk M, Cowling R, Horan M, Knoesen D, Mills A, Powell M, Schulze R: Baviaanskloof-Tsitsikamma payments for ecosystem services: a feasibility assessment. 2010.
Matthews JH, Wickel BAJ, Freeman S. Converging currents in climate-relevant conservation: water, infrastructure, and institutions. PLoS Biol. 2011;9(9):1–4.
McConnachie MM, Cowling RM, van Wilgen BW, McConnachie DA: Evaluating the cost-effectiveness of invasive alien plant clearing: a case study from South Africa. Biological Conservation (In Press). 2012.
McGuinness JL, Harrold L. Reforestation influences on small watershed streamflow. Water Resour Res. 1971;7(4):845–52.
Michalková M, Piégay H, Kondolf GM, Greco SE: Lateral erosion of the Sacramento River, California (1942–1999), and responses of channel and floodplain lake o human influences. Earth Surf Process Land. 2011;36:257–72.

Middleton BJ, Bailey AK: Water resources of South Africa 2005 study (WR2005). WRC Report Number TT 380/08. 2008.

Midgley DC, Pitman WV, Middleton BJ: Surface water resources of South Africa 1990 volume III: Orange-Namaqualand. Report 1990 298/3.1/94. Water Research Commission. 1994.

Midgley GF, Chapman RA, Hewitson B, Johnston P, De Wit M, Ziervogel G, Mukheibir P, Van Niekerk I, Tadross M, Van Wilgen BW, Kgope B, Morant P, Theron A, Scholes RJ, Forsyth GG: A status quo, vulnerability and adaptation assessment of the physical and socio-economic effects of climate change in the Western Cape. Report to the Western Cape Government, Cape Town, South Africa. Report No: ENV-S-C 2005-073. Stellenbosch: CSIR. 2005.

Millennium Ecosystem Assessment. Millennium ecosystem assessment synthesis reports. 2006.

Mooney HA, Hamberg SP, Drake JA. The invasion of plants and animals. In: Mooney HA, Drake JA, editors. Ecology of biological invasions of North America and Hawaii. New York: Springer-Verlag; 1986.

Munro SL, Linder HP. The embryology and systematic relationships of *Prionium serratum* (Juncaceae: Juncales). Am J Bot. 1997;84(6):850–60.

Murray R, Goedhart M, Baron J: High-yielding groundwater areas around the Nelson Mandela Bay Municipality. Water Research Commission, Pretoria, South Africa. WRC Report No: TT 327/08. 2008.

Nel JL, Reyers B, Roux DJ, Cowling RM. Expanding protected areas beyond their terrestrial comfort zone: identifying spatial options for river conservation. Biol Conserv. 2009;142: 1605–16.

Nel JL, Roux DJ, Maree G, Kleynhans CJ, Moolman J, Reyers B, Rouget M, Cowling RM. Rivers in peril inside and outside protected areas: a systematic approach to conservation assessment of river ecosystems. Biodivers Res. 2007;13:341–52.

Pahl-Wostl C. The implications of complexity for integrated resources management. Environ Model Softw. 2007;22:561–9.

Pimm SL, Raven P. Extinction by numbers. Nature. 2000;403:843–5.

Prinsloo FW, Scott DF. Streamflow responses to the clearing of alien invasive trees from riparian zones at three sites in the Western Cape Province. South Afr For J. 1999;185:1–7.

Raimondo D, Von Staden L, Foden W, Victor JE, Helme NA, Turner RC, Kamundi DA, Manyama PA, editors. Red List of South African plants 2009. Strelitzia 25. Pretoria: South African National Biodiversity Institute; 2009.

Raymer D: Streams of life. David Raymer. 2008.

Rejmánek M, Randall R. Invasive alien plants in California: 1993 summary and comparison with other areas in North America. Madrono. 1994;41:161–77.

Richardson DM, Macdonald IAW, Hoffmann JH, Henderson L. Alien plant invasions. In: Cowling RM, Richardson DM, Pierce SM, editors. Vegetation of Southern Africa. Cambridge: Cambridge University Press; 1997. p. 534–70.

Rogers KH. Freshwater wetlands. In: Cowling RM, Richardson DM, Pierce SM, editors. Vegetation of Southern Africa. Cambridge: Cambridge University Press; 1997.

Rouget M, Reyers B, Jonas Z, Desmet P, Driver A, Maze K, Egoh B, and Cowling RM: South African National spatial biodiversity assessment. Technical Report, vol. 1: Terrestrial Component. 2004.

Rowntree K. An assessment of the potential impact of alien invasive vegetation on the geomorphology of river channels in South Africa. South Afr J Aquat Sci. 1991;17(1/2):28–43.

Scanlon BR, Jolly I, Sophocleous M, Zhang L. Global impacts of conversions from natural to agricultural ecosystems on water resources: quantity versus quality. Water Resour Res. 2007;43:1–18.

Schulze RE (Ed): Climate change and water resources in southern Africa: studies on scenarios, impacts, vulnerabilities and adaptation. Water research commission, Pretoria, South Africa. WRC Report 1430/1/05. 2005.

Scott DF. Managing riparian zone vegetation to sustain streamflow: results of paired catchment experiments in South Africa. Can J For Res. 1999;29:1149–57.

Scott DF, Lesch W. Streamflow responses to afforestation with *Eucalyptus grandis* and *Pinus patula* and to felling in the Mokobulaan experimental catchments, Mpumalanga province, South Africa. J Hydrol. 1997;199:360–77.

Scott DF, Bruijnzeel LA, Vertessy RA, Calder IR. Impacts of forest plantations on streamflow. Forest Hydrology. 2004;272:1–11.

Shiklomanov IA, Rodda JC, editors. World water resources at the beginning of the twenty-first century. Cambridge: Cambridge University Press; 2003.

Skead CJ. Historical plant incidence in southern Africa, *Strelitzia 24*. Pretoria: South African National Biodiversity Institute; 2009.

Smith RE, Scott DF. The effects of afforestation on low flows in various regions of South Africa. Water SA. 1992;18:185–94.

Toerien DK, Hill RS. The geology of the Port Elizabeth areas, explanation of sheet 3324. South Africa: Geological Survey; 1989. 35.

Van den Berg EC, Plarre C, Van den Berg HM, Thompson MW: The South African national land cover 2000. Agricultural Research Council (ARC) and Council for Scientific and Industrial Research (CSIR), Pretoria. Report no. GW/A/2008/86. 2008.

van Wilgen BW, Le Maitre DC, Wannenburgh A, Kotze IM, van den Berg L, Henderson L. An assessment of the effectiveness of a large, national-scale invasive alien plant control strategy in South Africa. Biol Conserv. 2012;148:28–38.

van Wilgen BW, Le Maitre DC, Cowling RM. Ecosystem services, efficiency, sustainability and equity: South Africa's working for water programme. Trends Ecol Evol. 1998;13:378.

van Wilgen BW, Reyers B, Le Maitre DC, Richardson DM, Schonegevel L. A biome scale assessment of the impact of invasive alien plants on ecosystem services in South Africa. J Environ Manage. 2008;89:336–49.

Vlok J, Euston-Brown D, Wolf T: A vegetation map for the garden route initiative. Unpublished 1:50,000 maps and reports supported by CAPE FSP task team. 2008.

World Water Assessment Program: the United Nations world water development report: water in a changing world. Paris: UNESCO and Earthscan. 2009, p. 318.

Erratum to: Discursive Relationships Between Landscape Science, Policy and Management Practice: Concepts, Issues and Examples

Simon Swaffield, Jorgen Primdahl and Mark Hoversten

Erratum to:
Chapter 12 in: B. Fu and K. B. Jones (eds.), *Landscape Ecology for Sustainable Environment and Culture*, DOI 10.1007/978-94-007-6530-6_12

Author name "Flyvberg" from two references in Chap. 12 was misspelt in few occurrences, both in in-text citations and reference list. The name should read as "Flyvbjerg".

The online version of the original chapter can be found under DOI 10.1007/978-94-007-6530-6_12

S. Swaffield (✉)
Landscape Architecture, Lincoln University, Lincoln, New Zealand
e-mail: swaffies@lincoln.ac.nz

J. Primdahl
GeoScience and Nature Management, Copenhagen University, Copenhagen, Denmark
e-mail: jpr@life.ku.dk

M. Hoversten
Landscape Architecture, University of Idaho, Idaho, USA
e-mail: hoverstm@uidaho.edu

B. Fu and K. B. Jones (eds.), *Landscape Ecology for Sustainable Environment and Culture*, DOI: 10.1007/978-94-007-6530-6_17,

Index

B. Fu and K. B. Jones (eds.), *Landscape Ecology for Sustainable Environment and Culture*, DOI: 10.1007/978-94-007-6530-6,

MIX
Papier aus verantwortungsvollen Quellen
Paper from responsible sources
FSC® C105338

If you have any concerns about our products,
you can contact us on
ProductSafety@springernature.com

In case Publisher is established outside the EU,
the EU authorized representative is:
Springer Nature Customer Service Center GmbH
Europaplatz 3, 69115 Heidelberg, Germany

Printed by Libri Plureos GmbH
in Hamburg, Germany